Lecture Notes in Computer Science 901

Edited by G. Goos, J. Hartmanis and J. van Leeuwen

Advisory Board: W. Brauer   D. Gries   J. Stoer

Ramayya Kumar   Thomas Kropf (Eds.)

Theorem Provers in Circuit Design

Theory, Practice and Experience

Second International Conference, TPCD '94
Bad Herrenalb, Germany, September 26-28, 1994
Proceedings

Springer

Series Editors

Gerhard Goos
Universität Karlsruhe
Vincenz-Priessnitz-Straße 3, D-76128 Karlsruhe, Germany

Juris Hartmanis
Department of Computer Science, Cornell University
4130 Upson Hall, Ithaca, NY 14853, USA

Jan van Leeuwen
Department of Computer Science, Utrecht University
Padualaan 14, 3584 CH Utrecht, The Netherlands

Volume Editors

Ramayya Kumar
Forschungszentrum Informatik
Haid-und-Neu-Straße 10-14, D-76131 Karlsruhe, Germany

Thomas Kropf
Institut für Rechnerentwurf und Fehlertoleranz, Universität Karlsruhe
Zirkel 2, D-76128 Karlsruhe, Germany

CR Subject Classification (1991): F.4.1, I.2.3, B.1.2

ISBN 3-540-59047-1 Springer-Verlag Berlin Heidelberg New York

CIP data applied for

© Springer-Verlag Berlin Heidelberg 1995
Printed in Germany

Typesetting: Camera-ready by author
SPIN: 10485414   45/3140-543210 - Printed on acid-free paper

Preface

This volume contains the final revised proceedings of the Second International Conference on Theorem Provers in Circuit Design, jointly organized by FZI (Forschungszentrum Informatik), Karlsruhe, and the University of Karlsruhe (Universität Karlsruhe, Institut für Rechnerentwurf und Fehlertoleranz), in cooperation with IFIP (International Federation of Information Processing) Working Group 10.2. The workshop took place in the Treff Hotel, Bad Herrenalb, Germany from 26 to 28 September 1994.

The conference was a sequel to the one held at Nijmegen in June 1992 and provided a forum for discussing the role of theorem provers in the design of digital systems. The topics of interest included original research as well as case studies and other practical experiments with new or established theorem proving tools, including tautology and model checkers.

The field of formal methods in hardware abounds with various kinds of formalisms, each of which have their advantages and disadvantages. Two of the popular theorem provers, with different underlying formalisms, were presented as tutorial talks.

Two invited papers highlighted the use of formal methods in circuit design from an academic and an industrial viewpoint. They were given by Tom Melham (*Inductive Reasoning about Circuit Design*) and Pasupati Subrahmanyam (*Compositionality, Hierarchical Verification and the Principle of Transparency*).

An interesting panel discussion on the *Use of Formal Methods in Industry* was conducted with the following participants: Massimo Bombana, Holger Busch, Alberto Camilleri, Pasupati Subrahmanyam, and John van Tassel.

All submitted research papers were reviewed by at least three independent reviewers, who are all experts in the field. The emphasis of the conference was laid on an in-depth presentation of the approaches instead of accepting many papers, hence only 50% of the submitted papers were accepted as full papers, each of which was given 40 minutes of presentation time. Four papers were acknowledged for their interesting ideas and accepted as short papers.

The research papers were complemented by the demonstration of verification systems — *Fancy, MEPHISTO, PVS, Prevail,* and *Synchronized Transitions.*

At this conference also a set of benchmark circuits for hardware-verification was presented. These circuits are the basis for an international standardization effort of IFIP WG10.2 in order to provide a common basis for evaluating and comparing different approaches of hardware verification and formal synthesis.

We thank all people who actively supported us in organizing this conference, all members of the programme committee, and especially Hilke Kloss for solving many organizational details and Frank Imhoff and Dirk Eisenbiegler for their work in setting up the hardware for the system demonstrations. We are also grateful to Michael Berthold for his help in publishing these proceedings.

January 1995   *Ramayya Kumar*
                                                                                      Thomas Kropf

Programme Committee

Dominique Borrione (IMAG, France)
Holger Busch (Siemens AG, Germany)
Luc Claesen (IMEC, Belgium)
David Dill (Stanford University, USA)
Hans Eveking (University of Frankfurt, Germany)
Simon Finn (Abstract Hardware Ltd., UK)
Mike Gordon (University of Cambridge, UK)
Warren A. Hunt Jr. (CL Inc., USA)
Paul Loewenstein (Sun, USA)
Miriam Leeser (Cornell University, USA)
Tom Melham (University of Glasgow, UK)
Tobias Nipkow (TU München, Germany)
Jørgen Staunstrup (Lyngby University, Denmark)
Victoria Stavridou (University of London, UK)
Pasupati Subrahmanyam (AT&T Bell Labs, USA)

Conference Organization

Conference Chair
Ramayya Kumar
Forschungszentrum Informatik
Haid-und-Neu Strasse 10-14
D-76131 Karlsruhe
Germany

Conference Co-Chair
Thomas Kropf
Institut für Rechnerentwurf und Fehlertoleranz
Universität Karlsruhe
Zirkel 2
D-76128 Karlsruhe
Germany

Local Arrangements
Hilke Kloss
Forschungszentrum Informatik
Haid-und-Neu Strasse 10-14
D-76131 Karlsruhe
Germany

Technical Arrangements
Frank Imhoff
Institut für Rechnerentwurf und Fehlertoleranz
Universität Karlsruhe
Zirkel 2
D-76128 Karlsruhe
Germany

Contents

Benchmark-Circuits for Hardware -Verification

Thomas Kropf

Institut für Rechnerentwurf und Fehlertoleranz (Prof. D. Schmid)
Universität Karlsruhe, Kaiserstr. 12, 76128 Karlsruhe, Germany
email: kropf@informatik.uni-karlsruhe.de
WWW: http://goethe.ira.uka.de/hvg/

Abstract. This document describes the IFIP WG10.2 hardware-verification benchmark circuits, intended for evaluating different approaches and algorithms for hardware verification. The paper presents the rationale behind the circuits, describes them briefly and indicates how to get access to the verification benchmark set.

1 Introduction

1.1 Motivation and State-of-the-Art

Although having many drawbacks, benchmark circuits allow a more succinct and direct comparison of different approaches for solving a certain problem. This has lead to sets of widely accepted circuits e.g. in the area of testing [1, 2] or high-level synthesis [3].

In the area of hardware verification, this lead e.g. to the suggestion of "interesting" circuits, like Paillet's set of seven sequential circuits [4]. One of the first efforts to provide circuits for a broader community has been done by Luc Claesen for the 1990 International Workshop on Applied Formal Methods for VLSI Design [5]. The most prominent circuit evolving from this effort was the "Min_Max-Circuit".

The lack of additional, generally available verification benchmark circuits got aware in the preparation of the 2nd International Conference on Theorem Provers in Circuit Design (TPCD94). The motivation to provide additional benchmark circuits together with already ongoing standardization efforts of IFIP, coordinated by Jørgen Staunstrup [6], has led to an enhanced set of circuits. Thanks to J. Staunstrup, in the meantime these circuits have become the official "IFIP WG10.2 Hardware Verification Benchmark Circuit Set". It will be maintained and enhanced on a long term basis to promote a standardized benchmark set in the hardware verification community.

For the circuits a complete and self-contained implementation is provided, done in a commercial design system [7] (leading e.g. to additional timing diagrams for clarification) as well as a clear specification using the standardized hardware description language VHDL (see Section 2.3). This puts a comparison of different verification approaches on a sound basis, since - if the given implementations are used - identical circuits are verified instead of different designs implemented in a way especially suited for a certain approach. Moreover, people are not forced to tediously design the circuits before they can be verified - the latter being the main interest of people looking into these circuits. Naturally, when dealing with formal synthesis, the implementations provided here are less interesting and the specifications are the main thing to deal with.

1.2 Requirements for Verification Benchmark Circuits

A set of verification benchmark circuits has to provide easily usable circuits to evaluate and to compare different approaches to hardware verification. This comprises:

- availability via the World Wide Web and anonymous ftp,
- a high degree of diversity with regard to the underlying verification task (see Section 2.4),
- circuit descriptions without ambiguity, which are succinct and self-contained and
- circuits which span a wide range from introductory examples to real verification challenges [6].

2 The Benchmark Circuits

2.1 Scope

Currently, there is the set of benchmark circuits given in table 2-1. The main sources for these circuits have been the previous IFIP benchmark set [6] and various textbooks on circuit design.

Circuit Name	Short Circuit Description
Single Pulser	cuts input pulses to a fixed length
Traffic Light Controller	simplified controller for a traffic light
N-bit Adder	sum of two bitvectors of length N
Min_Max	the mean value of incoming integers
Black-Jack Dealer	the dealer's hand of a card game
Arbiter	access to shared resources for N clients
Rollback Chip	coprocessor for distributed simulation
Tamarack Processor	simplified microprocessor
Stop-Watch	digital stopwatch with 3 digits and 2 buttons
GCD	Greatest Common Divisor
Multiplier	N-bit Multiplier
Divider	N-bit Divider
FIFO	asynchronous FIFO queue with N places
Assotiative Memory	simple assotiative $N \times M$ memory

Table 2-1: Current Benchmark Circuits

Circuit Name	Short Circuit Description
1dim Systolic Array	unidimensional systolic filter array
2dim Systolic Array	two-dimensional systolic array for matrix multiplication

Table 2-1: Current Benchmark Circuits

2.2 Releases

Each release of the circuit is labeled with a version number, release number and patch level in the form v<Version>.<Release>.<Patchlevel>.

A version number is provided to make the inclusion of new circuits more explicit. Unfortunately we did not succeed in providing full implementation descriptions for all circuits in the first release. As these are provided, the release number is incremented.

Although the circuits have been designed with much care, reality-driven pessimism suggests that there will be the necessity for "patches" (i.e. bug fixes) at least in the first stage of the release process.

All changes are documented in a history file, patches are also explained in the documentation of the relevant circuits. You may want to be added to an emailing list dedicated to inform people about the actual verification benchmark status (see Section 3.5).

2.3 Verification Problem Presentation

Unambiguously specifying circuits in a general way without being forced to a certain description philosophy by the underlying notation is a challenge itself. The only way to circumvent a fixed description formalism is to use only informal descriptions (natural language, drawings[1], timing diagrams, etc.). However, then the descriptions are often not as crisp and exact as it is necessary especially for formal verification, where the information given here has to be translated into formal description languages like predicate logics, temporal logics, process algebras, Z and so forth.

Besides all drawbacks, we decided to provide more (and more formal) information for each circuit. This comprises for each circuit of:

- a description of the specification and the implementation in plain English,
- schematic diagrams of the implementation[2],
- a netlist of the implementation based on structural VHDL and
- a specification of the circuit in VHDL.

---

1. For some circuits, original schematics of a realization in the semi-custom design system CADENCE have been added. However, to fit these drawings on a single page they had to be shrunk in many cases so that labels etc. may not be well readable anymore. Nevertheless, the remaining circuit documentation without these figures is completely sufficient as an implementation description.

2. Schematics are based on an implementation of each circuit using a commercial semi-custom design system.

2.3.1 VHDL

The decision to use VHDL as a formalization means especially for specifications is probably the most disputable one for obvious reasons: the lack of a clean VHDL semantics, the danger of imposing a certain specification style, impossibility for expressing nondeterminism and so forth. However, in our opinion the advantages of providing a standardized (and simulatable) specification outweighs the disadvantages. Moreover, we tried to avoid ambiguous VHDL constructs and VHDL specifications may further encourage VHDL based verification (or at least the discussion about the "right" specification language is stimulated).

We use a restricted set of VHDL which should be sufficiently simple so that no semantic ambiguities occur.

2.3.2 General Conventions for Implementation Descriptions

The naming of inputs and outputs should indicate their functionality (see table 2-2). Internal lines without special meaning are labeled a, b, c, ... in the following.

Abbreviation	Semantics
CIN	carry in
COUT	carry out
S	sum (output)
S	select or set (input)
R	reset
P	product
I	input
O	output
Q, ¬Q	state

Table 2-2: Abbreviations for inputs and outputs

2.3.3 Storage Elements and Multiplexers

A register is treated as a *base module* (see Section 2.3.4). It is used as follows: The inputs S is connected to input lines named **Store(Name)**. A signal is stored, if, **Store(Name)** =1 and there is a $0 \rightarrow 1$ transition at the clock input.

Lines named **SelectXY** are controlling a multiplexer or a demultiplexer: **SelectXY** = 0 connects input 0 of a multiplexer to the output (output 0 of a demultiplexer) and **SelectXY** = 1 connects input 1 to the output (output 1 of a demultiplexer).

Busses are provided with their names and number of signal lines: **DataIn<31:0>** denotes a 32 bit bus for reading data.

2.3.4 Base Module Library

Most circuit implementations use a predefined base module library which contains parameterized modules. Parameters are delay time and, for certain elements, the bit-width. The library contains mainly simple gates and storages elements, listed in table 2-3.

Module Name	Description	Variable Bitwidth
INV	inverter	no
BUF	buffer	no
NAND2	two input nand	no
NAND3	three input nand	no
AND2	two input and	no
AND3	three input and	no
AND4	four input and	no
NOR2	two input nor	no
NOR3	three input nor	no
OR2	two input or	no
OR3	three input or	no
NXOR2	two input equal	no
XOR2	two input exor	no
MUX	two input, one output, one select multiplexer	no
DMUX	one input two output one select demultiplexer	no
DFF	D-flipflop	no
RSFFR	resetable RS-flipflop	no
nMUX	generic multiplexer	yes
nREG	generic register	yes
HA	half adder	no
FA	full adder	no

Table 2-3: Elements of the Base Module Library

2.3.5 Graphical Notation

The schematic diagrams consist of modules which are drawn according to the notation given in table 2-4.

Module type	Graphical Notation
Inverterc	
AND gates	
EXOR gate	
NAND gates	
NOR gates	
OR gates	
Demultiplexer 1:2	
Register	
Arithmetic Logic Unit	
RS-Flipflop	
Register with n Flipflops	

Table 2-4: Schematic Drawing Symbols

2.4 Classification of the Verification Tasks and the Circuits

2.4.1 Verification Tasks

There are mainly three verification tasks to be distinguished when talking about hardware verification:

1. verifying that a circuit specification is what it should be,

2. verifying that a given implementation behaves identically to a given specification and

3. verifying important (e.g. safety critical) properties of a given implementation.

According to [6], the first is called *requirements capture*, the second *implementation verification* and the third *design verification*.

The three tasks are often expressed in terms of a specification S and an implementation I, where a *complete* verification denotes some form of equivalence between S and I ($S = I$, $S \approx I$, $S \Leftrightarrow I$) and a *partial* verification denotes some form of implication ($I \Rightarrow S$, $I \supseteq S$, $I \vdash S$):

The first task is a partial verification (with S describing properties of the circuit specification and I being the circuit specification), the second is a complete verification (with S being the specification and I being the implementation) and the third is again a partial verification (with S describing the properties of the circuit implementation and I being the circuit implementation).

It is to be noted that if e.g. exact computation times are not stated in a specification then we have to cope with a verification problem of the third kind, since in that case an equivalence proof is not possible.

All three verification tasks are covered by the benchmark circuits.

2.4.2 Circuit Classification

Every classification scheme has its drawbacks, but we found the following circuit properties especially useful for classification purposes.

A circuit may be classified in several dimensions as depicted in table 2-5. Most of the criteria are self-explaining, besides complexity. In the area of testing usually the number of internal lines is directly used as a complexity measure [1, 2], motivated by the designated application of the circuits: test pattern generation. Using a classical stuck-at fault model, the set of faults to be treated equals the number of internal lines. Hence a circuit s713 denotes a sequential circuit with 713 internal lines [2].

In the area of hardware verification a similar complexity measure is not as obvious (at least we did not found any meaningful). Hence we use the coarse measure, proposed by J. Staunstrup: an example is either introductory, illustrative or a real challenge [6].

The abbreviations given in table 2-5 may be used to characterize each circuit using the following "signature":

    <Name>:<Spec>-<Imp>.<Sync>.<Hier>.<Det>.<Gener>.<Type>.Compl>

A circuit GCD.a-r.s.h.d.g.m.i denotes the Greatest Common Divisor circuit, which has a specification on algorithmic level and an implementation on register-trans-

Classification Criterion	Value Set	
Abstraction level	system	(s)
	algorithmic	(a)
	register-transfer	(r)
	gate	(g)
	transistor	(t)
Synchronicity of the implementation	synchronous	(s)
	asynchronous	(a)
	combinational	(c)
Hierarchy of the implementation	hierarchical	(h)
	flat	(f)
Determinism	deterministic	(d)
	nondeterministic	(n)
Genericity	generic	(g)
	concrete with (optional) bitwidth n	(cn)
Type	controller	(c)
	data path	(d)
	mixed (both)	(m)
Complexity	introductory	(i)
	standard illustrative	(s)
	challenge	(c)

Table 2-5: Possible Circuit Classifications

fer level. It is a synchronous, hierarchical and deterministic design with arbitrary bit width. Consisting of a controller and a data path it is a small, i.e. an introductory example.

Using this classification scheme, we can characterize all circuits as shown in table 2-6.

Circuit Name	Classification
Single Pulser	Pulser.g-g.a.f.d.c1.c.i
Traffic Light Controller	TLC.r-g.s.f.d.c1.c.i
N-bit Adder	Adder.a-g.c.h.d.g.d.i
Min_Max	Min_Max.a-g.s.h.d.c8/g.d.s
Black-Jack Dealer	Dealer.a-g.a.f.n.c.c.s
Arbiter	Arbiter.r-g.a.f.n.g.c.s
Rollback Chip	Rollback.?-?.?.h.d.?.m.s

Table 2-6: Classification of the Benchmark Circuits

Circuit Name	Classification
Tamarack Processor	Tamarack.a-g.s.h.d.?.m.s
Stop-Watch	Stopwatch.r-g.s.h.d.c.m.s
GCD	GCD.a-g.s.h.d.g.m.i
Multiplier	Mult.a-g.c.h.d.g.d.s
Divider	Div.a-g.c.h.d.g.d.s
FIFO	FIFO.a-g.a.h.d.g.m.s
Assotiative Memory	Assoc.r-g.s.h.d.g.d.s
1dim Systolic Array	1Syst.a-g.s.h.d.g.d.s
2dim Systolic Array	2Syst.a-g.s.h.d.g.d.s

Table 2-6: Classification of the Benchmark Circuits

3 You and the Verification Benchmarks

3.1 How to get the benchmark circuits

The most convenient way to a get access to the benchmark suite is via the World Wide Web. Using a WWW browser like Mosaic you can get the latest informations using the URL http://goethe.ira.uka.de/benchmarks/.

You can also directly use an anonymous FTP-server. All benchmark circuits as well as PostScript versions of the documents (including this paper) have been made available there in the directory pub/benchmarks. The FTP-server is reachable as goethe.ira.uka.de (current IP address: 129.13.18.22). The server will always hold the newest benchmark version (see Section 2.2).

Simple ASCII-files end in .txt, Postscript files always end in .ps, .ps.Z or ps.gz. The files with endings .gz (.Z) has been compressed using the UNIX gzip (compress) command. They must be transferred using binary ftp mode and must be expanded using the UNIX gunzip (uncompress) command before they are readable or printable. Tar-Files end in .tar (compressed .tar.gz or .tar.Z) and contain a whole directory in one file. The directory content may be rebuild by executing tar -xfv <name>.tar after having expanded the respective file.

3.2 Physical Organization

The main directory /pub/benchmarks contains:

- README — how to use and retrieve the benchmarks
- It_is_version_x.y.z — an empty dummy file indicating the current version, release and patchlevel
- Introduction.ps — this document as a PostScipt file
- Whole_documentation.ps this document plus all (currently available) circuit descriptions as a single PostScript File
- History.txt — the version history of the benchmarks

- `Library.vhdl`   the base module library (see Section 2.3.4)
- `<circuit>`   a directory for each benchmark circuit
- `documentation`   a directory containing the whole documentation in its original FrameMaker 4 format (for those who want to get the document "sources")
- `pending`   a directory containing circuits which will be probably be included in future releases of the benchmark circuits.[1]

Each circuit directory `<circuit>` contains:

- `<circuit>.ps`   a PostScript file describing the circuit
- `<circuit>_schem.ps`   a PostScript file containing schematic diagrams of the implementations (not for all circuits)
- `WAVE<circuit>.ps`   a PostScript file containing timing diagrams of the circuit implementation (not for all circuits)[2]
- `<circuit>.vhdl`   a VHDL specification and a (structural) VHDL implementation of the circuits

3.3 Using the benchmarks

You can use the benchmark circuits in any way which suits your needs. Especially when specifying the problems, you are in no way obliged to use VHDL. The VHDL specifications are mainly provided to clarify the intended proof goals.

However, if for example, you use an implementation completely different from the ones given here (e.g. a simplified version) you should state this clearly whenever you refer to the circuits provided here.

Some of the circuits have been designed hierarchically. If you are flattening the circuits in order to verify them, you should state this also.

3.4 Please Contact us if ...

Please contact us if you have any problems, especially

- if you have questions of any kind concerning the benchmarks,
- if you have comments or proposal for changes, additions or even new circuits,
- if you can provide "better" implementations for the circuits,
- if you have problems in printing the PostScript files,
- if you detect errors, inconsistencies or ambiguities, which should be fixed or
- if you have problems in accessing the files.

---

1. For these circuits the documentation may be incomplete, inconsistent or completely missing.

2. Note that this file resulted from a simulation of the VHDL sources <circuit>.vhdl. There may be additional simulations in the documentation of the respective circuit description. However, the latter have been the result of simulating the CADENCE designs of the circuits and there may be minor differences in the waveforms with regard to timing etc.

The easiest way is to send a brief email to Thomas Kropf or to Jørgen Staunstrup (kropf@informatik.uni-karlsruhe.de, jst@id.dth.dk).

3.5 Email list

To inform people about the latest release and patches of the benchmark circuits, we do maintain an informal emailing list. If you want to be added (or deleted) from this list, send a short note to kropf@informatik.uni-karlsruhe.de.

4 Present and Future Activity

4.1 The Present

At the moment, we are busy simply with completing all proposed circuits and — probably — by making the current set consistent.

4.2 The Future

The current set of circuits falls short of asynchronous verification examples. Moreover protocol verification problems are not covered, which may also be viewed at as important sub-aspects of circuit verification. To cover lower description levels, we also would like to add some switch-level or transistor level verification examples.

There is still a lack of "challenging" verification examples, i.e. circuits which are either of significant size or which reflect "real" commercial designs. As one of these circuits, we will probably provide a large RISC-processor: the DLX of Patterson and Hennesy [8].

4.3 Acknowledgments

We like to thank the following people, who have helped in creating this benchmark set and documentation:

• Ralf Reetz	for significant ideas and contributions like the parameterized base module library,
• Hans-Peter Eich	for implementing all circuits in a commercial design system and for providing all the figures and timing diagrams,
• Ramayya Kumar	for many valuable comments,
• Klaus Schneider	for pointing out interesting verification problems,
• Jørgen Staunstrup,	for supporting this activity via IFIP and suggesting the IFIP circuits and
• the Authors of TPTP	for the excellent documentation of their theorem proving benchmarks [9], which helped us in structuring this document

5 Literature

[1]   F. Brglez, P. Pownall, and R. Hum. Accelerating ATPG and fault grading via testability analysis. In *International Symposium on Circuits and Systems*, 1985.

[2]   F. Brglez, D. Bryan, and K. Kozminski. Combinational profiles of sequential benchmark circuits. In *International Symposium on Circuits and Systems*, May 1989.

[3]   R. Vemuri, J. Roy, P. Mamtora, and N. Kumar. Benchmarks for high-level synthesis. Technical Report ECE-DDE-91-11, Laboratory for Digital Design Environments, ECE Dept., University of Cincinnati, Ohio, USA, November 1991.

[4]   J.-L. Paillet. Un Modele de Fonctions Sequentielles pour la Verification Formelle de Systemes Digitaux. Technical Report 546, IMAG-ARTEMIS, Grenoble, June 1985.

[5]   L. Claesen, editor. *International Workshop on Applied Formal Methods for VLSI Design*, Leuven, Belgium, 1990.

[6]   J. Staunstrup. IFIP WG 10.2 collection of circuit verification examples, November 1993.

[7]   CADENCE Design Framework II version 4.2a. Reference Manual, February 1992.

[8]   J.L. Hennessy and D.A. Patterson. *Computer Architecture: A quantitative Approach*. Morgan Kaufmann Publishers Inc., San Mateo, CA, USA, 1990.

[9]   C. Suttner, G. Sutcliff, and T. Yemenis. *The TPTP (Thousands of Problems for Theorem Provers) Problem Library*. TU Muenchen, Germany and James Cook University, Australia, via anonymous ftp flop.informatik.tu-muenchen.de, tptpv1.0.0 edition, 1993.

Reasoning About
Pipelines with Structural Hazards

Mark Aagaard* and Miriam Leeser

School of Electrical Engineering
Cornell University
Ithaca, NY   14853
USA

Abstract. We have developed a formal definition of correctness for pipelines that ensures that transactions terminate and satisfy a functional specification. This definition separates the correctness criteria associated with the pipelining aspects of a design from the functional relationship between input and output transactions. Using this definition, we developed and formally verified a technique that divides the verification of a pipeline into two separate tasks: proving that the pipelining circuitry meets the pipelining correctness criteria and that the datapath and control circuitry meet the functional specification. The first proof is data independent (except for pipelines that use data-dependent control). The second proof is purely combinational: there is no notion of time and each possible input transaction can be dealt with independently. In addition, we have created a framework that structures and simplifies the proof of the pipelining circuitry.

1   Introduction

The work presented here is part of a larger effort to develop systematic techniques for the specification, design, and verification of large-scale, complex pipelined circuits [AL93]. We have concentrated on incorporating design features that are used in state-of-the-art high-performance microprocessors: super-scalar instruction issue, out-of-order completion and data-dependent control. These features are found in microprocessors such as the DEC Alpha [McL93], HP PA-Risc [AAD+93], IBM RS/6000 [Mis90], and Intel Pentium [AA93] and lead to implementations that contain structural hazards.

Structural hazards increase the difficulty of design and verification, because errors related to non-termination and interference between transactions may occur. They also complicate the specification of a pipeline, because transactions may have variable latencies (the transit time through a pipeline) and may exit in a different order than they entered. In pipelines without structural hazards, verifying termination and freedom from interference is trivial. Thus, the bulk

---

* Now at Department of Computer Science, University of British Columbia, Vancouver, BC, Canada

of the verification effort is concerned with verifying a datapath in which computations occur over some interval of time. When confronting pipelines with structural hazards, even specifying the intended behavior and verifying that all transactions will terminate can be a significant challenge.

1.1 Our Work

We often reason about hierarchical pipelines, where the datapaths for the stages may themselves be pipelines. We say that pipelines are composed of *segments* and that *stages* are segments that can contain at most one transaction (*i.e.* they are at the lowest level of the structural hierarchy).

Modern pipelined microprocessors contain: an instruction pipeline that fetches, decodes, and issues instructions; and several execution pipelines (*e.g.* integer, floating-point, and load/store). All of the floating-point pipelines, and most of the instruction and integer pipelines that we have found, contain structural hazards. In addition to structural hazards, instruction pipelines have data and control hazards. If a pipeline is free from *structural conflicts*[2] then *individual transactions* are able to proceed through the pipeline correctly. Control and data hazards cause problems with *dependencies between transactions*. Our belief is that before we can discuss dependencies between transactions, we should be able to reason effectively about individual transactions. The work presented here supports reasoning about structural hazards. Our preliminary efforts to extend this work to data and control hazards appears promising: we believe that the extensions can be added by building on our work with structural hazards.

We separate the general idea of correctness into two properties: *pipelining*, every transaction that wishes to enter the pipeline is eventually able to do so and there is a one-to-one mapping between transactions that enter and exit the pipeline; and *functionality*, every output transaction has the correct value. This definition of correctness allows us to separate the verification of a pipeline into a pipelining and a functionality proof. The pipelining proof is data-independent for almost all pipelines and the functionality proof is purely combinational.

The pipelining correctness criteria are the same for every pipeline. Because of this, even for pipelines which appear to be quite different, there is a great deal of similarity in how we verify that there are no structural conflicts. We have organized these similarities into a framework that captures reasoning common to these proofs. Using the framework allows the bulk of the verification of an individual pipeline to concentrate on the datapath.

The framework defines four parameters that are instantiated for individual pipelines: *Protocol* schemes describe how transactions are transferred between stages in the pipeline; *Arbitration* schemes specify how to handle collisions in the pipeline; *Control* schemes determine how transactions are routed through the pipeline and how stages know what operation to perform; and *Ordering* schemes

---

[2] Following the terminology of Tahar [TK94], which is based on that of Milutinovic [Mil89], we say that a pipeline is free from structural conflicts if it handles its structural hazards correctly or has no structural hazards.

describe a method for matching up a transaction as it leaves the pipeline with the corresponding input transaction.

We have derived specifications for the parameters and proved that any pipeline that meets the specifications is guaranteed to meet our pipelining correctness criteria. This proof was done on paper using a strongly-typed higher-order logic in a style that is compatible with interactive tactic-oriented proof development systems such as Nuprl, HOL, PVS, and Coq. Our syntax is based most closely on that used in Nuprl and SML. Using the framework divides the proof that a pipeline is free from structural conflicts into separate and largely independent tasks. The specifications for the parameters are very general, so as to allow for innovative design solutions and evolving technology.

In order to reason about pipelines in general and then apply the results to specific pipelines, we developed a model of a pipeline that is composed of generic segments. We defined a set of virtual functional units and signals in a generic segment and then derived behavioral specifications for these units. Using the framework to verify a pipeline is done by defining the correspondence between the hardware in the real pipeline and the virtual functional units and then proving that the circuits in the real pipeline meet the specifications for the corresponding virtual functional units.

We say the functional units and signals are "virtual," because a given pipeline may not implement all of the units in hardware. For example, a uniform-uni-functional pipeline (all transactions follow the same path and each segment can perform only one operation) does not need any control hardware. The verification of such a pipeline would define the virtual functional units for control to be constants that produce outputs of type *unit*, a type that has only one member.

1.2   Related Work

Previous research on verifying pipelined microprocessors using proof-development systems includes Srivas's verification of the Mini-Cayuga using Clio [SB90] and Saxe's verification of a very small microprocessor using the Larch Proof Assistant [SGGH92]. Saxe's example has also been done by Alex Bronstein using the Boyer-Moore theorem prover Nqthm [BT90]. Lessons learned from these efforts include Srivas's and Bronstein's use of equivalence mappings between the behavior of pipelined and unpipelined microprocessors.

Windley has developed a methodology for verifying microprocessors based on generic interpreters [Win91] and is extending this methodology to handle pipelined designs [WC94]. This methodology decomposes verifying a pipeline into a series of abstraction mappings between the gate level and the programmer's model. Incorporating pipelining into generic interpreters provides a systematic approach for specifying and verifying pipelined microprocessors, but complicates the abstraction mappings.

Tahar is using HOL to verify Hennessy and Patterson's academic microprocessor, the DLX [TK93]. As in the work presented in this paper, Tahar also separates the verification effort into pipelining and functionality concerns, but we have a stronger definition of correctness and have proved that the pipelining and functionality concerns can be separated.

Model checking systems have also been used to verify pipelines. Beatty has developed the theory of "marked strings", which provides a basis for the automated verification of circuits, including pipelines [Bea93]. Seger has used the Voss system to verify an integer pipeline that is reflective of those in a modern RISC microprocessor [Seg93]. The limiting factor in using model-checking techniques to verify a pipeline is the size of the state-space. To mitigate this, Burch [BD94] has separated the pipelining concerns from the functionality concerns. His work provides a highly automated approach for verifying pipelining circuitry, but he does not reason about composing pipelining and functionality correctness results.

A number of the efforts listed here have dealt with control and/or data hazards, but none of the pipelines have significant structural hazards. The only other formal work we are aware of that deals with realistic structural hazards is Harcourt's use of SCCS to formally specify microprocessors and derive instruction schedulers [HMC94]. This work is complementary to ours, in that Harcourt works upward from formal specifications of microprocessors to schedulers and we verify implementations against specifications.

1.3 Outline of Paper

In Section 2 we discuss our formal definition of correctness for pipelines. Section 3 describes how we use the framework to verify pipelines. Sections 4 and 5 include an introduction to the four parameters to the framework, their specifications, and a set of commonly used instantiations of the parameters. Section 6 contains an overview of how the framework can be used to characterize and verify the floating-point pipeline of the Adirondack, a fictitious super-scalar RISC microprocessor.

2 Correctness

We have defined what it means for a pipeline to be free from structural conflicts and to meet a functional specification. This definition ensures that every transaction that wishes to enter the pipeline is able to do so, every transaction that enters the pipeline eventually exits, and every transaction that exits the pipeline has the correct data. This definition of correctness is captures in the predicate *satisfies*, shown in Definition 1 and displayed using the symbol ⊐.

Definition 1: *Pipeline satisfies a specification*
  (Pipe, Match, Constraints) ⊐ *Spec* $\hat{=}$
    $\forall I, O$. *(Pipe I O)* & *(Constraints I O)* & *(envOk I O)* $\implies$
    PipeliningOk Match I O &
    $\forall t_i, t_o$. *Match I O t_i t_o* $\implies$ *Spec (I t_i) (O t_o)*

We parameterize *satisfies* by the implementation of a pipeline (*Pipe*), a relation for matching corresponding input and output transactions (*Match*), a possible set of constraints on the environment (*Constraints*), and a functional specification (*Spec*). The first part of the definition uses *PipeliningOk* (Definition 2) to ensure that the pipelining aspects of correctness are met. This definition is the same for every pipeline, which, as shown in Section 3.2, allows us to greatly simplify the pipelining part of the verification. The second part ensures that the functionality aspects of correctness are met. The correctness criteria for functionality says that when *Match* finds a matching input and output time (t_i and t_o), then the input transaction at t_i and the output transaction at t_o must satisfy the functional specification (*Spec*).

We parameterize *satisfies* by *Match*, because the temporal relationship between input and output transactions varies from pipeline to pipeline. Some pipelines work correctly only if their environment obeys some constraints, such as being always able to receive output transactions. We support this by parameterizing *satisfies* with *Constraints*. A functional specification (*Spec*) is a relation over an input transaction and an output transaction. It defines the computation that the pipeline is meant to perform. It is purely combinational and is not concerned with pipelining aspects of correctness.

In the definition of *satisfies*, the pipeline (*Pipe*) interacts with an environment through an input stream (*I*) and an output stream (*O*). The predicate *envOk* ensures that environment conforms to several requirements, such as every transaction that wants to exit the pipeline is eventually able to do so.

A pipeline is free from structural conflicts if: every transaction that wants to enter the pipeline is eventually able to do so, every transaction that enters eventually exits, and every transaction that exits the pipeline has entered. These criteria are captured in *PipeliningOk* (Definition 2). The relation *canEnter* guarantees that transactions that wish to enter the pipeline are eventually able to do so. In the second clause of *PipeliningOk* we use the matching relation for the pipeline (*Match*) to simplify the second and third correctness criteria to: *Match* defines a one-to-one mapping between input and output transactions (*isOneToOne*. These properties guarantee that the pipeline is free from deadlock and livelock and that transactions are not created inside the pipeline.

Definition 2: *Pipelining correctness criteria*
  PipeliningOk Match I O $\hat{=}$
      (*canEnter I O*) & (*isOneToOne Match I O*)

3 Verification of Pipelines

In this section we introduce our techniques for formally verifying pipelines using *satisfies*. The process is the same for all pipelines, so we illustrate it with a canonical pipeline (*Pipe*), matching relation (*Match*), constraints (*Constraints*), and specification (*Spec*). We begin with the proof goal that the pipeline, matching relation, and constraints satisfy the specification (Equation 1).

$$(Pipe, \ Match, \ Constraints) \ \sqsupset \ Spec \tag{1}$$

In a naive proof of Equation 1, we would unfold *satisfies* to produce two goals: one to show that the pipeline is free of structural conflicts and one to show that the pipeline and matching relation imply the functional specification (Figure 1). These two goals separate the *pipelining* and *functionality* aspects of the verification, but the implementation of the pipeline (*Pipe*) appears in the functionality goal. This means that the functionality proof must deal with pipelining concerns, such as potential collisions and out-of-order termination.

$\vdash$ (*Pipe, Match, Constraints*) $\sqsupset$ *Spec*
BY Unfold *satisfies* (* Definition 1 *)

- (*Pipe I O*) & (*Constraints I O*) & (*envOk I O*)
  $\vdash$ *PipeliningOk Match I O*

- (*Pipe I O*) & (*Constraints I O*) & (*envOk I O*)
  $\vdash$ *Match I O t_i t_o* $\implies$ *Spec* (*I t_i*) (*O t_o*)

Fig. 1. Direct verification of canonical example; unfold *satisfies*

Rather than following the naive approach just described, we have proved two theorems that completely separate the pipelining and functionality proofs and then greatly simplify the pipelining proof. In Section 3.1 we introduce the *combinational representation*, which leaves us with a functionality proof that is purely combinational and a pipelining proof that is almost always data-independent. In Section 3.2 we show how to simplify the pipelining proof using the specifications for the four parameters to the framework (protocol, arbitration, control, and ordering).

3.1 Using A Combinational Representation To Verify A Pipeline

To remove all aspects of pipelining from the functionality proof we use Theorem 1 to introduce a *combinational-logic representation* (*Comb*) of the pipeline. A combinational representation captures the input/output functionality but not the pipelining aspects of the implementation. This is because it reflects the data-path and control circuitry of the implementation, but not the pipelining circuitry. As an abbreviation, we package the implementation of our example pipeline (*Pipe*), matching relation (*Match*), and constraints (*Constraints*) as *PipeRecord*

When using Theorem 1 to verify a pipeline, we introduce the combinational representation and have the two goals shown in Figure 2, rather than those shown in Figure 1. The first goal, which we refer to as the pipelining goal, is to

Theorem 1: *Transitivity of satisfies*
⊢ ∀*PipeRecord, Comb, Spec .*
 PipeRecord ⊐ *Comb &*
 (∀T_i, T_o . *Comb* T_i T_o ⟹ *Spec* T_i T_o) ⟹
 PipeRecord ⊐ *Spec*

prove that the pipeline *satisfies* the combinational representation. This proof is much easier than showing that the pipeline *satisfies* the high-level specification (*Spec*), as in Figure 1. The combinational representation is closely related to the pipeline and the proof is data-independent (except for pipelines that use data-dependent control). In contrast, specifications are generally unrelated to the structure of the pipeline and reasoning about them is highly data-dependent. The second goal in Figure 2, which we refer to as the functionality goal, is to show that the combinational representation implies the specification.[3] This goal is purely combinational and can be solved using standard hardware verification techniques.

 ⊢ *PipeRecord* ⊐ *Spec*
BY Theorem 1

  ⊢ *PipeRecord* ⊐ *Comb* (* pipelining *)

  ⊢ ∀T_i, T_o . *Comb* T_i T_o ⟹ *Spec* T_i T_o (* functionality *)

Fig. 2. Verification of canonical example; introduce combinational representation

3.2   Using the Framework to Verify a Pipeline

In this section we introduce the framework to simplify the pipelining proof (the first goal in Figure 2). We use Theorem 2 to divide the proof that a pipeline *satisfies* its combinational representation into four subgoals (Figure 3): the core (datapath) for each segment is valid (*CoresOk*), the constraints of the segments are met (*ConstraintsOk*), the combinational representation is an accurate representation of the pipeline (*CombOk*), and the pipelining circuitry that glues the segments together meets the specifications of the framework parameters (*Fwork-ParamsOk*).

 Because we work with hierarchical pipelines, the core of a segment may itself be a pipeline. The core of each segment has an associated matching relation and combinational representation. The definition of *CoresOk* says that the core of

---

[3] We use the convention that capital Ts (*e.g.* T_i and T_o) are for transactions and lowercase Ts (*e.g.* t_i and t_o) are for times.

Theorem 2: *Pipelines that meet framework specifications satisfy their combinational representations*
$\vdash \forall$ *Pipe, Match, Constraints, Comb .*
  (CoresOk Pipe) &
  (ConstraintsOk Pipe Constraints) &
  (CombOk Pipe Comb) &
  (FworkParamsOk (Pipe, Match, Constraints)) $\implies$
   (Pipe, Match, Constraints) $\sqsupset$ *Comb*

 $\vdash$ *(Pipe, Match, Constraints)* $\sqsupset$ *Comb*
BY Theorem 2

  $\vdash$ *CoresOk Pipe*

  $\vdash$ *ConstraintsOk Pipe Constraints*

  $\vdash$ *CombOk Pipe Comb*

  $\vdash$ *FworkParamsOk (Pipe, Match, Constraints)*

Fig. 3. Verification of canonical example; use framework to show that pipeline satisfies combinational representation

each segment *satisfies* its combinational representation, the data in an output transaction does not change while waiting for its request to be accepted, and the combinational representation of the core is functional (has equal outputs for equal inputs).

 The second goal in Figure 3 is to prove that the constraints for each segment are met. This condition arises because in *satisfies* a pipeline may put constraints on its environment and the core of a segment may be a pipeline. Thus, using a segment in a pipeline requires showing that the pipeline meets the constraints of the segment. In the third goal, the relation *CombOk* relates the combinational representation *(Comb)* to *Pipe*. It requires that *Comb* is equivalent to composing the combinational representations of the segments. For the fourth goal, the framework defines four parameters (protocol, arbitration, control, and ordering) that characterize pipelining circuitry. Each of these parameters has an associated set of specifications and *FworkParamsOk* says that the pipeline meets these specifications.

3.3 Summary of Verifying a Pipeline

When verifying a pipeline we use Theorem 1 to introduce a combinational representation and separate the pipelining and functionality aspects of verification. We then use Theorem 2 to simplify the pipelining proof. This leaves us with the five proof obligations listed in Table 1.

Table 1. Proof obligations when verifying a pipeline

CoresOk PipeRecord
    Datapaths and control circuitry of segments are valid
ConstraintsOk PipeRecord
    Constraints of segments are met
CombOk Pipe Comb
    Combinational representation accurately represents pipeline
FworkParamsOk PipeRecord Comb
    Pipelining circuitry meets the specifications of the framework parameters
$\forall\ T_i,\ T_o\ .\ Comb\ T_i\ T_o\ \implies\ Spec\ T_i\ T_o$
    Combinational representation implies specification

For a stage (a segment that can contain at most one transaction), the pipelining circuitry is so simple that it is almost always trivial to prove that it conforms to *CoresOk*. For a segment, we use a theorem (not shown, because it is almost identical to Theorem 2) that says that if a pipeline meets the four antecedents of Theorem 2, then the pipeline meets all of the requirements in *CoresOk*. This means that the pipeline can be used as the core of a segment in a larger pipeline. Thus, the first goal is solved in the normal progression through the structural hierarchy of a large pipeline (*e.g.* a microprocessor composed of instruction, integer, floating-point, and load/store pipelines).

The second obligation, that the constraints of the segments are met, is usually quite easy to solve. Many segments do not put any constraints on their environment. Every constraint that we have found affects only one segment or the environment, and the constraints are direct consequences of the behavior of the segment or environment.

The third goal in Table 1 is to prove that the combinational representation of the pipeline is equivalent to the composition of the combinational representations of the segments. We systematically build the combinational representation of a pipeline by composing the combinational representations of the segments according to the paths that transactions follow. Thus, just as with the first goal, the third goal is solved simply by following our standard techniques.

The fourth goal is to prove that the specifications of the framework parameters are met. The informal specifications of the parameters are given in Section 4. In addition, we have found a set of instantiations of the parameters that is sufficient to characterize the pipelining circuitry of many microprocessors (Section 5). These instantiations guide the proofs that a pipeline meets the specifications of the parameters.

The fifth goal is for the functionality proof. It requires showing that the combinational representation of the pipeline implies the specification. This goal is purely combinational and can be solved without any pipeline-related reasoning.

4 Specifications for Framework Parameters

The specifications for a protocol scheme ensure that when two segments agree to transfer a transaction, the transaction is transferred correctly. In the arbitration specifications we check that when one segment wants to transfer a transaction to another, the second segment eventually agrees to receive the transaction. By combining the protocol and arbitration specifications, we prove that transactions make progress and flow through the pipeline. The control specifications require that all paths through the pipeline lead to an exit and that transactions follow the same path through the implementation and the combinational representation. From this we know that transactions will traverse the correct path and then exit. The ordering specification says that when a transaction exits, we can match it up with its corresponding input transaction. This allows us to check that corresponding input and output transactions meet the functional specification. There are a total of eighteen specifications, of which only six require significant effort to prove for most pipelines. To give a flavor of the specifications, we describe them textually and show several lemmas that were proved using the specifications and the requirements for cores of segments. At the end of this section, Table 2 summarizes the specifications.

4.1 Protocol

The purpose of a protocol scheme is to move transactions from one segment to the next in a pipeline. In order to transfer a transaction between segments, the segment that is to receive the transaction needs to know that the other segment wants to send it a transaction. Conversely, the sending segment needs to know if the receiving segment is able to receive the transaction. We have named these two properties, the desire to send and the ability to receive, "request" and "accept." We use the protocol specifications to prove Lemma 1, which says that at time t, a transfer must occur from segment s_0 to s_1 if s_0 sends a request to s_1 and s_1 accepts the request.

Lemma 1: *Correctness of transfers between segments*
$$\vdash \forall s_0,\, s_1,\, t \,.\, (req\,(s_0,\, s_1)\, t)\, \&\, (acc\,(s_0,\, s_1)\, t) \implies xfr\,(s_0,\, s_1)\, t$$

4.2 Arbitration

The specifications for an arbitration scheme are concerned with ensuring that every request is eventually accepted, as shown in Lemma 2. The predicate *holdUntil* is used to say that if $req\,(s_0,\, s_1)$ is true at time t_r, then there is exactly one time t_a such that $req\,(s_0,\, s_1)$ is true from t_r to t_a and t_a is the first time after t_r that $acc\,(s_0,\, s_1)$ is true.

Lemma 2: *Every request is eventually accepted*

$\vdash \forall s_0, s_1, t_r . \exists t_a . holdUntil (req (s_0, s_1)) t_r (acc (s_0, s_1)) t_a$

We separate the arbitration specifications into two parts: the highest priority request to a segment is eventually accepted and every request to a segment becomes the highest priority request. These two properties are related to deadlock and livelock respectively. If the highest priority request to each segment is always accepted, then the pipeline can not deadlock. Adding the requirement that every request becomes the highest priority request ensures that every request is eventually accepted, and hence livelock is prevented.

4.3   Control

A control scheme determines how a segment decides what operation to perform on each transaction and where to send the transaction when it is done in the segment. The control specifications allow us to prove that every path through a pipeline leads to an exit and that transactions follow the same path through the implementation and combinational representation of a pipeline. We require that every path through a pipeline leads to an exit, because without this requirement it would be valid for a pipeline to contain paths that loop through the same set of segments an infinite number of times. Such paths would not produce any output transactions, despite the fact that the pipeline is working "correctly."

Lemma 3: *Transactions transfer to the correct next segment*

$\vdash \forall s_0, s_1, t_0, t_1, T .$
$\quad\quad match\ s_0\ (t_0,\ t_1)\ \&$
$\quad\quad segComb\ s_0\ (transP\ s_0\ t_0)\ (T,\ s_1) \implies$
$\quad\quad\quad xfr\ (s_0,\ s_1)\ t_1$

Lemma 3 says for each step in a path, a transaction will transfer to the same next segment and its combinational representation. Formally, the lemma says that if a transaction transfers into segment s_0 at time t_0 and the combinational representation of s_0 produces a transaction T and selects s_1 as the next segment, then in the implementation the transaction will transfer from s_0 to s_1. We relate the combinational representation of s_0 to its implementation by using the input transaction to s_0 at t_0 ($transP\ s_0\ t_0$) as the input transaction to the combinational representation and using the matching relation for s_0 to detect when the transaction exits from s_0.

4.4   Ordering

Ordering schemes are used to verify that output transactions contain the correct results. To do this, we match up an output transaction with the input transaction that caused it. For example, if a transaction contains an add instruction, we need to check that the data in the output transaction is the sum of the two operands in the input transaction. Each pipeline defines a matching relation (*Match*), which takes two times (t_0) and (t_n) and returns true if the input transaction at t_0 results in the output transaction at t_n.

The relation *isOneToOne*, which is part of the pipelining correctness criteria, says that a matching relation defines a one-to-one mapping between the input and output transactions of a pipeline. The framework uses the ordering specification to prove that a matching relation defines a one-to-one mapping between input and output transactions and to prove that if the matching relation matches an input and output transaction then they satisfy the combinational representation.

Table 2. Summary of major specifications

Protocol	Transactions are transferred between segments correctly
Arbitration	The highest priority request to each segment is eventually accepted
	Every request to a segment becomes the highest priority request
Control	All paths through the pipeline lead to an exit
	The implementations of the control circuitry imply their specifications
Ordering	The matching relation for the pipeline finds corresponding input/output transactions

5 Common Instantiations of Framework Parameters

We have found a number of commonly used instantiations of the framework parameters. These instantiations are sufficient to characterize the instruction, integer, and floating-point pipelines of the DEC Alpha 21064, HP PA-Risc 7100, IBM RS/6000, Intel Pentium, MIPS R4000, Motorola 88110, and PowerPC 601. At the end of this section, Table 3 summarizes the instantiations.

5.1 Protocol

For some pipelines, we can guarantee that every request will be immediately accepted. These pipelines are free of structural hazards and use a *transit* protocol scheme, so called because transactions can "transit" through the pipeline. In the second scheme, called *general* because it matches the general specification of protocol correctness, transactions are allowed to proceed until just before a collision. When a transaction cannot proceed any further without colliding, it stalls until the potential collision is cleared.

5.2 Arbitration

In a *degenerate* arbitration scheme segments receive input transactions from only one segment. If a segment is connected to multiple input segments, but we can show that only request will be active at a time, then we have an *exclusive* arbitration scheme. The name is derived from the observation that each request is guaranteed to be the exclusive request to the segment. In this arbitration scheme the segment simply selects whichever request is active.

For pipelines that allow multiple simultaneous requests to the same segment, we need to provide a method for prioritizing the requests. In all pipelines that we have found, we can assign *static* priorities to requests based upon the segment that the request comes from.

5.3 Control

If all transactions use each segment in a pipeline at most once, and every segment sends transactions to another segment or to an exit port, then we know that all paths are finite and reach an exit. We refer to this as a *no-loops* control scheme. The only control schemes that are non-trivial to verify are those in which transactions may use segments multiple times. We need to prove that none of these loops can be repeated an infinite number of times. In all of the pipelines that we surveyed, there is fixed upper bound on the number of iterations that transactions can pass through each loop. So, in practice, control schemes are straightforward to verify.

5.4 Ordering

The ordering schemes described here are listed in order of increasing generality. In most cases, it is easiest to use the most specific ordering scheme that is applicable to a pipeline. The simplest way to match input and output transactions is if the pipeline has a *uniform latency*. In these pipelines all transactions will exit the pipeline a given number of cycles after they enter the pipeline. The next simplest case is pipelines where transactions exit in the same order that they enter the pipeline. This is an *in-order* scheme.

In all pipelines that we have found where transactions may exit out of order, we can assign tags to transactions in such a way that there is only one transaction with a given tag in the pipeline at a time (*tags-unique*) or transactions with the same tag exit in-order (*tags-in-order*). In the pipelines that we surveyed, either the opcode of the transaction or the destination of the result of the operation (*e.g.* a register or memory address) is used as a tag.

Table 3. Instantiations of framework parameters for commercial microprocessors

		Instruction							Integer							Floating Point						
		DECchip 21064	HP PA-Risc 7100	IBM RS/6000	Intel Pentium	MIPS R4000	M88110	PowerPC 601	DECchip 21064	HP PA-Risc 7100	IBM RS/6000	Intel Pentium	MIPS R4000	M88110	PowerPC 601	DECchip 21064	HP PA-Risc 7100	IBM RS/6000	Intel Pentium	MIPS R4000	M88110	PowerPC 601
Proto	Transit					●	●	●	●	●	●	●	●	●	●	●	●	●	●	●	●	●
	General	●	●	●	●	●	●	●			●												
Arb	Degenerate		●		●	●			●		●		●					●					
	Exclusive			●					●	●		●		●		●	●	●	●		●	●	●
	Static	●		●																			
Ctrl	No loops	●	●	●	●	●	●	●	●		●		●				●						
	Loops									●		●		●	●	●		●	●	●		●	●
Ordering	In order		●		●				●		●		●	●	●						●		●
	Tags: Unique								●		●		●			●	●		●	●			
	Tags: In order	●		●	●		●	●															

6 Floating Point Pipeline

This section describes the floating-point pipeline from a fictitious RISC super-scalar microprocessor, the Adirondack (ADK). The ADK is based primarily on the DEC Alpha 21064 [McL93], except for the the floating-point pipeline, which is based on the VAX 8600 Model 200 [BBC+89]. We show how the pipeline instantiates the framework parameters and discuss how the framework can be used to guide the verification of the pipeline. Because we have concentrated on developing general techniques for complex pipelines and not on verifying a specific circuit, we do not yet have rigorous proofs for this example.

The floating-point pipeline in the ADK consists of five main stages (AddRecode, MultShift, AddNorm, AddRound, and WriteBack) plus the Divide stage. Most transactions go through each of the main stages once and go through the stages in the same order. The two classes of transactions that are exceptions are: multiplies, which use the MultShift stage twice, and divides, which use the Divide stage between eighteen and fifty-four cycles and skip the MultShift stage.

6.1 Instantiation of Framework Parameters

This pipeline uses a transit protocol scheme. In a transit protocol scheme, when a transaction enters a pipeline it is guaranteed not to encounter any collisions. This pipeline contains a circuit at the entrance that only accepts requests from the environment if it can guarantee that they will not collide with any transaction already in the pipeline. The circuit knows what stages will be available in the future. Davidson's shift-register based circuit [Dav71], which is described in

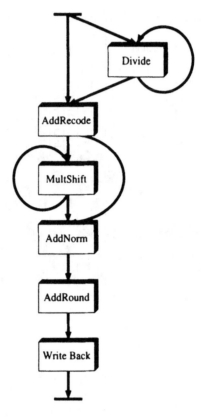

Fig. 4. Block diagram of floating-point pipeline of Adirondack microprocessor

many computer architecture textbooks, is the standard implementation for this protocol scheme.

Because divide transactions have such variable latencies, it would be very inefficient to prevent a transaction from entering the pipeline whenever there is a chance that it may collide with a divide transaction. The Divide stage sends a request to the protocol unit just before a transaction is ready to transfer from the Divide stage to the AddRecode stage. These requests are always accepted and the protocol unit updates its state to ensure that it does not accept any incoming transactions that will collide with the divide transaction.

Because the pipeline uses a transit protocol scheme, we know that there will never be more than one active request per stage, thus the pipeline uses an exclusive arbitration scheme.

For the control scheme, we need to ensure that the control logic will eventually send all transactions to the exit of the pipeline. Most transactions use each of the main stages in the pipeline once, and so follow the no-loops scheme, but multiplies and divides each contain loops, so in the proof we must show that these loops terminate.

The microprocessor *locks* the destination register for each transaction when its operands are fetched in the instruction pipeline and then unlocks the register when the transaction writes its results to the register file. This is an implementation of a tags-unique ordering scheme, in that only one transaction with a given tag (destination register) is in the pipeline at a time.

6.2 Preparing the Floating-Point Pipeline for Verification

Before we verify the pipeline, we need to define the combinational-representation, specification, matching function, and constraints.

Definition 3: *Combinational representation of ADK floating-point pipeline*
  FloatComb T_i T_o $\hat{=}$
    $\exists T_d,\ T_s,\ T_1,\ T_2,\ T_s,\ T_3,\ T_4$.
      case (*opcode* T_i)
        of AddF $\Rightarrow$
          ((*AddRecode subExps* T_i T_1) & (*MultShift align* T_1 T_2)   &
          (*AddNorm add* T_2 T_3)   & (*AddRound round* T_3 T_4) &
          (*WriteBack* T_4 T_o))
        | *DivF* $\Rightarrow$
          ((*varloop* 54 *decVal Divide* T_i T_d) &
          (*AddRecode add* T_d T_s)   & (*AddNorm add* T_s T_3)   &
          (*AddRound round* T_3 T_4)   & (*WriteBack* T_4 T_o))
        | *SubF* $\Rightarrow$ ...
        | *MulF* $\Rightarrow$ ...

The combinational representation of the pipeline (*FloatComb*) is shown in Definition 3, which for conciseness only includes the addition and division operations. We begin by existentially quantifying on the internal signals in the pipeline and then do a case split on the opcode of the input transaction. Each path describes the stages that the transaction goes through and the operation that the stages perform. Addition transactions go through the five main stages, performing the *subExps* operation in the *AddRecode* stage, etc.. The path for division transactions includes a data-dependent loop with the *Divide* stage. The *varloop* function initializes the loop counter to fifty-four and decrements it a data-dependent amount (one, two or three — corresponding to the number of bits of the quotient that were calculated) each iteration, until it reaches zero. The amount is calculated by applying the function *decVal* to the transaction in the stage.

The specification for the pipeline (*FloatSpec*, Definition 4) checks that the destination register of the output transaction is the same as that for the input transaction, does a case split on the opcode of the input transaction and then checks that the result (*getResult* T_o) is correct for the input operands (*getIOps* T_i).

Definition 4: *Specification of ADK floating-point pipeline*
$FloatSpec\ T_i\ T_o \doteq$
    $getDestReg\ T_i = getDestReg\ T_o$ &
    $case\ (opcode\ T_i)$
        $of\ AddF$ $\Rightarrow$ $AddFSpec\ (getIOps\ T_i)\ (getResult\ T_o)$
        $|\ \ SubF$ $\Rightarrow$ $SubFSpec\ (getIOps\ T_i)\ (getResult\ T_o)$
        $|\ \ MulF$ $\Rightarrow$ $MulFSpec\ (getIOps\ T_i)\ (getResult\ T_o)$
        $|\ \ DivF$ $\Rightarrow$ $DivFSpec\ (getIOps\ T_i)\ (getResult\ T_o)$

The pipeline uses a tags-unique ordering scheme, where transactions are tagged with their destination register. Each ordering scheme has an associated generic relation that is used as the basis of the matching relations for pipelines. The generic matching relation for a tags-unique scheme is *tagsUnique*, shown in Definition 5. It is a relation over an input stream and an output stream of tags, requests and accepts. Tags are only significant when a transaction transfers into or out of the pipeline, so we use the request and accept signals to detect transfers into and out of the pipeline.

In Definition 7 the matching relation for the floating-point pipeline (*Float-Match*) is defined in terms of *tagsUnique* and *mkTagFloat*, which extracts the destination register for a transaction.

Definition 5: *Matching with unique active tags*
$tagsUnique\ I\ O\ t_i\ t_o \doteq$
    $let\ (TagP,\ ReqP,\ AccP) \doteq map\ split3\ I$
        $(TagN,\ ReqN,\ AccN) \doteq map\ split3\ O$
    in
        $(ReqP\ t_i\ \&\ AccP\ t_i)$ &
        $nextTimeTrue\ (\lambda t\ .\ (TagN\ t = TagP\ t_i)\ \&\ (ReqN\ t)\ \&\ (AccN\ t))\ t_i\ t_o$
    end

Definition 6: *Function to calculate tags for ADK floating-point pipeline*
$mkTagFloat\ (Trans,\ Req,\ Acc) \doteq (getDestReg\ Trans,\ Req,\ Acc)$

Definition 7: *Matching relation for ADK floating-point pipeline*
$FloatMatch\ I\ O\ t_i\ t_o \doteq$
    $tagsUnique\ (map\ mkTagFloat\ I)\ (map\ mkTagFloat\ O)\ t_i\ t_o$

To guarantee that the tags in the pipeline are unique, the pipeline imposes the constraint on the the environment that a transaction does not begin to request the pipeline if the previous transaction with the same tag has not yet exited. This is captured in *FloatConst* (Definition 8), which says that if a transaction enters the pipeline at t_i and exits at t_o, then the next time (t_i') that a transaction with the same tag requests the pipeline must be after t_o.

Definition 8: *Constraints for ADK floating-point pipeline*
$FloatConst\ (I\ as\ (TagP,\ ReqP,\ AccP))\ O \doteq$
$\quad \forall\ t_i,\ t_o\ .$
$\qquad FloatMatch\ I\ O\ t_i\ t_o\ \&$
$\qquad (nextTimeTrue$
$\qquad\quad (\lambda t\ .\ (mkTagFloat\ (I\ t)) = (mkTagFloat\ (I\ t_i))\ \&\ (ReqP\ t))$
$\qquad\quad (t_i+1)\ t_i') \implies$
$\qquad\qquad t_o < t_i'$

6.3 Verification of Floating-Point Pipeline

Following the process described in Section 3 We begin the verification of a pipeline by saying that the implementation, matching relation, and constraints satisfy the specification (Equation 2).

$$(FloatPipe,\ FloatMatch,\ FloatConst)\ \sqsupset\ FloatSpec \qquad (2)$$

We use Theorem 1 to introduce the combinational representation of the pipeline (*FloatComb*) and separate the proof into pipelining and functionality goals. We then apply Theorem 2 and simplify the pipelining proof. This leaves us with the five proof goals that were listed in Table 1: the cores of the segments are valid, the constraints of the segments are met, the combinational representation accurately represents the pipeline, the specifications of the framework parameters are met, and the combinational representation implies the specification.

Because the segments in this pipeline are just stages and we derived the combinational representation using the standard algorithm, the first and third goals are easily solved. None of the stages impose any constraints on the pipeline, so the second proof obligation from Table 1 is also trivially solved.

The fourth obligation is to prove that the framework specifications are met. The pipeline uses Davidson's shift-register circuit to maintain the current state of the pipeline and guarantee that transactions do not collide. This is a valid implementation of a transit protocol scheme and so takes care of the protocol specifications. Because we are using a transit protocol scheme, we know that transactions will always be immediately accepted, and so the arbitration specifications are met.

For the control scheme, we need to prove that all paths through the pipeline lead to an exit. Multiply transactions use the MultShift stage exactly twice, and therefore these loops terminate. For divide transactions, we begin with a finite upper bound (fifty-four) on the loop counter for the divide stage and decrement the counter each iteration until we reach zero. For the tags-unique ordering scheme, the pipeline constrains the environment such that transactions do not request to enter the pipeline if the previous transaction with the same tag has not yet exited. We combine this with a lemma that a transaction's tag does not change as it traverses the pipeline and prove that the tag of each transaction in the pipeline is unique.

The fourth and final proof obligation is to show that *FloatComb* implies *FloatSpec*. This proof is still a significant challenge, but it is far simpler than it would have been without the framework. The floating-point datapath is complex and the abstraction gap between the implementation and specification is very large, but the proof is purely combinational and does not need to reason about pipelining concerns.

7 Conclusion

The work presented here is aimed at developing general techniques for applying formal methods to pipelined circuits. We began with a formal definition of correctness for pipelines that guarantees that transactions are able to traverse through the pipeline and will exit with the correct data. This definition allows us to separate the pipelining and functionality related parts of the verification into two proofs. The pipelining proof is data-independent (except for pipelines that use data-dependent control) and the functionality proof is purely combinational. We simplified the pipelining proof by introducing a framework with four parameters (protocol, arbitration, control, and ordering) that characterize pipelines. Each parameter has an associated set of specifications that are used to verify the correctness of the pipelining circuitry. We have found a set of commonly used instantiations of the framework parameters that sufficient to characterize and the guide the verification of a number of commercial microprocessors.

Acknowledgments

We are deeply indebted to the Semiconductor Engineering Group at DEC, who have provided a great deal of useful information and feedback. Mark Aagaard is supported by a fellowship from the Digital Equipment Corporation. Miriam Leeser is supported in part by NSF National Young Investigator Award CCR-9257280.

References

[AA93]   D. Alpert and D. Avnon. Architecture of the Pentium microprocessor. *IEEE Micro*, 12:11–21, June 1993.

[AAD+93] T. Aspre, G. Averill, E. DeLano, R. Mason, B. Weiner, and J. Yetter. Performance features of the PA7100 microprocessor. *IEEE Micro*, pages 22–35, June 1993.

[AL93]   M. D. Aagaard and M. E. Leeser. A framework for specifying and designing pipelines. In *ICCD*, pages 548–551. IEEE Comp. Soc. Press, Washington D.C., October 1993.

[BBC+89] B. J. Benschneider, W. J. Bowhill, E. M. Cooper, M. N. Gavrelov, P. E. Gronowski, V. K. Maheshwari, V. Peng, J. D. Pickholtz, and S. Samudrala. A pipelined 50-MHz CMOS 64-bit floating-point arithmetic processor. *IEEE Jour. of Solid-State Circuits*, 24(5):1317–1323, October 1989.

[BD94]   J. R. Burch and D. L. Dill. Automatic verification of pipelined micropro-
           cessor control. In *CAV*, July 1994.

[Bea93]   D. L. Beatty. *A methodology for formal hardware verification, with appli-
           cation to microprocessors.* PhD thesis, Computer Science Dept, Carnegie
           Mellon Univeristy, 1993.

[BT90]   A. Bronstein and C. L. Talcott. Formal verification of pipelines. In L. J. M.
           Claesen, editor, *Formal VLSI Specification and Synthesis*, pages 349-366.
           Elsevier, 1990.

[Dav71]   E. Davidson. The design and control of pipelined function generators. In
           *Proceedings 1971 International IEEE Conference on Systems, Networks and
           Computers*, pages 19-21, January 1971.

[HMC94]   E. Harcourt, J. Mauney, and T. Cook. From processor timing specifications
           to static instruction scheduling. In *Static Analysis Symposium*, September
           1994.

[McL93]   E. McLellan. Alpha AXP architecture and 21064 processor. *IEEE Micro*,
           13(3):36-47, June 1993.

[Mil89]   V. Milutinovic. *High Level Language Computer Architecture.* Comp. Sci.
           Press Inc., 1989.

[Mis90]   M. Misra. *IBM RISC System/6000 Technology.* IBM, 1990.

[SB90]   M. Srivas and M. Bickford. Formal verification of a pipelined microproces-
           sor. *IEEE Software*, pages 52-64, September 1990.

[Seg93]   C.-J. Seger. Voss — A formal hardware verification system user's guide.
           Technical Report 93-45, Dept. of Comp. Sci, Univ. of British Columbia,
           1993.

[SGGH92]   J. B. Saxe, S. J. Garland, J. V. Guttag, and J. J. Horning. Using trans-
           formations and verification in circuit design. In *Designing Correct Circuits,
           Lyngby 1992*, 1992.

[TK93]   S. Tahar and R. Kumar. Implementing a methodology for formally verifying
           RISC processors in HOL. In J. Joyce and C. Seger, editors, *Higher Order
           Logic Theorem Proving and Its Applications*, pages 283-296, August 1993.

[TK94]   S. Tahar and R. Kumar. Implementation issues for verifying RISC-pipeline
           conflicts in HOL. In J. Camelleri and T. Melham, editors, *Higher Order
           Logic Theorem Proving and Its Applications*, August 1994.

[WC94]   P. J. Windley and M. Coe. A correctness model for pipelined microproces-
           sors. In *Theorem Provers in Circuit Design*. Springer Verlag; New York,
           1994.

[Win91]   P. J. Windley. The practical verification of microprocessor designs. In *IEEE
           COMPCON*, pages 462-467. IEEE Comp. Soc. Press, Washington D.C.,
           February 1991.

A Correctness Model for Pipelined Microprocessors

Phillip J. Windley[1] and Michael L. Coe[2]

[1] Laboratory for Applied Logic, Brigham Young University,Provo, UT 84602-6576
[2] Laboratory for Applied Logic, University of Idaho, Moscow, ID 84843-1010

Abstract. What does it mean for an instruction pipeline to be correct? We recently completed the specification and verification of a pipelined microprocessor called UINTA. Our proof makes no simplifying assumptions about data and control hazards. This paper presents the specification, describes the verification, and discusses the effect of pipelining on the correctness model. The most significant effect on the pipeline is that data and temporal abstractions in the correctness model are *not* orthogonal as they are in non-pipelined implementations.

1 Introduction

Much has been written over the years regarding the formal specification and verification of microprocessors. Most of these efforts have been directed at non-pipelined microprocessors. See [Gor83, Bow87, CCLO88, Coh88, Joy88, Hun89, Win90, Her92, SWL93, Win94a] for examples.

The verification of pipelined microprocessors presents unique challenges. The correctness model is somewhat different than the standard correctness models used previously (see Section 7.1). Besides the correctness model, the concurrent operations inherent in a pipeline lead to *hazards* which must be considered in the proof. There are three types of hazards:

- **structural hazards** which arise because of resource constraints (i.e. more than one operation needing the ALU at a time),
- **data hazards** which arise when data is needed before it has been calculated or, alternately when data is changed before it has been used, and
- **control hazards** which arise when instructions change the flow of control after some operations in the original flow of control have already begun.

Several papers have presented the verification of pipelined microprocessors:

In [SB90], the verification of a three stage pipelined machine name Mini-Cayuga is presented. The verification is the first, to our knowledge, of a pipelined microprocessor. Because the pipeline has only three stages, however, the verification did not have to deal with data and control hazards in the pipeline.

The verification of a machine similar to the DLX processor of [HP90] is presented in [TK93]. The machine has a five stage pipeline and encounters data and control hazards, but it is not clear from the presentation whether these are dealt with in the proof or in the assumptions.

This paper presents the verification of a pipelined microprocessor called UINTA. UINTA has a five stage pipeline which presents data and control hazards (there are no structural hazards). Mitigation of the data hazards is done using two levels of data forwarding; mitigation of the control hazards is accomplished using a delayed branch (2 stages). Our verification makes no assumptions about software constraints or the ordering of instructions.

Our work in microprocessor verification has been characterized by the development of formal models for microprocessor correctness and a standard model of microprocessor semantics [Win93]. In [Win94a] we present the verification of a non–pipelined microprocessor using our model, which we call the generic interpreter theory. The generic interpreter theory does several things:

1. The formalization provides a step–by–step approach to microprocessor specification by enumerating the important definitions that need to be made for any microprocessor specification.
2. Using the formalization, the verification tool can derive the lemmas that need to be verified from the specification.
3. After these lemmas have been established, the verification tool can use the formalization to automatically derive the final result from the lemmas.

Using the generic interpreter theory provides a standardized model that ensures that the theorems used can be put together in standard ways and used in other places in the proof. One of the goals of the effort presented here was to evaluate the use of the generic interpreter theory in verifying pipelined processors. We will see that while the generic interpreter theory provides the same benefits for most of the verification of UINTA, its fails in one important place. This is discussed in more detail in Section 7.1.

The specification and verification of UINTA is done hierarchically to reduce the abstraction distance between successive layers. As noted in [Mel88], there are four types of abstraction: structural, behavioral, data, and temporal. Where possible, we limit the types of abstraction between any two layers. The four specification models employed in the verification are:

- **Electronic Block Model.** This model is a structural description of register transfer level. The model states how the major components such as the register file and arithmetic logic unit (ALU) are connected together.
- **Phase Model.** This model is a behavioral abstraction of the electronic block model. There is no data or temporal abstraction between the electronic block model and the phase model.
- **Pipeline Model.** This model is a temporal abstraction of the phase model. The two phases of the phase model are combined in the pipeline model. Each time unit in the pipeline model represents one execution of each stage in the pipeline.
- **Architectural Model.** This model is a data and temporal abstraction of the pipeline model. the architectural model describes the instruction set semantics and is intended to represent the assembly language programmer's

view of the microprocessor. We will say more about why we perform the data and temporal abstract concurrently in Section 7.1.

The verification of UINTA shows that the resultant specifications and theorems need not be different from those used in non–pipelined microprocessor verification, but that the correctness model and the important lemmas change considerably. We will briefly present the specifications of each level (in a slightly different order than that above) and concentrate on the parts of the verification that differ significantly from previous microprocessor verifications.

2   A Brief Introduction to HOL

To ensure the accuracy of our specifications and proofs, we developed them using a mechanical verification system. The mechanical system performs syntax and type checking of the specifications and prevents the proofs from containing logical mistakes. The HOL system was selected for this project because is has higher-order logic, generic specifications and polymorphic type constructs. These features directly affect the expressibility of the specification language. Furthermore HOL is widely available, robust, and has a growing world-wide user base. However, there is nothing our work that requires the HOL theorem proving system.

HOL is a general theorem proving system developed at the University of Cambridge [CGM87, Gor88] that is based on Church's theory of simple types, or higher-order logic [Chu40]. Similar to predicate logic in allowing quantification over variables, higher-order logic also allows quantification over predicates and functions thus permitting more general systems to be described.

For the most part, the notation of HOL is that of standard logic: $\forall$, $\exists$, $\wedge$, $\vee$, etc. have their usual meanings. There are a few constructs that deserve special attention due to their use in the remainder of the paper:

- HOL types are identified by a prefixed colon. Built–in types include :**bool** and :**num**. Function types are constructed using $\longrightarrow$. HOL is polymorphic; type variables are indicated by a type names beginning with an asterisk.
- The HOL conditional statement, written a $\rightarrow$ b | c, means "if a, then b, else c." A statement that would read "if a, then b, else if c then d else if …else e" would appear in HOL as

  a $\rightarrow$ b |
  c $\rightarrow$ d |
  ... | e

- The construct **let v1 = expr1 and v2 = expr2 and ...** in defines local variables **v1**, **v2**, etc. with values **expr1**, **expr2**, etc.simultaneously.
- Comments in HOL are enclosed in percent signs, %

3 Architectural Specification

Our intent is to present just enough of the specification of the architectural level to show that it is unchanged from the standard model and to support the discussion of the verification. Our presentation follows that of any denotational semantics: we discuss the syntax, the semantic domain, and the denotations, in that order. We conclude by showing the specification developed from the denotations using the generic interpreter theory. A more complete discussion of the use of HOL for specifying architectures is available in [Win94b].

3.1 Instruction Set Syntax

The instruction set for UINTA contains 27 instructions. The small number is not an issue since, as we show later, the verification would not change significantly with the addition of new instructions and the proof time is $O(n)$ in the size of the instruction set.

The instruction set contains instructions from most of the important classes of instructions one would find in any instruction set: ALU instructions, immediate instructions, branch instructions, jump instructions, load instructions, and store instructions. The following is the abstract syntax for part of the instruction set:

```
Instruction =
        LDI    *ri *ri *short   |
        STI    *ri *ri *short   |
        ADD    *ri *ri *ri      |
        ADDI   *ri *ri *short   |
        JMP    *word26          |
        BEQ    *ri *short       |
        NOOP                    |
        ...
```

3.2 Semantic Domain

The semantic domain is a record containing the state variables that the assembly language programmer would see. The name of the record and the name of each field is given in backquotes and the type of each field is enclosed in double quotes:

```
create_record 'State'
  ['Reg', ":*ri->*wordn";     % register file %
   'Pc',  ":*wordn";          % program counter %
   'NPc', ":*wordn";          % next program counter %
   'NNPc', ":*wordn";         % next next program counter %
   'Imem', ":*memory";        % instruction memory %
   'Dmem', ":*memory";        % data memory %
  ];;
```

The register file is modeled as a function from register indices to n-bit words, the program counters are n-bit words. **Imem** and **Dmem** are both memories. The legal operations on n-bit words and memories are specified algebraically. We do not present those specifications here. Interested readers are referred to [Win94b].

The three instances of the program counter in the semantic domain are an artifact of the delayed branches. Because delayed branches appear to the assembly language programmer, they are visible at the architectural level. We will see that in lower level of the specification hierarchy, there is only one program counter and the three program counters of the architectural level are merely temporal projections of the single program counter.

The separation of the memory into instruction and data memory is a convenience that allows us to ignore self modifying programs. Self modifying programs do not cause much concern in a non-pipelined machine, but when instructions are pipelined, an instruction in the pipeline can modify another instruction that has already been loaded and started to execute. This kind of behavior hardly seems worth the trouble it causes, so we disallow it.

3.3 Instruction Denotations

Instruction denotation can be given for classes of instructions. We call these specifications *semantic frameworks* since they specify a framework for the semantics of an entire class of instructions. They are similar to the class level specifications of [TK93]. For example, here is the semantic framework for the ALU instructions in UINTA. Notice that it is parameterized by the ALU operation to be performed, **op**:

```
1  ⊢_def ALU_FM op Rd Ra Rb s e =
2         let reg      = Reg s and
3             pc       = Pc s and
4             nextpc   = NPc s and
5             nextnextpc  = NNPc s and
6             imem     = Imem s and
7             dmem     = Dmem s in
8         let a        = INDEX_REG Ra reg and
9             b        = INDEX_REG Rb reg in
10        let result   = op (a, b) in
11        let new_reg  = UPDATE_REG Rd reg result and
12            new_pc   = nextpc and
13            new_nextpc  = nextnextpc and
14            new_nextnextpc  = inc nextnextpc in
15        (State new_reg new_pc new_nextpc new_nextnextpc imem dmem)
```

The framework is also parameterized by the destination register index, **Rd** and the source register indices, **Ra** and **Rb**. Because the function is curried, applying **ALU_FM** to an operation and the register indices like so:

```
(ALU_FM add Rd Ra Rb)
```

returns a state transition function (i.e., a function that takes a state, **s**, and environment. **e**, and returns a new state).

Lines 2–7 of the preceding definition bind local names to the contents of the fields of the state **s**. Lines 8–9 bind **a** and **b** to the contents of the register file, **reg**, at indices **Ra** and **Rb** respectively. The **op** parameter is used to calculate the result in line 10. Lines 11–14 calculate new values for those members of the state that change in this framework. For example, in line 11, a new register file is calculated by updating the old register file at location **Rd** with the result calculated in line 10. Line 15 creates the new state record that is returned as the result of the function.

We create a denotation for the instruction set by relating the instruction syntax to the semantic frameworks using the following definition:

```
⊢def (M_INST (LDI Rd Ra imm) =
           LOAD_FM Rd Ra imm)              ∧
      (M_INST (ADD Rd Ra Rb) =
           ALU_FM add Rd Ra Rb)            ∧
      (M_INST (SUB Rd Ra Rb) =
           ALU_FM sub Rd Ra Rb)            ∧
      (M_INST (ADDI Rd Ra imm) =
           ALUI_FM add Rd Ra imm)          ∧
      (M_INST (BNOT Rd Ra) =
           UNARY_FM bnot Rd Ra)            ∧
      (M_INST (JALI Rd imm) =
           JALI_FM Rd imm)                 ∧
      (M_INST (BEQ Ra imm) =
           BRA_FM eqzp Ra imm)             ∧
      ...
```

M_INST maps a valid instruction, given syntactically, to a state transition function denoting the meaning of that instruction.

3.4 Interpreter Specification

The architectural level specification is created by the generic interpreter theory from the preceding definitions:

```
⊢ Arch_Interp s e =
       (∀t.
          let k = Opcode s e in
          (s (t + 1)) = M_INST k (s t) (e t))
```

The definition, in classic form, declares that the state of the architecture, s, at time $t + 1$ is a function, **M_INST**, of the state at time t.

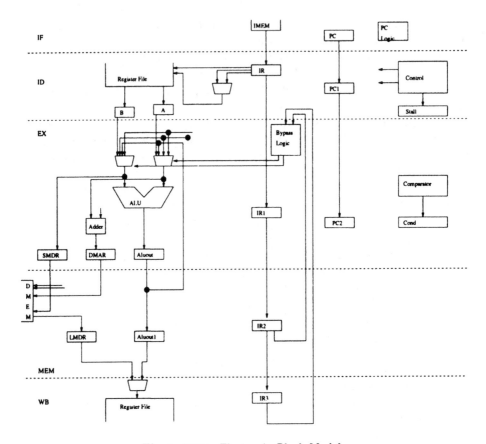

Fig. 1. UINTA Electronic Block Model

4 Electronic Block Model

The electronic block model, EBM, is a structural model of the register transfer level and is shown in Figure 1. The model describes the connections between the major components of the microprocessor. The EBM is the lowest level in the verification hierarchy. For the most part there is a recognizable correspondence between the EBM and synthesizable statements in a hardware description language such as VHDL.

The EBM state record is shown in Figure The state of the EBM is, obviously, larger than the state of the architectural level. Comparing the state record with Figure 1 shows that the EBM state record contains a field for each register and flipflop in the implementation. The EBM state record contains a field for each component of the architectural state record as well as all of the state invisible at the architectural level. Note that the next program counter, **npc**, and next next program counter, **nnpc** are not present in the electronic block model; we will discuss the disappearance of these later. The stage markers (in the comments) indicate the stage in which the register is set, not the stage in which it is used.

```
create_record 'EbmState'
   ['EbmReg',       ":*ri->*wordn";      % register file %
    'EbmPc',        ":*wordn";           % program counter %
    'EbmIMem',      ":*memory";          % instruction memory %
    'EbmDMem',      ":*memory";          % data memory %
    'EbmIr',        ":*wordn";           % instruction register, fetch %
    'EbmIr1',       ":*wordn";           % instruction register, decode %
    'EbmIr2',       ":*wordn";           % instruction register, execute %
    'EbmIr3',       ":*wordn";           % instruction register, memory %
    'EbmA',         ":*wordn";           % ALU input latch A %
    'EbmB',         ":*wordn";           % ALU input latch B %
    'EbmPc1',       ":*wordn";           % program counter, decode %
    'EbmPc2',       ":*wordn";           % program counter, execute %
    'EbmALUout',    ":*wordn";           % ALU output latch %
    'EbmALUout1',   ":*wordn";           % ALU output latch, memory %
    'EbmDMAR',      ":*wordn";           % data memory address register %
    'EbmSMDR',      ":*wordn";           % store memory data register %
    'EbmLMDR',      ":*wordn";           % load memory data register %
    'EbmCond',      ":bool";             % branch condition flipflop %
    'EbmStall',     ":bool";             % stall flipflop %
    'clk',          ":bool";             % 2 phase clock %
   ];;
```

Fig. 2. UINTA Electronic Block Model State Record

The top–level description of the EBM connects three large blocks; the control block, the clock, and the data path; together. The structure is modeled in the usual existentially quantified conjunction of predicates format. Each of the predicates is itself an existentially quantified conjunction of predicates. When fully expanded, the structural definition of UINTA is approximately four pages of text.

```
⊢_def uintaEBM s e p =
    ∃ (clk_1 clk_2 wrsig rsig newstall:time->bool).
    (CONTROL_BLOCK (EbmIr o s, EbmIr1 o s, EbmIr2 o s,
                    clk_1, EbmStall o s, newstall, rsig, wrsig))  ∧
    (CLOCK_SPEC (clk o s, clk_1, clk_2 ))  ∧
    (DATA_PATH (EbmReg o s, EbmPc o s, EbmIMem o s,
        EbmDMem o s, EbmIr o s, EbmIr1 o s, EbmIr2 o s, EbmIr3 o s,
        EbmA o s, EbmB o s, EbmPc1 o s, EbmPc2 o s, EbmALUout o s,
        EbmALUout1 o s, EbmDMAR o s, EbmSMDR o s, EbmLMDR o s,
        EbmCond o s, EbmStall o s, clk o s, clk_1, clk_2,
        wrsig ,rsig, newstall))
```

The arguments to the predicates are not just values, but signals (time dependent values). Thus, most of the arguments to the predicates are some function name, such as **EbmDMem**, composed with the EBM state stream **s**. Because the state stream, **s**, is a function with type "**:time** →**EbmState**", composing a field selector function with **s** returns a function with type "**:time** →**f**" where **f** is the type of the field. For example, **EbmDMem o s** has type "**:time** →***memory**".

5 Phase Model Specification

The phase model provides a behavioral abstraction of the EBM. The behavior of the phase model is equivalent to the EBM (i.e. there is not data or temporal abstraction). The phase model can be specified and verified from the EBM using the generic interpreter theory.

We do not give the details of the specification or verification here. Like the verification of most behavioral abstractions, the proof is quite irregular, but not technically difficult.

6 Pipeline Model Specification

The pipeline model is a temporal abstraction of the phase model. The behavior of the two phases is collapsed into one behavior for the pipeline. The data abstraction that takes place is mostly related to the temporal abstraction. Thus, the state description is largely the same for the pipeline model as for the EBM.

The specification of the state transition in the pipeline model is given as one large function since all of the changes take place concurrently. We will present the behavior by stages, but keep in mind that the stages make their state changes concurrently.

Fetch Stage. The primary state change in the fetch stage is loading the instruction register:

```
% stalls occur when ... %
let stall = STALL ir ir1 in
% fetch stage %
let new_ir = stall →   ir
              | decode_word (
                     fetch (imem, address pc)) in
```

The instruction register is unchanged in the event of a stall and gets the current instruction from memory otherwise.

Decode Stage. The primary state change in the decode stage is loading the ALU input latches, **A** and **B** from the register file:

```
% decode stage, use ir and pc %
let new_ir1 = stall →  NOOP
                    | ir and
   new_pc1 = stall →  pc1 | pc and
   new_a = INDEX_REG (sel_Ra ir) new_reg and
   new_b = ((CLASSIFY ir) = STORE) →
                   INDEX_REG (sel_Rd ir) new_reg |
                   INDEX_REG (sel_Rb ir) new_reg in
```

In the event of a stall, the decode stage instruction register, ir1 gets a NOOP rather than the instruction in ir.

Execute Stage. The execute stage can be broken into two large blocks. The first block describes the state changes associated with the ALU:

```
% execute stage, use ir1 and pc1 %
 1   let bsrc = ((CLASSIFY ir1) = STORE) →  (sel_Rd ir1) |
 2                                          (sel_Rb ir1) in
 3   let ra = (sel_Ra ir1) and
 4       imm = (sel_Imm ir1) in
 5   let amux = 5
 6        (¬(ra = R0)) →
 7          (((ra = (sel_Rd ir2)) ∧
 8           (CLASSIFY ir2 = JUMPLINK)) →  pc2     |
 9          ((ra = (sel_Rd ir2)) ∧
10           (IS_REG_WRITE ir2))          →  aluout |
11          ((ra = (sel_Rd ir3)) ∧
12           (IS_REG_WRITE ir3))          →  aluout1 |
13          ((ra = (sel_Rd ir3)) ∧
14           (CLASSIFY ir3 = LOAD))       →  lmdr   |
15                                        a)     |
16                                        a     . and
17       bmux = ...   in
18   let new_ir2 = ir1 and
19       new_pc2 = pc1 and
20       new_dmar = add (amux, shl (sel_Imm ir1, 2)) and
21       new_smdr = bmux and
22       new_aluout =
23          ((CLASSIFY ir1) = ALU)     →  ((BINOP ir1) (amux, bmux)) |
24          ((CLASSIFY ir1) = ALUI)    →  ((BINOP ir1) (amux, imm)) |
25          ((CLASSIFY ir1) = UNARY)   →  ((UNOP ir1) amux) |
26                                        ARB in
```

The new value of the aluout latch is calculated from the value given by the amux and bmux. These multiplexors supply the correct value for the respective inputs to the ALU based on the data forwarding conditions. For example, in lines 9–10, if the destination of the previous instruction in the pipe is the same as the A source for the current instruction and that instruction writes to the register file, then we supply the value in **aluout** to the **A** input to the ALU rather than the value in the **A** latch.

The second major state change in the execute stage is the calculation of the new value for the program counter:

```
% execute stage, use ir1 and pc1 %
 1  let cond = (CMPOP ir1 amux) in
 2  let new_pc = stall                              →  pc |
 3                (((CLASSIFY ir1) = BRA) ∧ cond) →
 4                     (add (pc2, shl (sel_Imm ir1, 2)))   |
 5                ((CLASSIFY ir1) = JUMPLONG)      →
 6                     (add (pc2, shl (sel_Imm26 ir1, 2))) |
 7                ((CLASSIFY ir1) = JUMPREG)       →
 8                     (add (amux, shl (sel_Imm ir1, 2)))  |
 9                ((CLASSIFY ir1) = JUMPLINK)      →
10                     (add (pc2, shl (sel_Imm ir1, 2)))   |
11                (inc pc)   in
```

Note that in the event of a stall, the program counter does not change (line 2). The new value of the program counter if the current instruction is a branch and the condition is true, for example, is the sum of **pc2** and the immediate portion of the instruction shifted left twice for word boundary alignment (lines 3-4).

Memory Stage. The memory stage calculates values for the load memory data register if the current instruction is a load or the stores a value in the data memory if the current instruction is a store:

```
let new_ir3 = ir2 and
    new_lmdr = ((CLASSIFY ir2) = LOAD) →
                     (fetch (dmem, address dmar)) |
                     lmdr and
    new_dmem = ((CLASSIFY ir2) = STORE) →
                     (store (dmem, address dmar, smdr)) |
                     dmem and
    new_aluout1 =((CLASSIFY ir2) = JUMPLINK) →  pc2 |
                                                aluout in
```

Write Back Stage. The write-back stage updates the register file if necessary:

```
let new_reg =
    (IS_REG_WRITE ir3)     →   UPDATE_REG (sel_Rd ir3) reg aluout1 |
    (CLASSIFY ir3 = LOAD)  →   UPDATE_REG (sel_Rd ir3) reg lmdr |
                     reg in
```

The New Pipeline State. The new values calculated in the preceding code fragments comprise the updated pipeline state. The state returned from the pipeline model is a new pipeline state record created be the following expression:

```
(PipelineState new_reg new_pc imem new_dmem
               new_ir new_ir1 new_ir2 new_ir3
               new_a new_b new_pc1 new_pc2
               new_aluout new_aluout1
               new_dmar new_smdr new_lmdr)
```

Verifying the Pipeline Correct. The proof that the pipeline model is correctly implemented by the phase model is done within the generic interpreter theory. There are no special considerations or exceptions.

7 Verifying UINTA

In this section we will concentrate on the proof that the pipeline model implements the architectural model since this is the part of the proof that differs from previous microprocessor verifications. We first discuss why the correctness model at this level in the proof differs from our previous notion of correctness and then discuss the proof itself in three crucial areas: important lemmas, the proof tactic, and the correctness theorem.

7.1 The Correctness Model

The correctness model presented by the generic interpreter theory is based a notion of state stream abstraction. In the model, state stream u is an abstraction of state stream u' (written $u \preceq u'$) if and only if

1. each member of the range of u is a state abstraction of some member of the range of u' and
2. there is a temporal mapping from time in u to time in u'.

There are two distinct kinds of abstraction going on: the first is a data abstraction and the second is a temporal abstraction. Thus, using a state abstraction function, $\mathcal{S}$, and a temporal abstraction function, $\mathcal{F}$, we define stream abstraction as follows

$$u \preceq u' \equiv \exists (\mathcal{S} : \mathbf{S}' \to \mathbf{S}). \, \exists (\mathcal{F} : \mathbf{N} \to \mathbf{N}). \, \mathcal{S} \circ u' \circ \mathcal{F} = u$$

where $\circ$ denotes function composition. The important part of this model for our purposes is the orthogonality of the temporal and data abstractions (indicated by function composition).

This is the crux of the problem with using the generic interpreter theory for verifying pipelined microprocessors: the data and temporal abstractions in

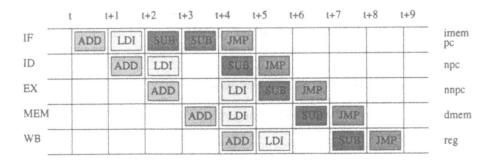

Fig. 3. Uinta Pipeline Execution

a pipeline are *not* orthogonal. Indeed, the correctness model depends on being able to mix the temporal and data abstractions.

The following discussion gives some idea of what we mean when we say a pipeline is correct and shows how the temporal and data abstractions are mixed. Suppose that we execute the following program fragment:

PC	Instruction
y	ADD x1 x2 x3
y + 1	LDI x4 x5 imm1
y + 2	SUB x6 x4 x7
y + 3	JMP z

We can picture the execution of the pipe as shown in Figure 3. On the left of the figure are the names of the pipeline stages. On the top is the time relative to the start of the execution of this code fragment. The square labeled **ADD** in the line labeled with **EX** indicates that the execute stage of the pipeline performs the state transitions for the **ADD** instruction between time $t + 2$ and $t + 3$. The labels on the right side of the figure indicate the architectural model states that are updated in the corresponding stage.

There are several points of interest regarding Figure 3:

- The **ADD** instruction updates the data memory between times $t + 3$ and $t + 4$, for example, and the register file between times $t + 4$ and $t + 5$. The state transitions indicated by the architectural model are spread out over time. We call this *skew*.
- Instruction execution can be delayed by hazards. The **SUB** instruction must wait to be executed because one of its arguments, **x4**, depends on the value being loaded by the **LDI** instruction. The value will not be ready for forwarding to the ALU in time to avoid a delay. This is called a *stall*. The hardware automatically injects a **NOOP** instruction, represented by blank cells, into the pipeline between the **LDI** and **SUB** instruction.
- The skew is not uniform in the presence of stalls. For example, the effect of the **LDI** instruction on the **nnpc** state variable does not occur between times

$t+3$ and $t+4$, as we would expect, but occurs between times $t+4$ and $t+5$. We call this *shifting*.

Because of skewing, stalling, and shifting, the data and temporal abstraction cannot be separated. Indeed, the architectural state variables, pc, npc, and nnpc are all the same variable in the pipeline model, they are merely different temporal views of the same data.

7.2 State Stream Abstraction

The discussion in the last section suggests that a function mapping the state stream of a pipeline into a state stream for a non–pipelined model must collect different pieces of the pipelined state stream at different times and package them into a state record to appear in the non-pipelined state stream at a particular time. This section describes such an abstraction function for UINTA.

The correctness model for UINTA depends on a function, called abs that maps a pipeline model state stream into an architectural model state stream. This function allows us to maintain the illusion in the architectural model that the state changes occur in a single time step between times t and $t + 1$.

```
⊢def abs s t =
    let t' = Temp_Abs (λ t. ¬(STALL (Ir (s t)) (Ir1 (s t)))) t in
    let j = (STALL (Ir (s (t' + 1))) (Ir1 (s (t' + 1)))) → 1 | 0 in
    let reg   = PipelineReg (s (t' + 4)) and
        pc    = PipelinePc (s t') and
        npc   = PipelinePc (s (t'+1)) and
        nnpc  = PipelinePc (s ((t' + 2) + j)) and
        imem  = PipelineIMem (s t') and
        dmem  = PipelineDMem (s (t' + 3)) in
    State reg pc npc nnpc imem dmem
```

The abs function is a curried function of two arguments. The first argument is the pipeline level state stream; the second argument is the architectural level time. Thus, abs ps represents the architectural level state stream that is an abstract of the pipeline level state stream ps.

In the function, the variable t' is defined using the temporal abstraction function Temp_Abs [Win94a] as the next time after t when there is not a stall. The variable j gets the value 1 if the next instruction stalls and 0 otherwise.

An architectural level state record has five components 3.2: the register file, the program counter, the next program counter, the next next program counter, the instruction memory, and the data memory.

If we examine Figure 3, we see that, for example, the register file in the architectural model at time t, is the register file in the pipeline model skewed by 4 from the next time after t there is not a stall. Thus, we define reg, the register file component of the architectural level state record as:

```
      let reg      = PipelineReg (s (t' + 4)) and
```

Other components of the architectural level state are defined similarly. Notice
that the pc, npc, and nnpc are all defined in terms of the pipeline model program counter with differing skews. The value of nnpc is shifted when the next
instruction stalls as indicated in Figure 3.

7.3 Important Lemmas

There are two important lemmas in the verification of UINTA. The first, the data
forwarding lemma, proves that the data forwarding behavior of the pipeline is
correct:

```
1  ⊢ Pipeline_Interp
2      (λ t. PipelineState (reg t) (pc t) (imem t) (dmem t) (ir t)
3                          (ir1 t) (ir2 t) (ir3 t) (a t) (b t) (pc1 t)
4                          (pc2 t) (aluout t) (aluout1 t) (dmar t)
5                          (smdr t) (lmdr t))
6      (λ t. Env(ivec t)(int t)(reset t)) (λ t. p t) ⇒
7  ∀ t.
8  let new_reg =
9      INDEX_REG x1
10         (IS_REG_WRITE(ir3(t+1)) →
11             UPDATE_REG (sel_Rd(ir3(t+1)))
12                        (reg(t+1))
13                        (aluout1(t+1)) |
14         (CLASSIFY(ir3(t+1)) = LOAD) →
15             UPDATE_REG (sel_Rd(ir3(t+1)))
16                        (reg(t+1))
17                        (lmdr(t+1)) |
18         reg(t+1)) in
19  ¬((CLASSIFY(ir1(t + 1)) = LOAD) ∧  (x1 = sel_Rd(ir1(t + 1)))) ⇒
20  ((INDEX_REG x1(reg(t+4))) =
21  ((¬(x1 = R0)) →  (((x1 = sel_Rd(ir2(t+2))) ∧
22                     (CLASSIFY(ir2(t+2))=JUMPLINK)) →  pc2(t+2) |
23                     ((x1 = sel_Rd(ir2(t+2))) ∧
24                     IS_REG_WRITE(ir2(t+2)))    →  aluout(t+2) |
25                     ((x1 = sel_Rd(ir3(t+2))) ∧
26                     IS_REG_WRITE(ir3(t+2)))    →  aluout1(t+2) |
27                     ((x1 = sel_Rd(ir3(t+2))) ∧
28                     (CLASSIFY(ir3(t+2))=LOAD)) →  lmdr(t+2) |
29                     new_reg |
30                     new_reg))
```

The theorem states that the pipeline model (lines 1–6) implies that reading, x1,
from the register file at time $t + 4$ (line 20) is equivalent to the data forwarding
behavior of the pipeline (lines 21–30) which is reading values at time $t + 2$.

The other important lemma is used to limit the case analysis in the proof. In the proof that any given instruction works, we want to be very general about the instruction sequencing. Thus we do not want to make any assumptions about whether the instructions before the current instruction stalls or not. This means that as we do the proof, we have to do a case analysis on stalling. Whenever we don't stall, we can proceed with the the symbolic execution of the instruction and complete the proof for that case, but there is always the left–over stall case to consider. The following lemma, however, limits the proof to one stall case split by showing that the pipeline cannot stall twice in a row (because the stall inserts NOOP which cannot cause a stall):

```
⊢ Pipeline_Interp
     (λ t. PipelineState (reg t) (pc t) (imem t) (dmem t) (ir t)
                         (ir1 t) (ir2 t) (ir3 t) (a t) (b t) (pc1 t)
                         (pc2 t) (aluout t) (aluout1 t) (dmar t)
                         (smdr t) (lmdr t))
     (λ t. Env(ivec t)(int t)(reset t)) (λ t. p t) ⇒
  (∀t . (STALL (ir t) (ir1 t)) ⇒
        ¬(STALL (ir (t+1)) (ir1 (t+1)))))
```

Without this lemma, the symbolic execution would lead to an infinite number of case splits on stalling. Every pipeline, to be correct, must have a similar lemma showing that it cannot issue an unlimited number of stalls. If it could, of course, the processor could go into an infinite loop of sorts under certain conditions. This is not usually the behavior one wants from a processor.

7.4   The Instruction Tactic

The proof of UINTA breaks into 27 cases, one for each of the instructions in the instruction set. [3] Each of these instructions can be solved using the same tactic. The tactic considers each instruction under the case that it stalls and the case that is does not. In either case, the method of proof is symbolic execution of the pipeline model using a general purpose symbolic execution tactic that we developed for microprocessor proofs.

A single /uinta/ instruction can be verified in less than 10 minutes on an HP 735 running HOL88 version 2.01 compiled with AKCL. Increasing the size of the instruction set thus increases the overall verification time linearly. Unless the instruction differs significantly from the instructions already in the instruction set, it is likely that the tactic used to perform the verification will not change, so the human effort to add additional instructions is probably small.

When there is an error, either in the design or the specification, the tactic fails, leaving a symbolic record of what the pipeline computed in each stage. We

---

[3] For larger instruction sets, we could have done case analysis on instruction classes (which correspond to the semantic frameworks) to limit the case explosion.

have found that this record is very helpful for debugging the implementation and specification as it is usually quite easy to see where what the pipeline computed differs from what the designer expected.

7.5 The Correctness Theorem

The overall correctness theorem (for this level in the proof hierarchy) shows that the pipeline model implies the architectural model when it is applied to an abstraction of the pipeline state stream:

$$\vdash \texttt{Pipeline\_Interp s e p} \Rightarrow \texttt{Arch\_Interp (abs s) e p}$$

This result can be combined with the proofs of the other levels (using the generic interpreter theory) to get a result that states the architectural model follows from the EBM. The overall form of the goal is familiar and has not changed from the goals commonly used in non-pipelined verification.

8 The Correctness Model and Hazards

The correctness model we have developed handles read after write (RAW) data hazards in the general case as shown in the data forwarding lemma of Section 7.3. The same cannot be said of control and structural hazards because of the variety of design techniques for mitigating them. RAW data hazards are almost always mitigated by forwarding the needed information from later stages of the pipeline to earlier stages. Because a single technique suffices, a single model suffices.

UINTA mitigates control hazards using delayed branching. As the discussion in Section 3 shows, delayed branching is visible at the architectural level. Another popular technique for mitigating control hazards, branch prediction, would *not* be visible at the architectural level. We have verified a processor that uses a simple form of branch prediction. As expected, changing the architectural model to that extent has significant effects on the abstraction function. For now, as the techniques for mitigating control hazards change, so will our verification methodology.

While we have not used our methodology for reasoning about a pipelined architecture with structural hazards, we believe that the techniques we have outlined are sufficient with some minor changes. The most important change relates to skew in the abstraction function: the abstraction function of Section 7.2 uses a specific number for the skew. A structural hazard could make the skew non-deterministic. In that case, the skew would have to be determined from a signal indicating when the structural hazard had cleared in the same way that asynchronous memory is specified now (see [Win94a] for more information). Aagaard and Leeser have described a very general methodology for reasoning about pipelines with structural hazards [AL94].

9 Conclusion

We have completed the specification and verification of a pipelined microprocessor called UINTA. UINTA has a five stage pipeline with two levels of data forwarding and delayed branching. This paper has presented techniques for verifying that the pipeline model correctly implements the architectural model.

Because we have developed tools for dealing with specifications and correctness theorems in a standard format, one of our goals in the verification of UINTA was to ensure that the specification and correctness theorem were stated using that format. The standard format has proven useful in proving properties about the architecture [Win91] and for extending the verification hierarchy [Win93]. The verification in this paper was completed using specifications exactly like those used for non-pipelined microprocessors. The correctness result is exactly like the correctness results for non-pipelined microprocessors.

We have also presented the abstraction function at the heart of the correctness model and shown how and why it differs from the abstractions commonly used to verify non-pipelined microprocessors. This function is the essence of what it means to say that the UINTA pipeline correctly implements the UINTA architecture. The function allows us to preserve the illusion that instructions execute sequentially in the architectural model even though the pipelined implementation performs operations in parallel.

We are presently verifying a pipelined microprocessor called SAWTOOTH. SAWTOOTH has a different set of features that UINTA including a simple form of branch prediction, user and multi–level system interrupts, a windowed register file, and supervisory mode. We plan to use the experience of verifying UINTA and SAWTOOTH to modify the generic interpreter theory to include pipeline semantics.

References

[AL94]   Mark D. Aagaard and Miriam E. Leeser. Reasoning about pipelines with structural hazards. In Ramayya Kumar and Thomas Kropf, editors, *Proceedings of the 1994 Conference on Theorem Provers in Circuit Design*. Springer–Verlag, September 1994.

[Bow87]   Jonathan P. Bowen. Formal specificaiton and documentation of microprocessor instruction sets. In *Microprocessing and Microprogramming 21*, pages 223–230, 1987.

[CCLO88] S. D. Crocker, E. Cohen, S. Landauer, and H. Orman. Reverification of a microprocessor. In *Proceedings of the IEEE Symposium on Security and Privacy*, pages 166–176, April 1988.

[CGM87] Albert Camilleri, Mike Gordon, and Tom Melham. Hardware verification using higher order logic. In D. Borrione, editor, *From HDL Descriptions to Guaranteed Correct Circuit Designs*. Elsevier Scientific Publishers, 1987.

[Chu40]   Alonzo Church. A formulation of the simple theory of types. *Journal of Symbolic Logic*, 5, 1940.

[Coh88]   Avra Cohn. A proof of correctness of the VIPER microprocessor: The first level. In G. Birtwhistle and P. Subrahmanyam, editors, *VLSI Specification, Verification, and Synthesis*, pages 27–72. Kluwer Academic Publishers, 1988.

[Gor83]   Michael J.C. Gordon. Proving a computer correct. Technical Report 41, Computer Lab, University of Cambridge, 1983.

[Gor88]   Michael J.C. Gordon. HOL: A proof generating system for higher-order logic. In G. Birtwhistle and P.A Subrahmanyam, editors, *VLSI Specification, Verification, and Synthesis*. Kluwer Academic Press, 1988.

[Her92]   John Herbert. Incremental design and formal verification of microcoded microprocessors. In V. Stavridou, T. F. Melham, and R. T. Boute, editors, *Theorem Provers in Circuit Design, Proceedings of the IFIP WG 10.2 International Working Conference, Nijmegen, The Netherlands*. North–Holland, June 1992.

[HP90]   John L. Hennesy and David A. Patterson. *Computer Architecture: A Quantitative Approach*. Morgan Kaufmann Publishers Inc., 1990.

[Hun89]   Warren A. Hunt. Microprocessor design verification. *Journal of Automated Reasoning*, 5:429–460, 1989.

[Joy88]   Jeffrey J. Joyce. Formal verification and implementation of a microprocessor. In G. Birtwhistle and P.A Subrahmanyam, editors, *VLSI Specification, Verification, and Synthesis*. Kluwer Academic Press, 1988.

[Mel88]   Thomas Melham. Abstraction mechanisms for hardware verification. In G. Birtwhistle and P. A. Subrahmanyam, editors, *VLSI Specification, Verification and Synthesis*. Kluwer Academic Publishers, 1988.

[SB90]   M. Srivas and M. Bickford. Formal verification of a pipelined microprocessor. *IEEE Software*, 7(5):52–64, September 1990.

[SWL93]   E. Thomas Schubert, Phillip J. Windley, and Karl Levitt. Report on the ucd microcoded viper verification project. In Jeffery J. Joyce and Carl Seger, editors, *Proceedings of the 1993 International Workshop on the HOL Theorem Prover and its Applications.*, August 1993.

[TK93]   Sofiene Tahar and Ramayya Kumar. Implementing a methodology for formally verifying RISC processors in HOL. In Jeffery J. Joyce and Carl Seger, editors, *Proceedings of the 1993 International Workshop on the HOL Theorem Prover and its Applications.*, August 1993.

[Win90]   Phillip J. Windley.. *The Formal Verification of Generic Interpreters*. PhD thesis, University of California, Davis, Division of Computer Science, June 1990.

[Win91]   Phillip J. Windley. Using correctness results to verify behavioral properties of microprocessors. In *Proceedings of the IEEE Computer Assurance Conference*, June 1991.

[Win93]   Phillip J. Windley. A theory of generic interpreters. In George J. Milne and Laurence Pierre, editors, *Correct Hardware Design and Verification Methods*, number 683 in Lecture Notes in Computer Science, pages 122–134. Springer-Verlag, 1993.

[Win94a]   Phillip J. Windley. Formal modeling and verification of microprocessors. *IEEE Transactions on Computers*, 1994. (to appear).

[Win94b]   Phillip J. Windley. Specifying instruction set architectures in HOL: A primer. In Thomas Melham and Juanito Camilleri, editors, *Proceedings of the 1994 International Workshop on the HOL Theorem Prover and its Applications.* Sspringer Verlag, Spetember 1994.

Non-Restoring Integer Square Root:
A Case Study in Design by Principled
Optimization

John O'Leary[1], Miriam Leeser[1], Jason Hickey[2], Mark Aagaard[1]

[1] School Of Electrical Engineering
Cornell University
Ithaca, NY 14853
[2] Department of Computer Science
Cornell University
Ithaca, NY 14853

Abstract. Theorem proving techniques are particularly well suited for reasoning about arithmetic above the bit level and for relating different levels of abstraction. In this paper we show how a non-restoring integer square root algorithm can be transformed to a very efficient hardware implementation. The top level is a Standard ML function that operates on unbounded integers. The bottom level is a structural description of the hardware consisting of an adder/subtracter, simple combinational logic and some registers. Looking at the hardware, it is not at all obvious what function the circuit implements. At the top level, we prove that the algorithm correctly implements the square root function. We then show a series of optimizing transformations that refine the top level algorithm into the hardware implementation. Each transformation can be verified, and in places the transformations are motivated by knowledge about the operands that we can guarantee through verification. By decomposing the verification effort into these transformations, we can show that the hardware design implements a square root. We have implemented the algorithm in hardware both as an Altera programmable device and in full-custom CMOS.

1 Introduction

In this paper we describe the design, implementation and verification of a subtractive, non-restoring integer square root algorithm. The top level description is a Standard ML function that implements the algorithm for unbounded integers. The bottom level is a highly optimized structural description of the hardware implementation. Due to the optimizations that have been applied, it is very difficult to directly relate the circuit to the algorithmic description and to prove that the hardware implements the function correctly. We show how the proof can be done by a series of transformations from the SML code to the optimized structural description.

At the top level, we have used the Nuprl proof development system [Lee92] to verify that the SML function correctly produces the square root of the input.

We then use Nuprl to verify that transformations to the implementation preserve the correctness of the initial algorithm.

Intermediate levels use Hardware ML [OLLA93], a hardware description language based on Standard ML. Starting from a straightforward translation of the SML function into HML, a series of transformations are applied to obtain the hardware implementation. Some of these transformations are expressly concerned with optimization and rely on knowledge of the algorithm; these transformations can be justified by proving properties of the top-level description.

The hardware implementation is highly optimized: the core of the design is a single adder/subtracter. The rest of the datapath is registers, shift registers and combinational logic. The square root of a $2n$ bit wide number requires n cycles through the datapath. We have two implementations of square root chips based on this algorithm. The first is done as a full-custom CMOS implementation; the second uses Altera EPLD technology. Both are based on a design previously published by Bannur and Varma [BV85]. Implementing and verifying the design from the paper required clearing up a number of errors in the paper and clarifying many details.

This is a good case study for theorem proving techniques. At the top level, we reason about arithmetic operations on unbounded integers, a task theorem provers are especially well suited for. Relating this to lower levels is easy to do using theorem proving based techniques. Many of the optimizations used are applicable only if very specific conditions are satisfied by the operands. Verifying that the conditions hold allows us to safely apply optimizations.

Automated techniques such as those based on BDDs and model checking are not well-suited for verifying this and similar arithmetic circuits. It is difficult to come up with a Boolean statement for the correctness of the outputs as a function of the inputs and to argue that this specification correctly describes the intended behavior of the design. Similarly, specifications required for model checkers are difficult to define for arithmetic circuits.

There have been several verifications of hardware designs which lift the reasoning about hardware to the level of integers, including the Sobel Image processing chip [NS88], and the factorial function [CGM86]. Our work differs from these and similar efforts in that we justify the optimizations done in order to realize the square root design. The DDD system [BJP93] is based on the idea of design by verified transformation, and was used to derive an implementation of the FM9001 microprocessor. High level transformations in DDD are not verified by explicit use of theorem proving techniques.

The most similar research is Verkest's proof of a non-restoring division algorithm [VCH94]. This proof was also done by transforming a design description to an implementation. The top level of the division proof involves consideration of several cases, while our top level proof is done with a single loop invariant. The two implementations vary as well: the division algorithm was implemented on an ALU, and the square root on custom hardware. The algorithms and implementations are sufficiently similar that it would be interesting to develop a single verified implementation that performs both divide and square root based on the research in these two papers.

The remainder of this paper is organized as follows. In section 2 we describe the top-level non-restoring square root algorithm and its verification in the Nuprl proof development system. We then transform this algorithm down to a level suitable for modelling with a hardware description language. Section 3 presents a series of five optimizing transformations that refine the register transfer level description of the algorithm to the final hardware implementation. In section 4 we summarize the lessons learned and our plans for future research.

2 The Non-Restoring Square Root Algorithm

An integer square root calculates $y = \sqrt{x}$ where x is the radicand, y is the root, and both x and y are integers. We define the *precise* square root (p) to be the real valued square root and the *correct* integer square root to be the floor of the precise root. We can write the specification for the integer square root as shown in Definition 1.

Definition 1 *Correct integer square root*
    y is the *correct* integer square root of $x \hateq y^2 \leq x < (y+1)^2$

We have implemented a subtractive, non-restoring integer square root algorithm [BV85]. For radicands in the range $x = \{0..2^{2n} - 1\}$, subtractive methods begin with an initial guess of $y = 2^{(n-1)}$ and then iterate from $i = (n-1)\ldots0$. In each iteration we square the partial root (y), subtract the squared partial root from the radicand and revise the partial root based on the sign of the result. There are two major classes of algorithms: restoring and non-restoring [Flo63]. In restoring algorithms, we begin with a partial root for $y = 0$ and at the end of each iteration, y is never greater than the precise root (p). Within each iteration (i), we set the ith bit of y, and test if $x - y^2$ is negative; if it is, then setting the ith bit made y too big, so we reset the ith bit and proceed to the next iteration.

Non-restoring algorithms modify each bit position once rather than twice. Instead of setting the the ith bit of y, testing if $x - y^2$ is positive, and then possibly resetting the bit; the non-restoring algorithms add or subtract a 1 in the ith bit of y based on the sign of $x - y^2$ in the previous iteration. For binary arithmetic, the restoring algorithm is efficient to implement. However, most square root hardware implementations use a higher radix, non-restoring implementation. For higher radix implementations, non-restoring algorithms result in more efficient hardware implementations.

The results of the non-restoring algorithms do not satisfy our definition of *correct*, while restoring algorithms do satisfy our definition. The resulting value of y in the non-restoring algorithms may have an error in the last bit position. For the algorithm used here, we can show that the final value of y will always be either the precise root (for radicands which are perfect squares) or will be odd and be within one of the correct root. The error in non-restoring algorithms is easily be corrected in a cleanup phase following the algorithm.

Below we show how a binary, non-restoring algorithm runs on some values for n = 3. Note that the result is either exact or odd.

$x = 18_{10} = 10100_2$ iterate 1   y = 100 $x - y^2 = +$
   iterate 2   y = 110 $x - y^2 = -$
   iterate 3   y = 101

$x = 15_{10} = 01111_2$ iterate 1   y = 100 $x - y^2 = -$
   iterate 2   y = 010 $x - y^2 = +$
   iterate 3   y = 011

$x = 16_{10} = 10000_2$ iterate 1   y = 100 $x - y^2 = 0$

In our description of the non-restoring square root algorithm, we define a datatype **state** that contains the state variables. The algorithm works by initializing the state using the function **init**, then updating the state on each iteration of the loop by calling the function **update**. We refine our program through several sets of transformations. At each level the definition of **state**, **init** and **update** may change. The SML code that calls the square root at each level is:

```
fun sqrt n radicand =  iterate update (n-1) (init n radicand)
```

The **iterate** function performs iteration by applying the function argument to the state argument, decrementing the count, and repeating until the count is zero.

```
fun iterate f n state =
    if n = 0 then state else iterate f (n - 1) (f state)
```

The top level ($\mathcal{L}_0$) is a straightforward implementation of the non-restoring square root algorithm. We represent the state of an iteration by the triple State{x,y,i} where **x** is the radicand, **y** is the partial root, and **i** is the iteration number. Our initial guess is $y = 2^{(n-1)}$. In **update**, **x** never changes and **i** is decremented from n-2 to 1, since the initial values take care of the $(n-1)^{st}$ iteration. At $\mathcal{L}_0$, **init** and **update** are:

```
fun init (n,radicand) =
  State{x = radicand,  y = 2 ** (n-1),   i = n-2 }

fun update (State{x, y, i}) =
    let
      val diffx  = x - (y**2)
      val y'     = if        diffx = 0    then y
                   else if diffx > 0    then (y + (2**i))
                   else (* diffx < 0 *)      (y - (2**i))
    in
      State{x = x,  y = y',   i = i-1 }
    end
```

In the next section we discuss the proof that this algorithm calculates the square root. Then we show how it can be refined to an implementation that requires significantly less hardware. We show how to prove that the refined algorithm also calculates the square root; in the absence of such a proof it is not at all obvious that the algorithms have identical results.

2.1 Verification of Level Zero Algorithm

All of the theorems in this section were verified using the Nuprl proof development system. Theorem 1 is the overall correctness theorem for the non-restoring square root code shown above. It states that after iterating through **update** for n-1 times, the value of **y** is within one of the *correct* root of the radicand. We have proved this theorem by creating an invariant property and performing induction on the number of iterations of **update**. Remember that n is the number of bits in the result.

Theorem 1 *Correctness theorem for non-restoring square root algorithm*

$\vdash \forall n . \forall radicand : \{0..2^{2n} - 1\}$.

    let

        $\text{State}\{x, y, i\} = \text{iterate update } (n - 1) \ (\text{init } n \ radicand)$

    in

        $(y - 1)^2 \leq radicand < (y + 1)^2$

    end

The invariant states that in each iteration **y** increases in precision by 2^i, and the lower bits in **y** are always zero. The formal statement $(\mathcal{I})$ of the invariant is shown in Definition 2.

Definition 2 *Loop invariant*

$\mathcal{I}(\text{State}\{x, y, i\}) \doteq$

    $((y - 2^i)^2 \leq radicand < (y + 2^i)^2) \ \& \ (y \text{ rem } 2^i = 0) \ \& \ (x = radicand)$

In Theorems 2 and 3 we show that **init** and **update** are correct, in that for all legal values of n and **radicand**, **init** returns a legal state and for all legal input states, **update** will return a legal state and makes progress toward termination by decrementing **i** by 1. A legal state is one for which the loop invariant holds.

Theorem 2 *Correctness of initialization*

$\vdash \forall n . \forall radicand : \{0..2^{2n} - 1\}.\forall x, y, i$.

    $\text{State}\{x, y, i\} = \text{init } n \ radicand \implies$

        $\mathcal{I}(\text{State}\{x, y, i\}) \ \& \ (x = radicand) \ \& \ (y = 2^{n-1}) \ \& \ (i = n - 2)$

Theorem 3 *Correctness of update*

$\vdash \forall x, y, i . \mathcal{I}((\text{State}\{x, y, i\})) \implies$
$\quad \forall x', y', i' . \text{State}\{x', y', i'\} = \text{update} (\text{State}\{x, y, i\}) \implies$
$\quad \mathcal{I}(\text{State}\{x', y', i'\}) \,\&\, (i' = i - 1)$

The correctness of init is straightforward. The proof of Theorem 3 relies on Theorem 4, which describes the behavior of the update function. The body of update has three branches, so the proof of correctness of update has three parts, depending on whether $x - y^2$ is equal to zero, positive, or negative. Each case in Theorem 4 is straightforward to prove using ordinary arithmetic.

Theorem 4 *Update lemmas*

    Case $x - y^2 = 0$
      $\vdash \forall x, y, i . \mathcal{I}((\text{State}\{x, y, i\})) \implies$
        $(x - y^2 = 0) \implies \mathcal{I}(\text{State}\{x, y, i - 1\})$

    Case $x - y^2 > 0$
      $\vdash \forall x, y, i . \mathcal{I}((\text{State}\{x, y, i\})) \implies$
        $(x - y^2 > 0) \implies \mathcal{I}(\text{State}\{x, y + 2^i, i - 1\})$

    Case $x + y^2 > 0$
      $\vdash \forall x, y, i . \mathcal{I}((\text{State}\{x, y, i\})) \implies$
        $(x + y^2 > 0) \implies \mathcal{I}(\text{State}\{x, y - 2^i, i - 1\})$

We now prove that iterating update a total of n-1 times will produce the correct final result. The proof is done by induction on n and makes use of Theorem 5 to describe one call to iterate. This allows us to prove that after iterating update a total of n-1 times, our invariant holds and i is zero. This is sufficient to prove that square root is within one of the *correct* root.

Theorem 5 *Iterating a function*

    $\vdash \forall prop, n, f, s.$
      $prop \; n \; s \implies$
        $(\forall n', s'.prop \; n' \; s' \implies prop \; (n' - 1) \; (f \; s')) \implies$
          $prop \; 0 \; (\text{iterate} \; f \; n \; s)$

2.2 Description of Level One Algorithm

The $\mathcal{L}_0$ SML code would be very expensive to directly implement in hardware. If the state were stored in three registers, **x** would be stored but would never change; the variable i would need to be decremented every loop and we would need to calculate y^2, $x - y^2$, 2^i, and $y \pm 2^i$ in every iteration. All of these are expensive operations to implement in hardware. By restructuring the algorithm through a series of transformations, we preserve the correctness of our design and generate an implementation that uses very little hardware.

The key operations in each iteration are to compute $x - y^2$ and then update y using the new value $y' = y \pm 2^i$, where $\pm$ is $+$ if $x - y^2 \geq 0$ and $-$ if $x - y^2 < 0$. The variable x is only used in the computation of $x - y^2$. In the $\mathcal{L}_1$ code we introduce the variable $\mathtt{diffx}$, which stores the result of computing $x - y^2$. This has the advantage that we can incrementally update $\mathtt{diffx}$ based on its value in the previous iteration:

$$
\begin{aligned}
y' &= y \pm 2^i \\
y'^2 &= y^2 \pm 2*y*2^i + (2^i)^2 \\
&= y^2 \pm y*2^{i+1} + 2^{2*i}
\end{aligned}
\qquad
\begin{aligned}
\mathtt{diffx} &= x - y^2 \\
\mathtt{diffx'} &= x - y'^2 \\
&= x - (y^2 \pm y*2^{i+1} + 2^{2*i}) \\
&= (x - y^2) \mp y*2^{i+1} - 2^{2*i}
\end{aligned}
$$

The variable i is only used in the computations of 2^{2*i} and $y*2^{i+1}$, so we create a variable b that stores the value 2^{2*i} and a variable $\mathtt{yshift}$ that stores $y*2^{i+1}$. We update b as: $b' = b \ \mathtt{div}\ 4$. This results in the following equations to update $\mathtt{yshift}$ and $\mathtt{diffx}$:

$$
\begin{aligned}
y'*2^{i+1} &= (y \pm 2^i)*2^{i+1} \\
\mathtt{yshift'} &= y*2^{i+1} \pm 2^i*2^{i+1} \\
&= \mathtt{yshift} \pm 2*2^{2*i} \\
&= \mathtt{yshift} \pm 2*b
\end{aligned}
\qquad
\mathtt{diffx'} = \mathtt{diffx} \mp \mathtt{yshift} - b
$$

The transformations from $\mathcal{L}_0$ to $\mathcal{L}_1$ can be summarized:

$$
\mathtt{diffx} = x - y^2 \qquad \mathtt{yshift} = y*2^{i+1} \qquad b = 2^{2*i}
$$

The $\mathcal{L}_1$ versions of **init** and **update** are given below. Note that, although the optimizations are motivated by the fact that we are doing bit vector arithmetic, the algorithm is correct for unbounded integers. Also note that the most complex operations in the update loop are an addition and subtraction and only one of these two operations is executed each iteration. We have optimized away all exponentiation and any multiplication that cannot be implemented as a constant shift.

```
fun init1 (n,radicand) =
  let
    val b' = 2 ** (2*(n-1))
  in
  State{diffx  = radicand - b',
        yshift = b',
        b      = b' div 4
        }
  end
```

```
fun update1 (State{diffx, yshift, b}) =
  let
    val (diffx',yshift') =
        if      diffx > 0 then (diffx - yshift - b, yshift + 2*b)
        else if diffx = 0 then (diffx             , yshift      )
        else (* diffx < 0 *)   (diffx + yshift - b, yshift - 2*b)
  in
    State{diffx  = diffx',
          yshift = yshift' div 2,
          b      = b   div 4
         }
  end
```

We could verify the $\mathcal{L}_1$ algorithm from scratch, but since it is a transformation of the $\mathcal{L}_0$ algorithm, we use the results from the earlier verification. We do this by defining a mapping function between the state variables in the two levels and then proving that the two levels return equal values for equal input states. The transformation is expressed as follows:

Definition 3 *State transformation*

$$T(\text{State}\{x, y, i\}, \text{State}\{diffx, yshift, b\}) \doteq$$
$$(diffx = x - y^2 \ \& \ yshift = y * 2^{i+1} \ \& \ b = 2^{2*i})$$

All of the theorems in this section were proved with Nuprl. In Theorem 6 we say that for equal inputs init1 returns an equivalent state to init. In Theorem 7 we say that for equivalent input states, update1 and update return equivalent outputs.

Theorem 6 *Correctness of initialization*

$\vdash \forall n . \forall radicand : \{0..2^{2n} - 1\}.$
$\quad \forall x, y, i . \text{State}\{x, y, i\} = init \ n \ radicand \implies$
$\quad\quad \forall diffx, yshift, b . \text{State}\{diffx, yshift, b\} = init1 \ n \ radicand \implies$
$\quad\quad\quad T(\text{State}\{x, y, i\}, \text{State}\{diffx, yshift, b\})$

Theorem 7 *Correctness of update*

$\vdash \forall x, y, i, diffx, yshift, b .$
$\quad T(\text{State}\{x, y, i\}, \text{State}\{diffx, yshift, b\}) \implies$
$\quad\quad T(update(\text{State}\{x, y, i\}), update1(\text{State}\{diffx, yshift, b\}))$

Again, the initialization theorem has an easy proof, and the update1 theorem is a case split on each of the three cases in the body of the update function, followed by ordinary arithmetic.

2.3 Description of Level Two Algorithm

To go from $\mathcal{L}_1$ to $\mathcal{L}_2$, we recognize that the operations in init1 are very similar to those in update1. By carefully choosing our initial values for diffx and y, we increase the number of iterations from n-1 to n and fold the computation of radicand - b' in init into the first iteration of update. This eliminates the need for special initialization hardware. The new initialize function is:

```
fun init2 (n,radicand) =
  State{diffx  = radicand,
        yshift = 0,
        b      = 2 ** (2*(n-1))}
```

The update function is unchanged from update1. The new calling function is:

```
fun sqrt n radicand =  iterate update1 n (init2 n radicand)
```

Showing the equivalence between init2 and a loop that iterates n times and the $\mathcal{L}_1$ functions requires showing that the state in $\mathcal{L}_2$ has the same value after the first iteration that it did after init1. More formally, init1 = update1 o init2. We prove this using the observation that, after init2, diffx is guaranteed to be positive, so the first iteration of update1 in $\mathcal{L}_2$ always executes the diffx > 0 case. Using the state returned by init2 and performing some simple algebraic manipulations, we see that after the first iteration in $\mathcal{L}_2$, update1 stores the same values in the state variables as init1 did. Because both $\mathcal{L}_1$ and $\mathcal{L}_2$ use the same update function, all subsequent iterations in the $\mathcal{L}_2$ algorithm are identical to the $\mathcal{L}_1$ algorithm.

To begin the calculation of a square root, a hardware implementation of the $\mathcal{L}_2$ algorithm clears the yshift register, load the radicand into the diffx register, and initializes the b register with a 1 in the correct location. The transformation from $\mathcal{L}_0$ to $\mathcal{L}_1$ simplified the operations done in the loop to shifts and an addition/subtraction. Remaining transformations will further optimize the hardware.

3 Transforming Behavior to Structure with HML

The goal of this section is to produce an efficient hardware implementation of the $\mathcal{L}_2$ algorithm. The first subsection introduces Hardware ML, our language for specifying the behavior and structure of hardware. Taking an HML version of the $\mathcal{L}_2$ algorithm as our starting point, we obtain a hardware implementation through a sequence of transformation steps.

1. Translate the $\mathcal{L}_2$ algorithm into Hardware ML. Provision must be made to initialize and detect termination, which is not required at the algorithm level.
2. Transform the HML version of $\mathcal{L}_2$ to a set of register assignments using syntactic transformations.

3. Introduce an internal state register **Exact** to simplify the computation, and "factor out" the condition **DiffX >= %0**.
4. Partition into functional blocks, again using syntactic transformations.
5. Substitute lower level modules for register and combinational assignments. Further optimizations in the implementation of the lower level modules are possible.

Each step can be verified formally. Several of these must be justified by properties of the algorithm that we can establish through theorem proving.

3.1 Hardware ML

We have implemented extensions to Standard ML that can be used to describe the behavior of digital hardware at the register transfer level. Earlier work has illustrated how Hardware ML can be used to describe the structure of hardware [OLLA93, OLLA92]. HML is based on SML and supports higher-order, polymorphic functions, allowing the concise description of regular structures such as arrays and trees. SML's powerful module system aids in creating parameterized designs and component libraries.

Hardware is modelled as a set of concurrently executing *behaviors* communicating through objects called *signals*. Signals have semantics appropriate for hardware modelling: whereas a Standard ML reference variable simply contains a value, a Hardware ML signal contains a list of time-value pairs representing a waveform. The current value on signal **a**, written **$a**, is computed from its waveform and the current time.

Two kinds of signal assignment operators are supported. *Combinational assignment*, written s == v, is intended to model the behavior of combinational logic under the assumption that gate delays are negligible. s == v causes the current value of the target signal s to become v. For example, we could model an exclusive-or gate as a behavior which assigns **true** to its output **c** whenever the current values on its inputs **a** and **b** are not equal:

```
fun Xor (a,b,c) = behavior (fn () => (c == ($a <> $b)))
```

HML's **behavior** constructor creates objects of type **behavior**. Its argument is a function of type **unit -> unit** containing HML code – in this case, a combinational assignment.

Register assignment is intended to model the behavior of sequential circuit elements. If a register assignment s <- v is executed at time t the waveform of s is augmented with the pair $(v, t+1)$ indicating that s is to assume the value v at the next time step. For example, we could model a delay element as a behavior containing a register assignment:

```
fun Reg (a,b) = behavior (fn () => (b <- $a))
```

Behaviors can be composed to form more complex circuits. In the composition of b_1 and b_2, written b_1 || b_2, both behaviors execute concurrently. We compose our exclusive-or gate and register to build a parity circuit, which outputs **false** if and only if it has received an even number of **true** inputs:

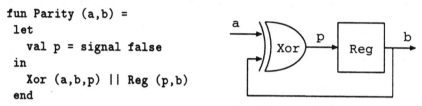

```
fun Parity (a,b) =
  let
    val p = signal false
  in
    Xor (a,b,p) || Reg (p,b)
  end
```

The **val p** = ... declaration introduces an internal signal whose initial value is **false**.

Table 1 summarizes the behavioral constructs of HML. We have implemented a simple simulator which allows HML behavioral descriptions to be executed interactively – indeed, all the descriptions in this section were developed with the aid of the simulator.

$\alpha^=$ signal	Type of signals having values of equality type $\alpha^=$
signal v	A signal initially having the value v
$ s	The current value of signal s
s == v	Combinational assignment
s <- v	Register assignment
behavior	Type of behaviors
behavior f	Behavior constructor
b_1 || b_2	Behavioral composition

Table 1. Hardware ML Constructs

3.2 Behavioral Description of the Level 2 Algorithm

It is straightforward to translate the $\mathcal{L}_2$ algorithm from Standard ML to Hardware ML. The state of the computation is maintained in the externally visible signal YShift and the internal signals DiffX and B. These signals correspond to yshift, diffx and b in the $\mathcal{L}_2$ algorithm and are capitalized to distinguish them as HML signals. For concreteness, we stipulate that we are computing an eight-bit root ($n = 8$ in the $\mathcal{L}_2$ algorithm), and so the YShift and B signals require 16 bits. DiffX requires 17 bits: 16 bits to hold the initial (positive) radicand and one bit to hold the sign, which is important in the intermediate calculations. We use HML's register assignment operator <- to explicitly update the state in each computation step. The % characters preceding numeric constants are constructors for an abstract data type num. The usual arithmetic operators are defined on this abstract type. We are thus able to simulate our description using unbounded integers, bounded integers, and bit vectors simply by providing appropriate implementations of the type and operators. The code below shows the HML representation of the $\mathcal{L}_2$ algorithm. Two internal signals are defined, and the behavior is packaged within a function declaration. In the future we will omit showing the function declaration and declarations of internal signals; it is understood that all signals except `Init`, `XIn`, `YShift`, and `Done` are internal.

```
val initb = %16384                    (* 0x4000 *)

fun sqrt (Init, XIn, YShift, Done) =
    let
        (* Internal registers *)
        val DiffX = signal (%0)
        val B = signal (%0)
    in
        behavior
        (fn () =>
        (if $Init then
              (DiffX <- $XIn;
               YShift <- %0;
               B <- initb;
               Done <- false)
          else
              (if $DiffX = %0 then
                    (DiffX <- $DiffX;
                     YShift <- $YShift div %2;
                     B <- $B div %4)
               else if $DiffX > %0 then
                    (DiffX <- $DiffX - $YShift - $B;
                     YShift <- ($YShift + %2 * $B) div %2;
                     B <- $B div %4)
               else (* $DiffX < %0 *)
                    (DiffX <- $DiffX + $YShift - $B;
                     YShift <- ($YShift - %2 * $B) div %2;
                     B <- $B div %4));
               Done <- $B sub 0))
    end
```

In the SML description of the algorithm, some elements of the control state are implicit. In particular, the initiation of the algorithm (calling `sqrt`) and its termination (it returning a value) are handled by the SML interpreter. Because hardware is free running there is no built-in notion of initiation or termination of the algorithm. It is therefore necessary to make explicit provision to initialize the state registers at the beginning of the algorithm and to detect when the algorithm has terminated.

Initialization is easy: the `diffx`, `yshift`, and b registers are assigned their initial values when the `init` signal is high.

In computing an eight-bit root, the $\mathcal{L}_2$ algorithm terminates after seven iterations of its loop. An efficient way to detect termination of the hardware algorithm makes use of some knowledge of the high level algorithm. An informal analysis of the $\mathcal{L}_2$ algorithm reveals that b contains a single bit, shifted right two places in each cycle, and that the least significant bit of b is set during the execution of the last iteration. Consequently, the **done** signal is generated by testing whether the least significant bit of **b** is set (the expression **$b sub 0** selects the lsb of b) and delaying the result of the test by one cycle. **done** is therefore set during the clock cycle following the final iteration. To justify this analysis, we must formally prove that the following is an invariant of the $\mathcal{L}_2$ algorithm:

$$(i = 1) \longleftrightarrow (b_0 = 1)$$

3.3 Partitioning into Register Assignments

Our second transformation step is a simple one: we transform the HML version of the $\mathcal{L}_2$ algorithm into a set of register assignments. The goal of this transformation is to ensure that the control state of the algorithm is made completely explicit in terms of HML signals.

We make use of theorems about the semantics of HML to justify transformations of the **behavior** construct. First, **if** distributes over the sequential composition of signal assignment statements: if P and Q are sequences of signal assignments, then

$$\boxed{\begin{array}{l} \text{if } e \text{ then} \\ \quad (s \text{ <- } a; \ P) \\ \text{else} \\ \quad (s \text{ <- } b; \ Q) \end{array}} \quad = \quad \boxed{\begin{array}{l} \text{if } e \text{ then} \\ \quad s \text{ <- } a \\ \text{else} \\ \quad s \text{ <- } b; \\ \text{if } e \text{ then } P \text{ else } Q \end{array}}$$

We also use a rule which allows us to push **if** into the right hand side of an assignment:

$$\boxed{\begin{array}{l} \text{if } e \text{ then} \\ \quad s \text{ <- } a \\ \text{else} \\ \quad s \text{ <- } b \end{array}} \quad = \quad \boxed{s \text{ <- if } e \text{ then } a \text{ else } b}$$

Repeatedly applying these two rules allows us to decompose our HML code into a set of assignments to individual registers. The register assignments after this transformation are

```
DiffX   <- (if $Init then
                $XIn
            else if $DiffX = %0 then
                $DiffX
            else if $DiffX > %0 then
                $DiffX - $YShift - $B
            else (* $DiffX < %0 *)
                $DiffX + $YShift - $B);
```

```
YShift <- (if $Init then
                %0
            else if $DiffX = %0 then
                $YShift div %2
            else if $DiffX > %0 then
                ($YShift + %2 * $B) div %2
            else (* $DiffX < %0 *)
                ($YShift - %2 * $B) div %2);

B       <- (if $Init then initb else $B div %4);
Done    <- (if $Init then false else $B sub 0);
```

3.4 Simplifying the Computation

Our third transformation step simplifies the computation of DiffX and YShift. We begin by observing two facts about the $\mathcal{L}_2$ algorithm: if DiffX ever becomes zero the radicand has an exact root, and once DiffX becomes zero it remains zero for the rest of the computation. To simplify the computation of DiffX we introduce an internal signal Exact which is set if DiffX becomes zero in the course of the computation.

If Exact becomes set, the value of DiffX is not used in the computation of YShift (subsequent updates of YShift involve only division by 2). DiffX becomes a don't care, and we can merge the $DiffX = %0 and $DiffX > %0 branches. We also replace $DiffX = %0 with $Exact in the assignment to YShift, and change the $DiffX > %0 comparisons to $DiffX >= %0 (note that this branch of the if is not executed when $DiffX = %0, because Exact is set in this case).

```
ExactReg <- (if $init then false else $Exact);
Exact    == ($DiffX = %0 orelse $ExactReg);

DiffX    <- (if $Init then
                $XIn
            else if $DiffX >= %0 then
                $DiffX - $YShift - $B
            else (* $DiffX < %0 *)
                $DiffX + $YShift - $B);

YShift   <- (if $Init then
                %0
            else if $Exact then
                $YShift div %2
            else if $DiffX >= %0 then
                ($YShift + %2 * $B) div %2
            else (* $DiffX < %0 *)
                ($YShift - %2 * $B) div %2);

B        <- (if $Init then initb else $B div %4);
Done     <- (if $Init then false else $B sub 0);
```

Simplifying the algorithm in this way requires proving a history property of the computation of DiffX. Using $DiffX(n)$ to denote the value of the signal DiffX in the n'th computation cycle, we state the property as:

$$\forall t : t \geq 0.(\$DiffX(t) = 0) \implies (\$DiffX(t+1) = 0)$$

Next, we note that the condition $DiffX >= 0$ can be detected by negating DiffX's sign bit. We introduce the negated sign bit as the intermediate signal ADD (to specify that we are adding to YShift in the current iteration) and rewrite the conditions in the assignments to DiffX and Yshift to obtain:

```
ADD        == not ($DiffX sub 16);

ExactReg <- (if $init then false else $Exact);
Exact    == ($DiffX = %0 orelse $ExactReg);

DiffX    <- (if $Init then
                 $XIn
             else if $ADD then
                 $DiffX - $YShift - $B
             else $DiffX + $YShift - $B);

YShift   <- (if $Init then
                 %0
             else if $Exact then
                 $YShift div %2
             else if $ADD then
                 ($YShift + %2 * $B) div %2
             else ($YShift - %2 * $B) div %2);

B        <- (if $Init then initb else $B div %4);

Done     <- (if $Init then false else $B sub 0);
```

3.5 Partitioning into Functional Blocks

The fourth transformation separates those computations which can be performed combinationally from those which require sequential elements, and partitions the computations into simple functional units. The transformation is motivated by our desire to implement the algorithm by an interconnection of lower level blocks; the transformation process is guided by what primitives we have available in our library. For example, our library contains such primitives as registers, multiplexers, and adders, so it is sensible to transform

```
| s <- if x then a else b+c |  ⟶  | s <- s';
                                   | s' == if x then a else d;
                                   | d == b + c |
```

This particular example is a consequence of some more general rules which are justified as before by appealing to the semantics of HML.

$$\boxed{s \; == \; \text{if } e \text{ then } a \text{ else } b} \quad = \quad \boxed{\begin{array}{l} s \; == \; \text{if } e \text{ then } s_1 \text{ else } s_2; \\ s_1 \; == \; a; \\ s_2 \; == \; b \end{array}}$$

The assignments resulting from this transformation are shown below. `DiffX'` can be computed by a multiplexer. `DiffXTmp` and `Delta` can be computed by adder/subtracters. The `YShift` register can be conveniently implemented as a shift register which shifts when `Exact`'s value is true, and loads otherwise. `YShift'` can be computed by a multiplexer; `YShiftTmp` by an adder/subtracter – multiplication and division by 2 are simply wired shifts.

```
ADD        == not ($DiffX sub 16);

ExactReg   <- (if $init then false else $Exact);
Exact      == ($EqZero orelse $ExactReg);
EqZero     == $DiffX = %0;

DiffX      <- $DiffX';
DiffX'     == (if $Init then $XIn else $DiffXTmp);
DiffXTmp   == (if $ADD then $DiffX - $Delta else $DiffX + $Delta);
Delta      == (if $ADD then $YShift + $B else $YShift - $B);

YShift     <- (if $Exact then $YShift div %2 else $YShift');
YShift'    == (if $Init then %0 else $YShiftTmp);
YShiftTmp  == (if $ADD then
                      ($YShift + %2 * $B) div %2
               else
                      ($YShift - %2 * $B) div %2);

B          <- (if $Init then initb else $B div %4);

Done       <- (if $Init then false else $B sub 0);
```

3.6   Structural Description of the Level 2 Algorithm

The fifth, and final, transformation step is the substitution of lower-level modules for the register and combinational assignments; the result is a structural description of the integer square root algorithm which can readily be implemented in hardware, as shown in Figure 1.

`ShiftReg4` is a shift register which shifts its contents two bits per clock cycle; the `Done` signal is simply its shift output. `ShiftReg2`, `Mux`, `AddSub` and `Reg` are a shift register, multiplexer, adder/subtracter and register, respectively. `SubAdd` is an adder/subtracter, but the sense of its mode bit makes it the opposite of `AddSub`. The `Hold` element has the following substructure:

```
Hold        {exact=Exact, zero=EqZero, rst=Init}                    ||
IsZero      {out=EqZero, inp=DiffX}                                 ||
Reg         {out=DiffX, inp=DiffX'}                                 ||
Mux         {out=DiffX', in0=XIn, in1=DiffXTmp, ctl=Init}           ||
AddSub      {sum=DiffXTmp, in0=DiffX, in1=Delta, sub=ADD}           ||
Neg         {out=ADD, inp=$DiffX sub 16}                            ||
SubAdd      {out=Delta, in0=YShift, in1=B, add=ADD}                 ||
ShiftReg2   {out=YShift, inp=YShift', shift=Exact}                  ||
Mux         {out=YShift', in0=signal (%0),
             in1=YShiftTmp div 2, ctl=Init}                         ||
SubAdd      {out=YShiftTmp, in0=YShift, in1=2 * B, sub=ADD}         ||
ShiftReg4   {out=B, shiftout=Done, in=signal initb, shiftbar=Init}
```

Fig. 1. Structural Description

```
fun Hold {exact=exact, zero=zero, rst=rst} =
    let
        val ExactReg = signal false
    in
        RegR {out=ExactReg, inp=exact, rst=rst}    ||
        Or2  {out=Exact, in0=zero, in1=ExactReg}
    end
```

There are further opportunities for performing optimization in the implementation of the lower level blocks. Analysis of the $\mathcal{L}_2$ algorithm reveals that Delta is always positive, so we do not need its sign bit. This property can be used to save one bit in the AddSub used to compute Delta – only 16 bits are now required. One bit can also be saved in the implementation of SubAdd; the value of the ADD signal can be shown to be identical to a latched version of the carry output of a 16-bit SubAdd, provided the latch initially holds the value true. Figure 2 shows a block diagram of the hardware to implement the square root algorithm, which includes this optimization.

A number of other optimizations are not visible in the figure. The ShiftReg4 can be implemented by a ShiftReg2 half its width if we note that every second bit of B is always zero. The two SubAdd blocks can each be implemented with only a few AND and OR gates per bit, rather than requiring subtract/add modules, if we make use of some results concerning the contents of the B and YShift registers.

To justify these optimizations we are obliged to prove that every second bit of B is always zero:

$$\forall i : 0 \le i < n : \neg B_{2i+1}$$

and that the corresponding bits of B and YShift cannot both be set:

$$\forall i : 0 \le i < 2n. \neg (B_i \wedge YShift_i)$$

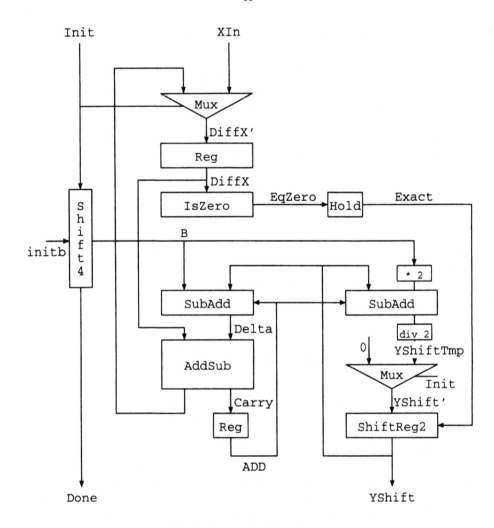

Fig. 2. Square Root Hardware

and that the corresponding bits of $B * \%2$ and YShift cannot both be set:

$$\forall i : 1 \leq i < 2n. \neg(B_{i-1} \wedge YShift_i)$$

We have produced two implementations of the square root algorithm which incorporate all these optimizations. The first was a full-custom CMOS layout fabricated by Mosis, the second used Altera programmable logic parts. In the latter case, the structural description was first translated to Altera's AHDL language and then passed through Altera's synthesis software.

4 Discussion

We have described how to design and verify a subtractive, non-restoring integer square root circuit by refining an abstract algorithmic specification through several intermediate levels to yield a highly optimized hardware implementation. We have proved using Nuprl that the $\mathcal{L}_0$ algorithm performs the square root function, and we have also used Nuprl to show how proving the first few levels of refinement ($\mathcal{L}_0$ to $\mathcal{L}_1$, $\mathcal{L}_1$ to $\mathcal{L}_2$) can be accomplished by transforming the top level proof in a way that preserves its validity.

This case study illustrates that rigorous reasoning about the high-level description of an algorithm can establish properties which are useful even for bit-level optimization. Theorem provers provide a means of formally proving the desired properties; a transformational approach to partitioning and optimization ensures that the properties remain relevant at the structural level. Each of the steps identified in this paper can be mechanized with reasonable effort. At the bottom level, we have a library of verified hardware modules that correspond to the modules in the HML structural description [AL94].

In many cases the transformations we applied depend for their justification upon non-trivial properties of the square root algorithm: we are currently working on formally proving these obligations. Some of our other transformations are purely syntactic in nature and rely upon HML's semantics for their justification. We have not considered semantic reasoning in this paper – this is a current research topic.

The algorithm we describe computes the integer square root. The algorithm and its implementation are of general interest because most of the algorithms used in hardware implementations of floating-point square root are based on the algorithm presented here. One difference is that most floating-point implementations use a higher radix representation of operators. In the future, we will investigate incorporating higher radix floating-point operations. We believe much of the reasoning presented here will be applicable to higher radix implementations of square root as well.

Many of the techniques demonstrated in this case study are applicable to hardware verification in general. Proof development systems are especially well suited for reasoning at high levels of abstraction and for relating multiple levels of abstraction. Both of these techniques must be exploited in order to make it feasible to apply formal methods to large scale highly optimized hardware systems. Top level specifications must be concise and intuitively capture the designers' natural notions of correctness (for example, arithmetic operations on unbounded integers), while the low level implementation must be easy to relate to the final implementation (for example, operations on bit-vectors). By applying a transformational style of verification as a design progresses from an abstract algorithm to a concrete implementation, theorem proving based verification can be integrated into existing design practices.

Acknowledgements

This research is supported in part by the National Science Foundation under contracts CCR-9257280 and CCR-9058180 and by donations from Altera Corporation. John O'Leary is supported by a Fellowship from Bell-Northern Research Ltd. Miriam Leeser is supported in part by an NSF Young Investigator Award. Jason Hickey is an employee of Bellcore. Mark Aagaard is supported by a Fellowship from Digital Equipment Corporation. We would like to thank Peter Soderquist for his help in understanding the algorithm and its implementations, Mark Hayden for his work on the proof of the algorithm, Shee Huat Chow and Ser Yen Lee for implementing the VLSI version of this chip, and Michael Bertone and Johanna deGroot for the Altera implementation.

References

[AL94]   Mark D. Aagaard and Miriam E. Leeser. A methodology for reusable hardware proofs. *Formal Methods in System Design*, 1994. To appear.

[BJP93]   Bhaskar Bose, Steve D. Johnson, and Shyamsundar Pullela. Integrating Boolean verification with formal derivation. In David Agnew, Luc Claesen, and Raul Camposano, editors, *Computer Hardware Description Languages and their Applications*, IFIP Transactions A-32. Elsevier, North-Holland, 1993.

[BV85]   J. Bannur and A. Varma. A VLSI implementation of a square root algorithm. In *IEEE Symp. on Comp. Arithmetic*, pages 159–165. IEEE Comp. Soc. Press, Washington D.C., 1985.

[CGM86]   Alberto Camilleri, Mike Gordon, and Tom Melham. Hardware verification using higher-order logic. In D. Borrione, editor, *From HDL Descriptions to Guaranteed Correct Circuit Designs*. Elsevier, September 1986.

[Flo63]   Ivan Flores. *The Logic of Computer Arithmetic*. Prentice Hall, Englewood Cliffs, NJ, 1963.

[Lee92]   Miriam E. Leeser. Using Nuprl for the verification and synthesis of hardware. In C. A. R. Hoare and M. J. C. Gordon, editors, *Mechanized Reasoning and Hardware Design*. Prentice-Hall International Series on Computer Science, 1992.

[NS88]   Paliath Narendran and Jonathan Stillman. Formal verification of the sobel image processing chip. In *DAC*, pages 211–217. IEEE Comp. Soc. Press, Washington D.C., 1988.

[OLLA92]   John O'Leary, Mark Linderman, Miriam Leeser, and Mark Aagaard. HML: A hardware description language based on Standard ML. Technical Report EE-CEG-92-7, Cornell School of Electrical Engineering, October 1992.

[OLLA93]   John O'Leary, Mark Linderman, Miriam Leeser, and Mark Aagaard. HML: A hardware description language based on SML. In David Agnew, Luc Claesen, and Raul Camposano, editors, *Computer Hardware Description Languages and their Applications*, IFIP Transactions A-32. Elsevier, North-Holland, 1993.

[VCH94]   D. Verkest, L. Claesen, and H. De Man A proof of the nonrestoring division algorithm and its implementation on an ALU. *Formal Methods in System Design*, 4(1):5–31, January 1994.

An automatic generalization method for the inductive proof of replicated and parallel architectures

Laurence PIERRE

Laboratoire d'Informatique de Marseille - URA CNRS 1787
CMI / Université de Provence
Technopôle de Château-Gombert, 39 rue Joliot-Curie
13453 Marseille Cedex 13 - FRANCE
e-mail : laurence@gyptis.univ-mrs.fr

Abstract : Our approach for verifying the equivalence of two VHDL architectures consists in translating these descriptions into functional forms and then in proving the equivalence of these functions. As far as replicated or parallel architectures are concerned, an induction-based method is used for verifying generic n-bit descriptions. This technique takes advantage of the regular structure of these devices and can give better results than the BDD-based approaches. However, induction requires complete specifications, whereas the designers usually supply partial specifications. Therefore, we propose a specialized automatic method for generalizing such incomplete statements, before the Boyer-Moore proof process.

I. INTRODUCTION

The concept of formal verification includes various aspects [9], this paper focuses on verifying that the function implemented by a given VLSI circuit implies or is equivalent to its specification. Our system PREVAIL™, that is being developed in cooperation with Imag/Artemis, is a prototype proof environment which includes a set of verification tools [5]. It takes as input VHDL descriptions [13] and verifies the equivalence (or the implication) of two different *architectures* of the same *entity* (two descriptions given at different abstraction levels, or a description and its optimization, etc...). The most appropriate verification tool is selected with the help of the designer, and the descriptions are automatically translated into the formalism of this proof system. One of the included proof tools is the Boyer-Moore system, Nqthm [3][4].

This system, which is essentially based on the induction principle, is particularly useful for the verification of generic replicated or parallel architectures. In effect, such a theorem prover bring several advantages :

- we do not have to care about the size of the device, we recursively describe generic n-bit architectures and we prove generic properties,

- the specification can be given at the arithmetic level, and we can use conversion functions between bit-vectors and integers.
  This feature is very important for the hierarchical verification of large arithmetic devices : after being verified, each elementary component can be replaced by its arithmetic specification. Thus, the whole system verification is performed at the arithmetic level, and the efficiency is significantly improved.

This approach circumvents some drawbacks of the well-known tautology-checking techniques based on BDD-like representations [6] (see for instance [10], [17]), where :

- the size of the circuit must be fixed to 16, 32, 64... bits, and the proof system verifies the correctness of each output independently,
- the "specification" of the circuit must be another implementation of the same arithmetical function, also considered at the boolean level. Usually, even if the specification is given at the arithmetical level, it has to be transformed into an equivalent bit-level implementation, using a library of simple basic modules (adders, multipliers,...).

Nevertheless, a correct recursive model cannot be deduced from a simple netlist of the circuit, and our approach requires a particular description methodology. The automatic translation task is feasible provided that some description rules are respected. In particular, the genericity of the architecture must be expressed by a declaration of the form generic(N:positive), and the regularity must be described using a for ... generate construct. Moreover, the theorems to be verified are usually expressed in terms of given values for the incoming boolean signals (input carries). It means that induction cannot be applied, unless we are able to provide a generalized version where carries are considered as symbolic variables. Generalization is a well-known problem in the framework of induction-based proofs, and no general-purpose method can be found. Conversely, it may be possible to mechanize some specialized methods. Here, we propose an automatic technique for generalizing the partial specifications of one-dimensional, as well as two-dimensional, replicated structures. This method extends a previous work on simple one-dimensional regular devices [19].

Finally, our methodology is able to give a positive answer to one of the questions that have been asked in the presentation of the N-bit adder TPCD benchmark [14] : "Is the verification done for arbitrary N?". We will use this simple ripple-carry adder as running example throughout this paper; its VHDL description is given by Figure 1. More elaborate examples can be found in paragraph V.

```
entity rca is
  generic(N:positive);
  port(a,b:in bit_vector(0 to N-1); c:in bit;
       z:out bit_vector(0 to N));
end rca;

architecture struct of rca is
component fulladd
  port(a,b,carry:in bit; fout,fcarry:out bit);
end component;
signal carry:bit_vector(0 to N);
begin
   addblock:block
   for all:fulladd use entity work.fulladd(stadd);
   begin      --- Regular part :
     fadd:for I in 0 to N-1 generate
       f1:fulladd port map(a(I),b(I),carry(I),z(I),carry(I+1));
     end generate;
     carry(0) <= c;
     z(N) <= carry(N);
   end block;
end struct;
```

FIGURE 1 : VHDL DESCRIPTION OF THE RIPPLE-CARRY ADDER

II. BRIEF OVERVIEW OF THE BOYER-MOORE SYSTEM

All along this paper, we only have to suppose that the proofs are performed in an induction-based environment. In fact, we use Nqthm and the reader will find some lines of Boyer-Moore code. Therefore, we give a brief overview of this system.

The Boyer-Moore theorem prover is based on a quantifier-free first order logic with equality. The three basic principles of this prover are [3][4] :

- **The "shell" principle** :
  Inductive abstract data types - called "shells" - can be built by means of a bottom object, a constructor, and one or more accessors. A boolean function, called a recognizer, recognizes if an object belongs to the shell.
  The type of "natural numbers" is a well-known example : the bottom is 0, the constructor is $\lambda x.\, x+1$, and the accessor is $\lambda x.\, x-1$; the predicate recognizer of this type in Boyer-Moore is *numberp*.

- **The definition principle** :
  Prior to accepting a recursive function definition, the system verifies that there exists a measure which decreases according to a "well-founded" relation.
  For example, we can define the following function "times" over natural numbers :
  $$times\,(i, j) =_{def} if\ i = 0\ then\ 0\ else\ j + times\,(i\text{-}1, j)$$
  The term i-1 decreases at each recursive call, and the recursion stops when i=0. Thus, this recursive definition is correct. Much more elaborate function templates can also be accepted.

- **The induction principle** :
  It allows to prove inductive theorems over recursive functions. Induction schemes are *automatically* generated according to the definitions of the recursive functions involved in the theorems. For example, let us assume that we want to verify the following proposition P(x,y) :
  $$numberp(x)\ and\ numberp(y)\ \Rightarrow\ times(x, y+1) = x + times(x, y)$$
  The induction scheme generated for the proof of P(x,y) is :
         1. $x = 0\ \Rightarrow\ P(x,y)$
         2. $(x \neq 0)$ and $P(x\text{-}1,y)\ \Rightarrow\ P(x,y)$

A brief state of the art about Nqthm and the formal proof of software/hardware can be found in [20].

Remark : In the following, we will have to use bit-vectors. The bit-vector "shell" has been proposed in [11]. It is defined by means of : the bottom (btm), the constructor **bitv**, the accessors **bit** (least significant bit) and **vec** (rest of the vector), and the recognizer **bitvp**.

III. MODELLING AND VERIFYING REPLICATED DEVICES

In this paragraph, we give recursive models for regular architectures. Recursion is well-suited for describing the duplication of a same hardware motif. Throughout this section, recursive functional forms are associated with the outputs of replicated structures :

- the parameters are the inputs of the system, and carry propagation is translated by the updating of the corresponding parameter(s) with the value(s) computed by the function(s) that express this propagation,

- if the output is vectorial, the function builds this vector by means of `bitv`, which appends a bit to a bit-vector.

III.1 A recursive model for one-dimensional replicated circuits

Here, we give recursive functional models for representing the output(s) of one-dimensional regular architectures. These devices usually take as inputs several bit-vectors of identical length N, which is given as a `generic` parameter of the description. Since we deal with bit-vectors of the same length, only one of them is used as recursion parameter in our definitions. In the following, we use the term "carry" for describing any boolean result computed in the ith module and passed to the $(i+1)^{th}$ one.

Notation : the vectorial inputs are denoted VI_1, ..., VI_j, and the scalar inputs (which include the propagated carries) are denoted SI_1, ..., SI_k.

First, we give the function which corresponds to a vectorial output vo_i, $1 \le i \le i_{vo}$, where i_{vo} is usually $\le j$. The function $REC\text{-}F_{VOi}$ builds the bit-vector formed by the result of F_{VOi} (the function associated with the corresponding boolean output of the basic cell), catenated to the result of the recursive call to $REC\text{-}F_{VOi}$ with the vectorial inputs of one element less and the scalar inputs updated by the functions F_{SI1}, ... , F_{SIk} (which express the carry propagation) :

```
REC-F_VOi (VI₁, …, VI_j, SI₁, …, SI_k) =def
   if bitvp(VI₁)        --- if VI₁ is a bit-vector
   then if VI₁ = (btm)           --- if VI₁ is empty, the result
        then F_SI(SI₁, …, SI_k)    --- is a function of SI₁, …, SI_k.
        --- if VI₁ is not empty, we recursively call REC-F_VOi :
        else bitv(F_VOi(bit(VI₁), …, bit(VI_j), SI₁, …, SI_k),
                  REC-F_VOi (vec(VI₁), …, vec(VI_j),
                        F_SI1(bit(VI₁), …, bit(VI_j), SI₁,…, SI_k),
                        …
                        F_SIk(bit(VI₁),…, bit(VI_j), SI₁,…, SI_k)))
   else F_SI(SI₁, …, SI_k)
```

The model which is associated with **a scalar output** so_i, $1 \le i \le i_{so}$, is similar, except that it does not build a vectorial result :

```
REC-F_SOi (VI₁, …, VI_j, SI₁, …, SI_k) =def
   if bitvp(VI₁)     --- if VI₁ is a bit-vector
   then if VI₁ = (btm)            --- if VI₁ is empty, the result
        then F_SI(SI_i)           --- is a function of SI_i
        --- if VI₁ is not empty, we recursively call REC-F_SOi :
        else REC-F_SOi (vec(VI₁), …, vec(VI_j),
                  F_SI1(bit(VI₁), …, bit(VI_j), SI₁, …, SI_k),
                  …
                  F_SIk(bit(VI₁), …, bit(VI_j), SI₁, …, SI_k))
   else F_SI(SI_i)
```

Example : The description of the ripple-carry adder (Figure 1) has one vectorial output z. It is expressed by the following recursive function RCA-z (given in the Boyer-Moore syntax), which is of the form of REC-F$_{VOi}$ where A and B are the vectorial inputs and c is a scalar input (the carry to be propagated) :

```
(defn RCA-z (A B c)
  (if (bitvp A)
      (if (equal A (btm))
          (bitv c (btm))
          (bitv (fulladd-fout (bit A) (bit B) c)
                (RCA-z (vec A)
                       (vec B)
                       (fulladd-fcarry (bit A) (bit B) c))))
      (bitv c (btm))))
```

where fulladd-fout and fulladd-fcarry are the non-recursive functions associated with the outputs fout and fcarry of the component fulladd.

III.2 A recursive model for two-dimensional replicated circuits

Now, let us focus on two-dimensional regular structures. More precisely, we are interested in parallel systems which are built from N instances of the same row, where each row is also a replicated device. Thus, the outputs of the basic row are represented by functional forms which correspond to the templates of § III.1.

Here, we give a recursive model for representing the output of the whole parallel architecture; this function will be defined in terms of the REC-F$_{VOi}$, $1 \le i \le i_{vo}$.

<u>Notation</u> : the vectorial inputs are denoted VI$_1$, ..., VI$_p$, and the scalar inputs are denoted SI$_1$, ..., SI$_q$. Typically, $p > j$ where j is the index used in § III.1. A necessary condition is $i_{vo} \le j$. Usually, the architecture is such that $i_{vo} = j-1$, $q = k-1$, and $p = j+1$ (i.e. VI$_p$ expresses the second dimension). In order to simplify the model given below, we will assume these equalities (but it can easily be transformed into a more general one).

The following function F$_{par}$ models the output of such a parallel circuit :

```
Fpar(VI1, ..., VIp, SI1, ..., SIq) =def
    if bitvp(VIp)       --- if VIp is a bit-vector
    then if VIp = (btm)            --- if VIp is empty, the result is
                                   --- output by the last component
            then Last-comp(VI1, ..., VIp-1, SI1, ..., SIq)
            --- if VIp is not empty, we recursively call Fpar :
         else bitv(Fb(bit(VI2), ..., bit(VIp-1))),
                    Fpar(VI1,
                         REC-FVO1(VI1, ..., bit(VIp), SI1, ..., SIq),
                         ...,
                         REC-FVOivo(VI1, ..., bit(VIp), SI1, ..., SIq),
                         vec(VIp), SI1, ..., SIq))
    else Last-comp(VI1, ..., VIp-1, SI1, ..., SIq)
```

This function builds the bit-vector formed by the result of F_b(bit(VI$_2$), ..., bit(VI$_{p-1}$)) where F_b is a boolean function (the function associated with the corresponding output of the basic row), catenated to the result of the recursive call to F_{par} with the vectorial input VI$_p$ of one element less and the vectorial inputs updated by the functions REC-F$_{VO1}$, ..., REC-F$_{VOivo}$ (which express, if needed, the carry propagation through the rows).

Remark : Last-comp is a vectorial function which represents the last row, in particular it can be different from the other ones.

An illustrative example of this modelling method can be found in § V.2.

III.3 The induction-based proof method

1. As far as **one-dimensional circuits** are concerned, the proof process consists in verifying the equivalence between each function REC-F$_{VOi}$ or REC-F$_{SOi}$ (or a combination of these functions) and a function which expresses the arithmetic specification. In order to (formally) compare the results computed by these functions, we have to abstract the first one to the arithmetic level, using the function bv-to-nat which converts bit-vectors into natural numbers, starting from the LSB.

The theorem to be proved generally states the equivalence between the result of the function REC-F$_{VOi}$ converted into a natural number and a specification function, denoted below SPEC$_{VOi}$. This property is generally given in terms of particular initial values for the scalar inputs, referred to as IV-SI$_1$, ..., IV-SI$_k$:

```
bitvp(VI₁) and ... and bitvp(VIⱼ)      --- the VIᵢ are bit-vectors
and size(VI₁) = size(VI₂)      --- the vectors have the same length
... and size(VI₁) = size(VIⱼ)
                        ⇓
bv-to-nat(REC-Fᵥₒᵢ(VI₁, ..., VIⱼ, IV-SI₁, ..., IV-SIₖ))
= SPECᵥₒᵢ(bv-to-nat(VI₁), ..., bv-to-nat(VIⱼ))
```

Th₁

Remarks : • In the case where we have to reason on REC-F$_{SOi}$, we use the same statement, except that we do not have to call on the function bv-to-nat since REC-F$_{SOi}$ gives a boolean result.

  • If the property to be proved is an equivalence between a combination of the results of the functions REC-F$_{VO1}$, ..., REC-F$_{VOj}$ and a specification expression, we use the same kind of theorem except that the term bv-to-nat(REC-F$_{VOi}$(VI$_1$, ..., VI$_j$, IV-SI$_1$,..., IV-SI$_k$)) is replaced by a term of the form F(bv-to-nat(REC-F$_{VO1}$(VI$_1$, ..., IV-SI$_k$)), ..., bv-to-nat(REC-F$_{VOj}$(VI$_1$, ..., IV-SI$_k$))).

Example : The correctness theorem for the circuit of Figure 1 is the theorem proof-of-RCA-z which verifies that this device outputs the sum of the two bit-vectors A and B, provided that the carry-in is set to false :

```
(prove-lemma proof-of-RCA-z (rewrite)
    (implies (and (bitvp A) (bitvp B) (equal (size A) (size B)))
            (equal (bv-to-nat (RCA-z A B f))
                    (plus (bv-to-nat A) (bv-to-nat B))))))
```

We will see in section V that other significant examples can be verified using the same modelling and proof methodology.

2. With respect to **two-dimensional architectures**, the verification process is hierarchically decomposed into sub-proofs : first we verify the correctness of the basic row, and then it is possible to validate the whole device.

The first necessary step consists in giving pieces of information about the size(s) of the bit-vector(s) output by the one-dimensional row. One theorem per row output indicates the size of this vectorial output in terms of the input size(s). Then, the correctness of this row is verified using one or several theorems of the form of Th_1, provided that they have previously been generalized. Finally, a theorem of the form of Th_2 below allows to validate the two-dimensional structure :

$$
\begin{array}{ll}
\texttt{bitvp(VI}_1\texttt{)} \ \textbf{and} \ \texttt{bitvp(VI}_p\texttt{)} & \texttt{--- VI}_1 \ \texttt{and VI}_p \ \texttt{are bit-vectors} \\
\textbf{and} \ \texttt{size(VI}_1\texttt{)} = \texttt{size(VI}_p\texttt{)} & \texttt{--- They have the same length} \\
\end{array}
$$

$$\Downarrow$$

$$
\begin{array}{l}
\texttt{bv-to-nat(F}_{par}\texttt{(VI}_1\texttt{, F}_{init1}\texttt{(VI}_1\texttt{, VI}_p\texttt{), ...,} \\
\qquad\qquad \texttt{F}_{initp-2}\texttt{(VI}_1\texttt{, VI}_p\texttt{), VI}_p\texttt{, IV-SI}_1\texttt{, ..., IV-SI}_q\texttt{))} \\
= \texttt{SPEC(bv-to-nat(VI}_1\texttt{), ..., bv-to-nat(VI}_p\texttt{))}
\end{array}
\qquad Th_2
$$

where - SPEC corresponds to the specification,

- F_{init1} is the function which computes the initial value of VI_2, ..., and
  $F_{initp-2}$ is the function which computes the initial value of VI_{p-1},
- and $IV-SI_i$, $1 \le i \le q$, is the initial value of SI_i.

Here, as well as in the case of one-dimensional devices, the presence of particular values for the incoming carries implies that an inductive proof is not feasible. Such a theorem must be generalized, this is the purpose of section IV.

IV. THE GENERALIZATION METHOD

IV.1 Motivation

In the fields of software as well as hardware verification, generalization is often necessary as soon as induction-based techniques are applied. For instance, Manna and Waldinger describe the problem of "generalization of specifications" in the framework of program synthesis [16]. Finding general-purpose generalization methods that can be fully mechanized is almost impossible. Some interesting heuristics, that require user-interaction, have been proposed. Among them, let us recall the following approaches :

- J Moore, one of the authors of the Boyer-Moore prover, proposed an interesting generalization heuristics for recursive functions with accumulating parameters [15]. However, there is a step where the user has to "guess" the term by which a certain variable must be replaced. In fact, this heuristics has not been implemented in the Boyer-Moore system.
- R.Aubin also worked on such generalization heuristics. With respect to the problem of verifying the equivalence between recursive functions and corresponding iterative ones, he gave a method called "indirect generalization" [2]. This method introduces, in the expression to be proved, a new function call

associated with its neutral element, and then generalizes this constant. Here also, the appropriate function has to be intuitively determined by the user.

Our goal is not to propose a general-purpose technique. We give a specialized method for our problem, and this special-purpose algorithm can be mechanized. A prototype implementation has been included in PREVAIL™.

IV.2 Generalization algorithm

The algorithm below is devoted to regular replicated architectures. Starting from a theorem of the form of Th_1 or Th_2, this method yields a more general lemma which is provable by induction. The generalization algorithm is :

Unfold once the definition associated with the implementation (i.e. REC-F_{VOi} or REC-F_{SOi} or F_{par}) in the left-hand side of the equality ;

While the new expression includes terms of the form bit(VI_n) (or bit(GVI_n)) **do**

    Perform a case analysis on the value of bit(VI_n) :   bit(VI_n) $= f$
                                                   bit(VI_n) $= t$

    For each case **do**

        Simplify the resulting equality ;
        Unfold once the definition of "bv-to-nat" ;
        Apply simplification rules ;
        Generalize the term vec(VI_n) into GVI_n ;
    EndFor ;
EndWhile ;

From the set of expressions obtained from the previous case analysis, deduce a single (possibly conditional) expression E ;

If the implementation function corresponds to a two-dimensional architecture

Then use theorems Th_1 to transform the left-hand side of this equality (and the right-hand side accordingly) :

E is of the form :

    bv-to-nat(F_{par}(VI_1, REC-F_{VO1}(VI_1, VI_2', —, vi_p', SI_1, —, SI_q),
                     —, REC-F_{VOivo}(VI_1, VI_2', —, vi_p', SI_1, —, SI_q),
                    VI_p, SI_1, —, SI_q))
  = G(bv-to-nat(VI_1),
      F_1(bv-to-nat(VI_1), —, bv-to-nat(VI_{p-1}'), vi_p', SI_1, —, SI_q),
      ...
      F_{ivo}(bv-to-nat(VI_1), —, bv-to-nat(VI_{p-1}'), vi_p', SI_1,—, SI_q),
      bv-to-nat(VI_p))

and theorems Th_1 give equalities of the form :

    bv-to-nat(REC-F_{VOi}(VI_1, —, VI_{p-1}, vi_p, SI_1, —, SI_q))
  = SPEC$_{VOi}$(bv-to-nat(VI_1), —, bv-to-nat(VI_{p-1}), vi_p, SI_1, —, SI_q)

where vi_p plays the role of bit(VI_p).

By replacing VI_2' by VI_2', ..., VI_{p-1} by VI_{p-1}', vi_p by vi_p', in these equalities, we get equalities such as :

```
bv-to-nat(REC-Fvoi(VI1, -, VIp-1', vip', SI1, -, SIq))
= SPECvoi(bv-to-nat(VI1), -,bv-to-nat(VIp-1'),vip',SI1, -,SIq)
```

which is precisely of the form of

```
Fi(bv-to-nat(VI1), -,bv-to-nat(VIp-1'),vip', SI1, -, SIq)
```

It allows to rewrite E into the following expression :

```
bv-to-nat(Fpar(VI1, REC-Fvo1(VI1, VI2', -, vip', SI1, -, SIq),
           -, REC-Fvoivo(VI1, VI2', -, vip', SI1, -, SIq), VIp,
           SI1, -, SIq))
= G(bv-to-nat(VI1),
    bv-to-nat(REC-Fvo1(VI1, -, VIp-1', vip', SI1, -, SIq))),
    -,
    bv-to-nat(REC-Fvoivo(VI1, -, VIp-1', vip', SI1, -, SIq))),
    bv-to-nat(VIp))
```

which is finally generalized into :

```
bv-to-nat(Fpar(VI1, X1, -, Xivo, VIp, SI1, -, SIq))
= G (bv-to-nat(VI1), bv-to-nat(X1), -, bv-to-nat(Xivo),
     bv-to-nat(VIp))
```

Result := this expression ;

Else   Result := E;

EndIf.

Example : As far as the ripple-carry adder is concerned, the algorithm stops at the end of the *While* loop. Starting from :

```
(prove-lemma proof-of-RCA-z (rewrite)
   (implies (and (bitvp A) (bitvp B) (equal (size A) (size B)))
            (equal (bv-to-nat (RCA-z A B f))
                   (plus (bv-to-nat A) (bv-to-nat B)))))
```

with

```
(defn bv-to-nat (x)
   (if (bitvp x)
       (if (equal x (btm))
           0
           (plus (if (bit x) 1 0) (times 2 (bv-to-nat (vec x)))))
   0))
```

the generalization process is :

Assuming that we are not in the basis case, and unfolding the definition of RCA-z *gives :*

```
(equal (bv-to-nat
              (bitv (xor f (xor (bit A) (bit B)))
                    (RCA-z (vec A) (vec B)
                           (or (and (bit A) (bit B))
                               (or (and (bit A) f) (and (bit B) f))))))
       (plus (bv-to-nat A) (bv-to-nat B)))
```

which simplifies into :

```
(equal (bv-to-nat
              (bitv (xor (bit A) (bit B))
                    (RCA-z (vec A) (vec B) (and (bit A) (bit B)))))
       (plus (bv-to-nat A) (bv-to-nat B)))
```

and the case analysis is :

1. If (bit a) = true

```
(equal (bv-to-nat
              (bitv (not (bit b)) (RCA-z (vec A) (vec B) (bit B))))
       (plus (add1 (times 2 (bv-to-nat (vec A)))) (bv-to-nat B)))
```

i.e.

```
(equal (bv-to-nat (bitv (not (bit b)) (RCA-z GA (vec B) (bit B))))
       (plus (add1 (times 2 (bv-to-nat GA))) (bv-to-nat B)))
```

1.1. If (bit b) = true

```
(equal (bv-to-nat (bitv f (RCA-z GA (vec B) t)))
       (plus (add1 (times 2 (bv-to-nat GA)))
             (add1 (times 2 (bv-to-nat (vec B))))))
```

i.e.

```
(equal (times 2 (bv-to-nat (RCA-z GA (vec B) t)))
       (times 2 (add1 (plus (bv-to-nat GA))
                             (bv-to-nat (vec B)))))
```

which is simplified and generalized into

```
(equal (bv-to-nat (RCA-z GA GB t))
       (add1 (plus (bv-to-nat GA)) (bv-to-nat GB)))
```

1.2. If (bit b) = false

```
(equal (bv-to-nat (bitv t (RCA-z GA (vec B) f)))
       (plus (add1 (times 2 (bv-to-nat GA)))
             (times 2 (bv-to-nat (vec B)))))
```

i.e.

```
(equal (add1 (times 2 (bv-to-nat (RCA-z GA (vec B) f))))
       (add1 (plus (times 2 (bv-to-nat GA))
                   (times 2 (bv-to-nat (vec B))))))
```

which is simplified and generalized into

```
(equal (bv-to-nat (RCA-z GA GB f))
       (plus (bv-to-nat GA) (bv-to-nat GB)))
```

2. If (bit a) = false

```
(equal (bv-to-nat (bitv (bit b) (RCA-z (vec A) (vec B) f)))
       (plus (times 2 (bv-to-nat (vec A))) (bv-to-nat B)))
```

i.e.

```
(equal (bv-to-nat (bitv (bit b) (RCA-z GA (vec B) f)))
       (plus (times 2 (bv-to-nat GA)) (bv-to-nat B)))
```

Similarly, we get in that case :

2.1. If (bit b) = true

```
(equal (bv-to-nat (RCA-z GA GB f))
       (plus (bv-to-nat GA) (bv-to-nat GB)))
```

2.2. If (bit b) = false

```
(equal (bv-to-nat (RCA-z GA GB f))
       (plus (bv-to-nat GA) (bv-to-nat GB)))
```

Then, from sub-cases 1.2, 2.1 and 2.2, we have :
```
(equal (bv-to-nat (RCA-z GA GB f))
       (plus (bv-to-nat GA) (bv-to-nat GB)))
```
and from sub-case 1.1, we have :
```
(equal (bv-to-nat (RCA-z GA GB t))
       (add1 (plus (bv-to-nat GA) (bv-to-nat GB))))
```

Thus, we deduce the complete generalized theorem :
```
(prove-lemma proof-of-RCA-z-gen (rewrite)
   (implies (and (bitvp GA) (bitvp GB) (boolp c)
                 (equal (size GA) (size GB)))
            (equal (bv-to-nat (RCA-z GA GB c))
                   (if c (add1 (plus (bv-to-nat GA) (bv-to-nat GB)))
                         (plus (bv-to-nat GA) (bv-to-nat GB)))))))
```

V. APPLICATION TO THE TPCD BENCHMARKS

Among the devices to which this technique applies, some of them have been proposed as *TPCD benchmarks*, or parts of these benchmarks [14]. Apart from the N-bit ripple-carry adder, some sub-modules of the Min-Max circuit, as well as the parallel multiplier can be given as illustrative examples.

V.1 Min-Max

A specification of this benchmark has been proposed for the IFIP WG 10.2 International Workshop on "Applied Formal Methods for Correct VLSI Design" [22]. We have designed an implementation (with arithmetic modules for unsigned bit-vectors only) and we have realized a hierarchical proof of this circuit [18].

The specification is that the Min-Max unit has 3 boolean control signals CLEAR, RESET and ENABLE. The unit produces an output sequence OUT at the same rate as IN :

- if CLEAR is *true*, then OUT equals 0, independent of the other control signals,
- if CLEAR is *false* and ENABLE is *false*, then OUT equals the last value of IN before ENABLE became *false*,
- if CLEAR is *false*, ENABLE is *true*, and RESET is *true*, then OUT follows IN,
- if RESET becomes *false*, then OUT holds, on each time point t, the mean value of the maximum and minimum value of IN until that time point, since RESET became *false*.

The most important component of our implementation is a sub-module "MeanValue" which is supposed to correspond to the last part of the specification : *OUT holds, on each time point t, the mean value of the maximum and minimum value of IN until that time point, since RESET became false.* This device is depicted on Figure 2. PASTMAX and PASTMIN are two registers, and the other components are N-bit arithmetic modules :

- GREAT_N and LESS_N implement the ">" and "<" comparisons on N-bit vectors,
- MUX_N multiplexes two N-bit vectors according to a control bit,
- RCA is the ripple-carry adder ,
- and RIGHTSHIFT shifts a N-bit vector to the right, i.e. divides the corresponding natural number by 2.

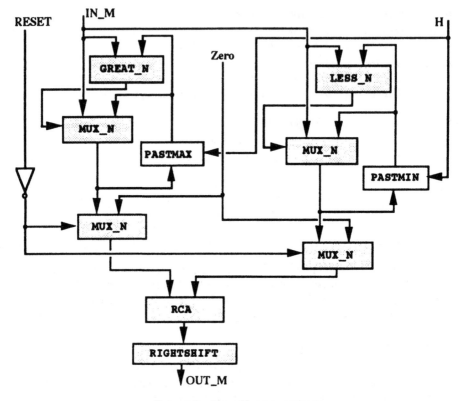

FIGURE 2 : MEANVALUE MODULE

Because of the lack of place, we give neither all the pictures of the sub-modules nor the VHDL descriptions. We will only describe the Boyer-Moore verification of the most significant components. The Boyer-Moore code is automatically generated from the VHDL descriptions, with a limited user interaction (through a menu-directed interface).

1. The rightshift component.

Here, we give a more elaborate version of this device than in [18]. This circuit inputs a N-bit vector A and a boolean signal c, and outputs a N-bit vector z which corresponds to A shifted to the right, and the most significant bit of which takes the value of c. Thus, the Boyer-Moore function associated with z is :

```
(defn rightshift-Z (A c)
  (if (bitvp A)
      (if (equal A (btm))
          (btm)
          (if (equal (vec A) (btm))
              (bitv c (btm))
              (bitv (bit (vec A)) (rightshift-Z (vec A) c))))
      (btm)))
```

We have to verify that, when c equals false, the result corresponds to (bv-to-nat A) divided by 2, i.e. :

```
(prove-lemma proof-of-rightshift-Z (rewrite)
  (implies (bitvp A)
           (equal (bv-to-nat (rightshift-Z A f))
                  (quotient (bv-to-nat A) 2))))
```

Since c is not propagated through the shifter cells, generalization is not mandatory. However, this lemma can be generalized into the following one :

```
(prove-lemma proof-of-rightshift-Z-gen (rewrite)
  (implies (and (bitvp A) (not (equal A (btm))) (boolp c))
           (equal (bv-to-nat (rightshift-Z A c))
                  (if c
                      (plus (quotient (bv-to-nat A) 2)
                            (exp 2 (sub1 (size A))))
                      (quotient (bv-to-nat A) 2)))))
```

2. The modules GREAT_N and LESS_N.

Each of these devices inputs two N-bit vectors A and B, and a carry-in X_IN, and compares the bit-vectors, starting from the least significant bit. The carry is propagated up to the last cell, where it finally represents the result of the comparison. Figure 3 depicts the component GREAT_N, the other one is rather similar. We give the proof of this module GREAT_N, the verification of LESS_N is processed similarly.

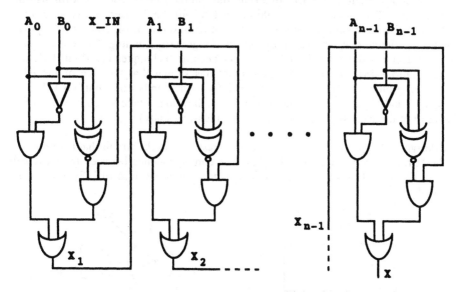

FIGURE 3 : GREAT_N MODULE

This circuit has a boolean output x, and the corresponding Boyer-Moore function is :

```
(defn great_n-X (A B X_IN)
  (if (bitvp A)
      (if (equal A (btm))
          X_IN
          (great_n-X (vec A) (vec B)
                     (or (and (bit A) (not (bit B)))
                         (and (eqv (bit A) (bit B)) X_IN))))
      X_IN))
```

Then, we have to verify that this device implements correctly the associated comparison, provided that the input carry equals `false`. Our system generates the following theorem :

```
(prove-lemma proof-of-great_n-X (rewrite)
    (implies (and (bitvp A) (bitvp B) (equal (size A) (size B)))
            (equal (great_n-X A B f)
                    (greaterp (bv-to-nat A) (bv-to-nat B))))))
```

Here, generalization is mandatory. The generalization algorithm generates 4 sub-cases and finally produces the following generalized lemma :

```
(prove-lemma proof-of-great_n-X-gen (rewrite)
    (implies (and (bitvp A) (bitvp B)
                (equal (size A) (size B)) (boolp x))
            (equal (great_n-X A B x)
                    (if x
                        (or (greaterp (bv-to-nat A) (bv-to-nat B))
                            (equal (bv-to-nat A) (bv-to-nat B)))
                        (greaterp (bv-to-nat A) (bv-to-nat B)))))))
```

V.2 Parallel multiplier

This second benchmark will illustrate our methodology for two-dimensional architectures. This device is a combinational parallel multiplier. Many significant methodologies have been developed for the verification of parallel array multipliers. The most efficient ones are based on constraint logic programming [21], or have been inspired from the development of tautology-checking techniques based on BDD-like representations [8], [1]. However, R.Bryant has shown that the size of the BDD representing multiplication grows exponentially in the number of input bits [7]. Even though [8] and [1] try to overcome this problem, they do not take advantage of the fact that most of these circuits are completely (or at least partially) regular.

The design that is proposed in [14] exactly corresponds to the circuit that is referred to as the "Braun's array multiplier" in [12]. The correspondence between the basic cell used in the TPCD design and the fulladder-based module used in [12] is given by Figure 4 below.

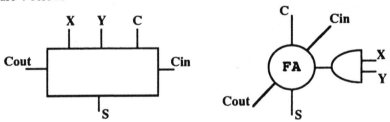

Figure 4a. TPCD module   *Figure 4b. FullAdder-based module*

FIGURE 4 : CORRESPONDENCE BETWEEN BASIC MODULES

A 4-bit version of the Braun multiplier is depicted by Figure 5. A generic N-bit VHDL description and a detailed Nqthm proof can be found in [20]. Here we propose an outline of this verification. This device is made of a succession of similar rows, and the last row is a ripple-carry adder. Its hierarchical proof consists in validating the basic row, the ripple-carry adder (proof already presented), and then the whole multiplier.

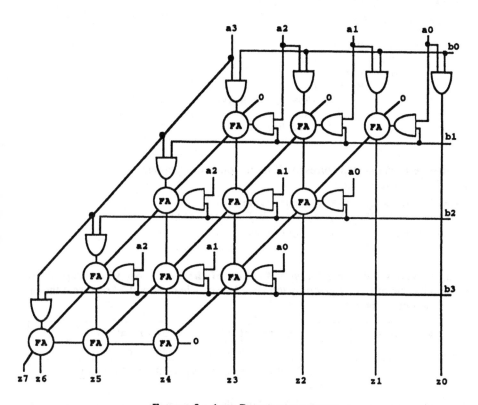

FIGURE 5 : 4-BIT BRAUN MULTIPLIER

Two functions are associated with the two vectorial outputs of the basic row. The case where (vec A) = (btm) must be processed separately (irregularity on the last cell) :

```
(defn Row-x_out (A X Y b)
  (if (bitvp A)
      (if (equal A (btm))
          (btm)
          (if (equal (vec A) (btm))
              (bitv (and (bit A) b) (btm))
              (bitv (fulladd-fout (bit X) (bit Y) (and (bit A) b))
                    (Row-x_out (vec A) (vec X) (vec Y) b))))
      (btm)))
(defn Row-y_out (A X Y b)
  (if (bitvp A)
      (if (equal A (btm))
          (btm)
          (if (equal (vec A) (btm))
              (btm)
              (bitv (fulladd-fcarry (bit X) (bit Y) (and (bit A) b))
                    (Row-y_out (vec A) (vec X) (vec Y) b))))
      (btm)))
```

The following function Mult-Z corresponds to the vectorial output of the multiplier, where left-shift is the function which specifies a left-shifter, i.e. a multiplication by two (in our system, its definition and associated properties are included in a library of pre-proven and re-usable basic components) :

```
(defn Mult-Z (A X Y B)
  (if (bitvp B)
      (if (equal B (btm))
          (RCA-z X (left-shift Y) f)
          (bitv (bit X)
                (Mult-Z A
                        (Row-x_out A (vec X) Y (bit B))
                        (Row-y_out A (vec X) Y (bit B))
                        (vec B))))
      X))
```

The first verification step corresponds to the following theorems size-of-Row-x_out and size-of-Row-y_out :

```
(prove-lemma size-of-Row-x_out (rewrite)
  (equal (size (Row-x_out A X Y b)) (size A)))))
```

```
(prove-lemma size-of-Row-y_out (rewrite)
  (equal (size (Row-y_out A X Y b)) (subl (size A))))
```

Then, we verify the functional **specification of the basic row** which depends on the initial value of b. It is given by the table below, where X' and Y' represent the bitvectors output by the row :

value of b	bv-to-nat (X') + 2 bv-to-nat (Y')
f	bv-to-nat (x) + bv-to-nat (y)
t	bv-to-nat (x) + bv-to-nat (y) + bv-to-nat (a)

The Boyer-Moore theorem which translates this specification is of the form of *Th1* where we consider the appropriate combination of the row outputs :

```
(prove-lemma proof-of-Row-x_out-y_out (rewrite)
  (implies (and (bitvp X) (bitvp Y) (bitvp A) (boolp b)
                (equal (size Y) (size X))
                (equal (size X) (subl (size A))))
           (equal (plus (bv-to-nat (Row-x_out A X Y b))
                        (times 2 (bv-to-nat (Row-y_out A X Y b))))
                  (plus (plus (bv-to-nat X) (bv-to-nat Y))
                        (if b (bv-to-nat A) 0))))))
```

After this proof, we should try to verify the **correctness of the multiplier**, i.e. the lemma proof-of-Mult-Z **below** :

```
(prove-lemma proof-of-Mult-Z (rewrite)
  (implies (and (bitvp A) (bitvp B)
                (equal (size B) (size A)))
           (equal (bv-to-nat (Mult-Z A
                                     (init_and A (bit B))
                                     (all-zeros (subl (size A)))
                                     (vec B)))
                  (times (bv-to-nat A) (bv-to-nat B)))))
```

where the functions init$_{and}$ and all-zeros are used to initialize the vectorial inputs X and Y : the initial value of X is the result of "and-ing" each bit of A with the first bit of B, and the initial value of Y is a vector each bit of which is equal to false.

The lemma proof-of-Mult-Z is generalized into the theorem proof-of-Mult-Z-gen :

```
(prove-lemma proof-of-Mult-Z-gen (rewrite)
   (implies (and (bitvp A) (bitvp B) (bitvp X) (bitvp Y)
                (equal (size Y) (sub1 (size A)))
                (equal (size X) (size A)) (lessp (size B) (size A)))
        (equal (bv-to-nat (Mult-Z A X Y B))
               (plus (times 2 (times (bv-to-nat A) (bv-to-nat B)))
                    (plus (times 2 (bv-to-nat Y))
                         (bv-to-nat X))))))
```

The first phase of the generalization algorithm generates 8 sub-cases, from which the following equalities are deduced :

```
(equal (bv-to-nat (Mult-Z A
                          (Row-x_out A (all-zeros (sub1 (size A)))
                                       (all-zeros (sub1 (size A))) c)
                          (Row-y_out A (all-zeros (sub1 (size A)))
                                       (all-zeros (sub1 (size A))) c)
                          B))
        (plus (times 2 (times (bv-to-nat A) (bv-to-nat B)))
             (if c (bv-to-nat A) 0)))
```
and
```
(equal (bv-to-nat (Mult-Z A
                          (Row-x_out A (vec A)
                                       (all-zeros (sub1 (size A))) c)
                          (Row-y_out A (vec A)
                                       (all-zeros (sub1 (size A))) c)
                          B))
        (plus (plus (times 2 (times (bv-to-nat A) (bv-to-nat B)))
                   (bv-to-nat (vec A)))
             (if c (bv-to-nat A) 0)))
```

With regard to the first one, an ad hoc instantiation of theorem proof-of-Row-x_out-y_out gives :

```
(equal (plus (bv-to-nat (Row-x_out A (all-zeros (sub1 (size A)))
                                     (all-zeros (sub1 (size A))) c))
            (times 2 (bv-to-nat
                         (Row-y_out A (all-zeros (sub1 (size A)))
                                      (all-zeros (sub1 (size A)))
                                      c))))
        (plus (plus (bv-to-nat (all-zeros (sub1 (size A))))
                   (bv-to-nat (all-zeros (sub1 (size A)))))
             (if c (bv-to-nat A) 0)))
```
i.e.
```
(equal (plus (bv-to-nat (Row-x_out A (all-zeros (sub1 (size A)))
                                     (all-zeros (sub1 (size A))) c))
            (times 2 (bv-to-nat
                         (Row-y_out A (all-zeros (sub1 (size A)))
                                      (all-zeros (sub1 (size A)))
                                      c))))
        (if c (bv-to-nat A) 0))
```

Thus, replacing `(if c (bv-to-nat A) 0)` **by the equivalent expression in the first equality above produces** :

```
(equal (bv-to-nat (Mult-Z A
                          (Row-x_out A (all-zeros (sub1 (size A)))
                                       (all-zeros (sub1 (size A))) c)
                          (Row-y_out A (all-zeros (sub1 (size A)))
                                       (all-zeros (sub1 (size A))) c)
                          B))
       (plus (times 2 (times (bv-to-nat A) (bv-to-nat B)))
             (plus (bv-to-nat (Row-x_out A (all-zeros (sub1 (size A)))
                                          (all-zeros (sub1 (size A))) c))
                   (times 2
                          (bv-to-nat (Row-y_out A
                                                (all-zeros (sub1 (size A)))
                                                (all-zeros (sub1 (size A)))
                                                c)))))
```

and generalizing the terms `(Row-x_out ...)` **and** `(Row-y_out ...)` **finally gives** :

```
(equal (bv-to-nat (Mult-Z A X Y B))
       (plus (times 2 (times (bv-to-nat A) (bv-to-nat B)))
             (plus (bv-to-nat X) (times 2 (bv-to-nat Y)))))
```

Similarly, for the second one, an ad hoc instantiation of theorem `proof-of-Row-x_out-y_out` **gives** :

```
(equal (plus (bv-to-nat (Row-x_out A (vec A)
                                     (all-zeros (sub1 (size A))) c))
             (times 2 (bv-to-nat (Row-y_out A (vec A)
                                             (all-zeros (sub1 (size A)))
                                             c))))
       (plus (plus (bv-to-nat (vec A))
                   (bv-to-nat (all-zeros (sub1 (size A)))))
             (if c (bv-to-nat A) 0)))
```

which finally produces, after generalization :

```
(equal (bv-to-nat (Mult-Z A X Y B))
       (plus (times 2 (times (bv-to-nat A) (bv-to-nat B)))
             (plus (bv-to-nat X) (times 2 (bv-to-nat Y)))))
```

Both equalities produce the same expression which becomes, with the appropriate hypotheses, the lemma `proof-of-Mult-Z-gen`.

Now, let us compare the efficiency of our approach with results given in significant articles referenced at the beginning of section V.2. The following table gives the number of theorems to be proved and the total CPU times (on a SUN Sparc2, with 32 Mb of memory) for the complete Boyer-Moore proof of this Braun multiplier, as well as for the verification of a simpler multiplier architecture given as example in [8].

	Simple N-bit multiplier	N-bit BRAUN multiplier
Number of theorems	3	5
Total CPU time	14.6 seconds	99.2 seconds

In the table below, we recall some of the experimental results presented in the referenced papers :

Paper	Multiplier size	Proof time
[21]	44-bit	~**100s** on a Sun 3/260
[1]	32-bit	output #31 in **533mn**, and output #15 in **25h**, on a IBM 6000/320 (with 256 Mb of memory)
[8]	16-bit (C6288, ISCAS'85 bench.)	~**40mn** on a Sun 3/60 (with 12 Mb of memory)

The comparison of these tables demonstrates that our technique allows the verification of N-bit architectures within satisfying CPU times.

VI. CONCLUSION

We have proposed an inductive proof methodology devoted to the formal verification of one-dimensional, as well as two-dimensional, replicated structures. This method has been mechanized in the Boyer-Moore logic. We have also described an associated generalization technique for transforming the properties to be verified, before the proof process. The automatic translator between VHDL and Boyer-Moore and the generalization algorithm are being implemented. The examples that illustrate this paper demonstrate the feasibility and the efficiency of the method for validating generic N-bit devices.

Our proof technique requires a particular design methodology; it is not applicable starting from a simple netlist of such a parallel circuit. The translation task between VHDL and Boyer-Moore is feasible if precise description rules have been respected. The illustrative examples show that the circuit descriptions must be hierarchically organized, and the regularity must be expressed by for ... generate and for ... loop statements. Future work will aim at defining a "Design For Verifiability" methodology for such parallel arrays.

In the case where the circuit under consideration consists of an interconnection of various components, this hierarchical aspect can be taken into account within PREVAIL™: several proof tools are integrated, and each component can be verified using the most appropriate system. The team of Flavio Wagner is developing a real framework-based environment for supporting hierarchical proof, library management, etc... [23].

REFERENCES

[1] P.ASHAR, A.GHOSH, S.DEVADAS, A.NEWTON : "Combinational and sequential logic verification using general binary decision diagrams". Proc. Int. Workshop on Logic Synthesis, Research Triangle Park (NC), May 1991.
[2] R.AUBIN : "Mechanizing structural induction-Part II : Strategies". Theoretical Computer Science 9. North-Holland,1979. pp. 347-362.
[3] R.S.BOYER, J S.MOORE : "A Computational Logic". ACM Monograph Series. Academic Press, Inc. 1979.
[4] R.S.BOYER, J S.MOORE : "A Computational Logic Handbook". Perspectives in Computing, Vol. 23. Academic Press, Inc. 1988.

[5] D.BORRIONE, L.PIERRE, A.SALEM : "Formal Verification of VHDL Descriptions in the PREVAIL Environment". IEEE Design&Test magazine, vol. 9, n°2, June 1992.

[6] R.E.BRYANT : "Graph-based Algorithms for Boolean Function Manipulation". IEEE Transactions on Computers, Vol. C-35, n°8, August 1986.

[7] R.E.BRYANT : "On the Complexity of VLSI Implementations and Graph Representations of Boolean Functions with Application to Integer Multiplication". IEEE Transactions on Computers, Vol. 40, n°2, February 1991.

[8] J.R.BURCH : "Using BDDs to verify multipliers". Proc. DAC'91, San Francisco (CA), June 1991.

[9] P.CAMURATI, P.PRINETTO : "Formal Verification of Hardware Correctness : Introduction and Survey of Current Research". IEEE Computer, Vol.21, n°7. July 1988.

[10] M.FUJITA, H.FUJISAWA, N.KAWATO : "Evaluation and Improvements of Boolean Comparison Method based on Binary Decision Diagrams". Proc. Int. Conference on Computer-Aided Design ICCAD'88, 1988.

[11] W.A.HUNT : "FM8501 : A verified microprocessor". Institute for Computing Science, University of Texas, Austin (USA). Technical Report n°47. February 1986.

[12] K.HWANG : "Computer arithmetic : principles, architecture and design", John Wiley & sons Inc., New-York, 1979.

[13] IEEE Standard VHDL Language Reference Manual, IEEE. 1988.

[14] T.KROPF : "Benchmark-Circuits for Hardware Verification, 2nd TPCD Conference". 2nd Conference on Theorem Proving in Circuit Design, Bad Herrenalb (Germany), 1994.

[15] J.S.MOORE : "Introducing Iteration into the Pure Lisp Theorem Prover". IEEE Transactions on Software Engineering, Vol. SE-1, n°3. September 1975.

[16] Z. MANNA, R. WALDINGER : "Knowledge and Reasoning in Program Synthesis". Artificial Intelligence Journal. Vol 6, 2. 1975.

[17] S.MALIK, A.R.WANG, R.K.BRAYTON, A.SANGIOVANNI-VINCENTELLI : "Logic Verification using Binary Decision Diagrams in a Logic Synthesis Environment". Proc. Int. Conference on Computer-Aided Design ICCAD'88, 1988.

[18] L.PIERRE : "The Formal Proof of the Min-Max sequential benchmark described in CASCADE using the Boyer-Moore Theorem Prover". Proc. IFIP WG 10.2 Int. Workshop Nov. 1989. In "Formal VLSI Correctness Verification", L.Claesen ed., North Holland, 1990.

[19] L.PIERRE : "One Aspect of Mechanizing Formal Proof of Hardware : the Generalization of Partial Specifications". Proc. ACM International Workshop on Formal Methods in VLSI Design. Miami (Fl). 9-11 January 1991.

[20] L.PIERRE : "VHDL Description and Formal Verification of Systolic Multipliers". In "CHDL and their Applications", D.Agnew, L.Claesen & R.Camposano Eds, North Holland, 1993.

[21] H.SIMONIS : "Formal verification of multipliers". Proc. IFIP WG 10.2 Int. Workshop Nov. 1989. In "Formal VLSI Correctness Verification", L.Claesen ed., North Holland, 1990.

[22] D.VERKEST, L.CLAESEN, H.DE MAN : "Special Benchmark Session on Formal System Design". Proc. IFIP WG 10.2 Int. Workshop Nov. 1989. In "Formal VLSI Correctness Verification", L.Claesen ed., North Holland, 1990.

[23] F.WAGNER : "Prevail-DM : a framework-based environment for formal hardware verification". In "CHDL and their Applications", D.Agnew, L.Claesen & R.Camposano Eds, North Holland, 1993.

ACKNOWLEDGEMENTS

This work has been supported by the EEC under Charme-II ESPRIT Basic Research Working Group n°6018.

The author is grateful to the anonymous reviewers for their fruitful comments.

A Compositional Circuit Model and Verification by Composition *

Zheng Zhu

Integrated System Design Laboratory
Department of Computer Science
The University of British Columbia
Vancouver, B.C. Canada V6T 1Z4
(zhu@cs.ubc.ca)

Abstract. To embrace the fast growth of circuit complexity, verification researchers are probing new verification methods. Verification by composition, among others, is regarded as a promising direction.

Symbolic Trajectory Evaluation (STE) is a theory for digital circuit verification. In the last a few years, STE has been used in proving practical digital circuits and has been proven a practical methodology with a mathematical foundation in circuit verification. However, the circuit model used in the existing STE verification systems is, in general, not compositional.

In this paper, we present a compositional circuit model. This model distinguishes two different types of unknown circuit values, *i.e.* driven undefined value and undriven undefined value. This treatment makes composition of circuit model possible. Major results of the paper are the following:

1. A language for describing finite state machines. This language is used to describe circuits behaviors. Expressions written in the language can be interpreted to the new model in this paper, as well as to the existing model. An operator in the language is designed for finite-state machine composition. The semantics of this operator is consistent to our intuitive understanding of "connecting two circuit node together". The major theorem concerning this operator is that it preserves the properties of the finite-state machines being composed.

2. The finite-state machine description can also be interpreted to the model which is used by the Voss system, a circuit verification system. A theorem shows that under certain conditions, two interpretations

* This research was supported, in part, by operating grants OGPO 109688 and OGPO O46196 from the Natural Sciences Research Council of Canada, fellowships from the Province of British Columbia Advanced Systems Institute, and by research contract 92-DJ-295 from the Semiconductor Research Corporation.

of the same finite-state machine description achieve the same verification results. This theorem allows us to perform circuit verification by using the well-developed STE verification system, and then to interpret the verification result in the model presented in this paper.

1 Introduction

To tackle fast growth of circuit complexity, verification researchers are probing new verification methods. Verification by composition, among others, is regarded as a promising and necessary direction.

To perform verification by composition, it is necessary that circuits are modeled in a compositional way. That is, composition of circuit models must preserve properties of each individual components. Therefore, properties which are verified in individual circuits do not need to be verified again in the composite.

The main focus of this paper is a circuit model which is compositional. It is an offspring of the circuit model used by Seger and Bryant [1] which regards a circuit node value as one of $\{0, 1, \bot, \top\}$ where 0, 1 are logical values, and $\bot$, $\top$ are under-defined and over-defined logical values respectively. As illustrated in Section 3 of this paper, an inadequacy of this quadruple circuit model is that it is not compositional.

The organization of this paper is the following: Section 2 briefly introduces the theory of symbolic trajectory evaluation (STE) and the quadruple circuit model used by STE systems. Section 3 first briefly discusses the inadequacy of the quadruple model with respect to model composition, and then introduces a circuit model, which is compositional. A language for describing finite state machines, named Set expressions, and its semantics are introduced. Section 4 discusses composition of finite state machines. Section 5 establishes a relationship between the proposed model and the one used in STE systems. Finally, Section 6 presents applications of this work.

2 STE and the $\mathcal{Q}$ Model

2.1 A Language for Trajectory Evaluation

In symbolic trajectory evaluation [1], system behaviors are given as trajectories over fixed length sequences of states. Each of these trajectories are described by a trajectory formula, called *trajectory formula*. A trajectory formula is in one the following forms:

1. unc. a constant in the language, which represents unconstrained circuit behavior.
2. Node specifications:
   (a) n is 0. The node "n" has value 0
   (b) n is 1. The node "n" has value 1.
3. $F_1 \wedge F_2$. Formulae F_1 and F_2 must both hold;
4. $E \rightarrow F$. The properties represented by the formula F need only hold the boolean expression E is evaluated to true;
5. $\mathbf{N}F$. $\mathbf{N}$ is the only temporal operator used in the language. $\mathbf{N}F$ specifies that F must hold in the following state.

A verification procedure checks *assertions* in the form of an implication $A \Rightarrow C$; the formula A (the antecedent) gives the stimulus and current state, and the formula C (the consequent) gives the desired response and state transition. Although the language has limited expressive power due to it's lack of such operators as disjunction and negation, along with temporal operators expressing properties of unbounded state sequences, it is designed as a compromise between expressive power and ease of evaluation. In practice, it is proven to be powerful enough to express timing and state transition behavior of circuits, while allowing assertions being verified efficiently.

2.2 Domain of Discourse

In symbolic trajectory evaluation, a circuit is modeled as operating over logic levels $0, 1$, a third value $\perp$ representing an indeterminate or unknown level, and a fourth value $\top$, representing an overly defined value (such as asserting value 0 and 1 to a circuit node at the same time). Let $Q = \{0, 1, \perp, \top\}$. Q is partially ordered as shown in Figure 1.

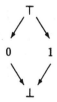

The ordering relation is: $\perp \sqsubseteq 0, 1 \sqsubseteq \top$

Fig. 1. Partial Order of Q

Intuitively, $\sqsubseteq$ orders the elements of Q according to the amount of information they carry: $\bot$ carries no information; 0, 1 carry fully defined circuit node values, and $\top$ is the overly defined, thus inconsistent value (too much information).

In lattice theory, a finite set S is a complete lattice under the partial ordering $\sqsubseteq$ if for every $a, b \in S$, there exist a unique smallest $c \in S$ (under the partial order $\sqsubseteq$) such that $a \sqsubseteq c$, $b \sqsubseteq c$, and a unique greatest $d \in S$ (under $\sqsubseteq$) such that $d \sqsubseteq a$, $d \sqsubseteq b$. Given $a, b \in S$, such unique c and d are denoted by $a \sqcup b$ (the least upper bound) and $a \sqcap b$ (the greatest lower bound) respectively. A finite set S has a least upper bound (greatest lower bound) under $\sqsubseteq$ if there exists a unique $l \in S$ ($g \in S$) such that for every $s \in S$, $l \sqsubseteq s$ ($s \sqsubseteq g$). By this definition, Q is a complete lattice under the partial ordering $\sqsubseteq$, where $\top$ and $\bot$ are the least upper bound and greatest lower bound of the lattice respectively. Furthermore, let $Q^n = Q \times Q \times \cdots \times Q$ be the cartesian product of n Qs. We can extend the relation $\sqsubseteq$ (of Q) to a relation of Q^n pair-wisely: for every $a, b \in Q^n$, $a \sqsubseteq b$ if and only if for every $i : 1 \leq i \leq n$, $a_i \sqsubseteq b_i$. It is easy to show that Q^n is a complete lattice under the extended relation $\sqsubseteq$, and for every $a, b \in Q^n$ and every $i : 1 \leq i \leq n$, $(a \sqcup b)_i = a_i \sqcup b_i$, $(a \sqcap b)_i = a_i \sqcap b_i$.

Our intention is to use Q^n as the set of all possible states of an n-node circuit. In practice, a circuit node is usually referenced by its name, or a character string, rather than by a natural number (as a subscript of a product). Therefore, it is often convenient to regard Q^n as a set of functions $N \to Q$ where N is a set of n node names. In particular, we use $\vec{\bot}$ to denote the function $N \to \{\bot\}$. That is, $\vec{\bot}(a) = \bot$ for every a.

To express the behavior of a system over time, we use *sequence* of circuit state, *e.g* sequence of elements in Q^n. Conceptually, these sequences are infinite, although the properties expressible in the language can be determined from some finite prefix of a sequence. Given two sequences (of elements of Q^n) $\sigma = \sigma^0 \sigma^1 \cdots$ and $\tau = \tau^0 \tau^1 \cdots$ we extend the relation $\sqsubseteq$ to the sequences pointwise: if $\sigma = \sigma^0 \sigma^1 \cdots$ and $\tau = \tau^0 \tau^1 \cdots$ are two sequences, then $\sigma \sqsubseteq \tau$ iff $\sigma^i \sqsubseteq \tau^i$ for every $i \geq 0$.

The definition of trajectory formulae can be extended to allow node specifications contain symbolic boolean expressions, rather than just 0 and 1. This extension makes specifications written in trajectory formulae very compact. Symbolic evaluation can be thought of as computing circuit behavior for many different operating conditions simultaneously, with each possible assignment of 0 or 1 to the variables in $\mathcal{V}$ indicating a different condition. Formally, this is expressed by defining an *assignment* φ to be a particular mapping from the elements of $\mathcal{V}$ to binary values. A formula F in the logic expresses some property of the circuit in terms of the symbolic variables. It may hold for only a subset of all possible assignments. Such a subset can be represented by a boolean domain function d over $\mathcal{V}$ yielding 1 for precisedly the assignments in the subset. For example, the constant functions 0 and 1, for example, represent the empty assignment set and the set of all possible assignments, respectively. However, allowing symbolic boolean expressions does not add any expressive power. Therefore, unless indicated explicitly, trajectory formulae in this paper are variable-free.

2.3 Use of Symbols

In this paper, we adopt the following convention of notations: every syntactic entity in a language (tern expressions and Set expressions, see Section 3.2) are in sans serif font, such as And, Not. Function symbols are represented by upper-case words, such as AND, NOT, UNION, plus conventional function symbols such as $\sqcup$ and $\sqcap$.

Uses of function $\sqcup$ and relation $\sqsubseteq$ are quite liberal in this paper. Although both are originally defined in a lattice such as Q, they are also used as binary function and relation on

1. functions such as elements of Q^n with the extension in a pair-wise manner; and
2. sequences of elements in Q^n with the extension: let $\sigma = \sigma^0 \sigma^1 \cdots$ and $\tau = \tau^0 \tau^1 \cdots$ by sequences such that for every $i \geq 0$, $\sigma^i \in Q^n$ and $\tau^i \in Q^n$, $\sigma \sqcup \tau$ is a sequence $\alpha = \alpha^0 \alpha^1 \cdots$ such that for every $i \geq 0$, $\alpha^i = \sigma^i \sqcup \tau^i$. $\sigma \sqsubseteq \tau$ if and only if for every $i \geq 0$, $\sigma^i \sqsubseteq \tau^i$.

2.4 Circuit Model Structures

A *circuit model structure* is $M = [(S, \sqsubseteq), Y]$ where S is the set of all functions from a set of nodes N to $\{0, 1, \perp, \top\}$, $\sqsubseteq$ is the ordering relation on S defined in the previous subsection, and Y is a monotone function $S \rightarrow S$. Let S^ω be the set of all (infinite) sequences of elements of S. In general, we are only interested in those sequences related to the behavior of a circuit model, namely, those sequences constrained by function Y in the model structure. We formalize this by introducing the concept of a trajectory. Given a model $M = [(S, \sqsubseteq), Y]$ and an arbitrary sequence $\sigma = \sigma^0 \sigma^1 \cdots \in S^\omega$, σ is a *trajectory* of M iff for every $i \geq 0$,

$$Y(\sigma^i) \sqsubseteq \sigma^{i+1}$$

We now assign a meaning to the specification language in terms of defining sequences. Let F be a trajectory formula, its *defining sequence*, denoted by $\delta(F)$, is defined as follows:

1. $\delta(\mathsf{unc}) = \vec{\mathcal{I}} \vec{\mathcal{I}} \cdots$.

2. Let $b \in \{0,1\}$, $\delta(n$ is $b)$ is a sequence $\sigma \vec{\bot} \cdots \vec{\bot} \cdots$ where σ a function $N \rightarrow \{0,1,\bot\}$ defined by: for every $x \in N$,

$$\sigma(x) = \begin{cases} b & x = n \\ \bot & x \neq n \end{cases}$$

3. $\delta(F_1 \wedge F_2) = \delta(F_1) \sqcup \delta(F_2)$.

4. $\delta(E \rightarrow F) = \begin{cases} \delta(F) & E \text{ is evaluated to 1} \\ \vec{\bot}\vec{\bot}\cdots & \text{Otherwise} \end{cases}$

5. $\delta(\mathbf{N}F) = \vec{\bot}\delta(F)$.

Assume that $\delta(F) = \delta^0 \delta^1 \cdots$ is the defining sequence of formula F, define the *defining trajectory* of F constrained by Y, denoted by $\tau_Y(F) = \tau^0 \tau^1 \cdots$, as follows:

$$\tau^i = \begin{cases} \delta_F^0 & i = 0 \\ \delta_F^i \sqcup Y(\tau^{i-1}) & i > 0 \end{cases}$$

2.5 Specification and Verification

The truth semantics of trajectory formulae is defined relative to circuit model and its defining trajectories. In particular, given a circuit model M and a trajectory σ, a trajectory formula F is true on the trajectory σ, written $\sigma \models F$, is defined as follows:

1. $\sigma \models$ unc for all σ.
2. $\sigma^0 \sigma^1 \cdots \models n$ is b iff $b \sqsubseteq \sigma^0(n)$.
3. $\sigma \models F_1 \wedge F_2$ iff $\sigma \models F_1$ and $\sigma \models F_2$.
4. (a) $\sigma \models E \rightarrow F$ if $\sigma \models F$ and E is evaluated to 1.
   (b) $\sigma \models E \rightarrow F$ for every σ if E is evaluated to 0.
5. $\sigma^0 \sigma^1 \cdots \models \mathbf{N}F$ iff $\sigma^1 \cdots \models F$.

A specification of a circuit is a pair of trajectory formulae A and C, denoted by $A \Rightarrow C$, where A and C are called *antecedent* and *consequent* respectively. $A \Rightarrow C$ is a specification of a circuit model M if for every trajectory σ of M, $\sigma \models A$ implies $\sigma \models C$.

A major theorem proved in [1] is the following:

Theorem 1. *Let A and C be two trajectory formulae and $M = [(S, \sqsubseteq), Y]$ be a circuit model. $\delta(C) \sqsubseteq \tau_Y(A)$, if and only if for every trajectory σ of M, $\sigma \models A$ implies $\sigma \models C$, namely, $A \Rightarrow C$ is a specification of M.*

Informally, this theorem can be interpreted as the following: the next-state function Y is a function from circuit states to circuit states. $\tau_Y(A)$ is the sequence of states when the input to the circuit is what specified by the formula A. Each state includes the values of input/output circuit nodes as well as internal state nodes. $\delta(C) \sqsubseteq \tau_Y(A)$ means that in every state, a circuit node value is either equal to what specified in the formula C, or its value is not mentioned in the formula C *i.e.* unrelated to the assertion $A \Rightarrow C$. By this theorem, in order to show that a circuit model has a property $A \Rightarrow C$, it is sufficient to show that $\delta(C) \sqsubseteq \tau_Y(A)$.

3 The Model $\mathcal{F}$

In this section, we introduce a slightly different lattice than $\mathcal{Q}$ for circuit modeling. The motivation of $\mathcal{F}$ can be illustrated by the example in Figure 2.

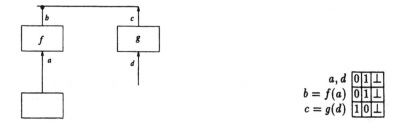

Fig. 2. Example: Composition of Circuits

In this example, functions f and g are defined in the truth table. An interesting situation is when the value on 'a' is $\bot$, and the value on 'd' is 0, which lead to $f(\bot) = \bot$ and $g(0) = 1$. Since the outputs of f and g are connected, how do we reconcile two different values $\bot$ and 1? There are two different interpretations to the fact $f(\bot) = \bot$, which lead to different answers to the question:

- The value on 'b' and 'c' can be $g(0) = 1$ if the node 'b' is not driven by any value, *i.e.* $\bot$ is interpreted as an undriven "unknown" value.
- Alternatively, the value on 'b' and 'c' can be an unknown value ($\top$) if 'b' is driven by some unknown but valid logical value, possibly be 0. *i.e.*, $\bot$ is interpreted as a driven but unknown value.

This example illustrate a need of differentiating two different type of $\bot$s, *i.e.* driven unknown value and undriven unknown value, when we consider composition of circuits. However, when modeling circuits by $\mathcal{Q}$, these two different unknown values are treated equally. Therefore, circuits modeled in $\mathcal{Q}$ are, in general, not compositional. This motivates our attempt to enrich $\mathcal{Q}$ in order to obtain a compositional model of circuits.

3.1 The Model $\mathcal{F}$

The example in Figure 2 reveals that, in order to be compositional, it is essential for a circuit model to distinguish two different types of unknown values. For this reason, we extend the lattice $(\mathcal{Q}, \sqsubseteq)$ to $(\mathcal{F}, \sqsubseteq)$ where $\mathcal{F} = \{0, 1, X, \bot, \top\}$ and the partial order $\sqsubseteq$ is shown in the following diagram.

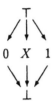

The ordering relation is: $\bot \sqsubseteq 0, X, 1 \sqsubseteq \top$

Apparently, $\mathcal{F}$ is a complete lattice under the ordering relation indicated by arrows in the picture. In the context of circuit modeling, the intuition behind these 5 values is the following: 0 and 1 have their conventional meaning. $\top$ is an over-constrained value. $\bot$ represents an unconstrained value. It could be used to model a don't-care input, or an undriven, unknown output value, such as the high-impedance state of a tri-state output. X is also an unconstrained value. The difference between X and $\bot$ is that X represents a driven unknown output value.

Conventional boolean functions, and $\sqcup$, $\sqcap$ can be extended to the values in $\mathcal{F}$. The following "truth-tables" are those for AND, NOT, $\sqcup$, and $\sqcap$:

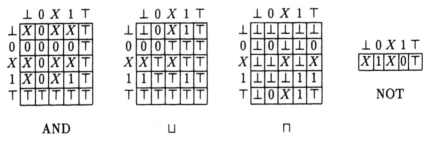

AND	$\bot$	0	X	1	$\top$
$\bot$	X	0	X	X	$\top$
0	0	0	0	0	$\top$
X	X	0	X	X	$\top$
1	X	0	X	1	$\top$
$\top$	$\top$	$\top$	$\top$	$\top$	$\top$

$\sqcup$	$\bot$	0	X	1	$\top$
$\bot$	$\bot$	0	X	1	$\top$
0	0	0	$\top$	$\top$	$\top$
X	X	$\top$	X	$\top$	$\top$
1	1	$\top$	$\top$	1	$\top$
$\top$	$\top$	$\top$	$\top$	$\top$	$\top$

$\sqcap$	$\bot$	0	X	1	$\top$
$\bot$	$\bot$	$\bot$	$\bot$	$\bot$	$\bot$
0	$\bot$	0	$\bot$	$\bot$	0
X	$\bot$	$\bot$	X	$\bot$	X
1	$\bot$	$\bot$	$\bot$	1	1
$\top$	$\bot$	0	X	1	$\top$

$\bot$	0	X	1	$\top$
X	1	X	0	$\top$

NOT

Similar to $\mathcal{Q}$-model structures, a $\mathcal{F}$-model structure is $((S, \sqsubseteq), Y)$ where S is the set of all functions from a set of nodes (names) to $\mathcal{F}$, and Y, the next-state function of the model structure, is a function $S \to S$.

3.2 Representation of Next-state Functions

We now introduce a notation for describing next-state functions of finite state machines: tern and Set expressions. A tern expression can be regarded as an extension of boolean expressions in $\mathcal{F}$, and a Set expression can be regarded as a description of next-state function of a finite-state machine.

3.2.1 tern Expressions and Set Expressions

A tern expression is defined as:

tern ::= One | Zero | X | Val str | And tern tern | Not tern

where One, Zero, X, And, and Not are the syntactical representations of 0, 1, X, and functions AND, and NOT respectively, and

Val str

is used to refer to the value on the node which is named by the string str. It plays a role similar to that of boolean variables in boolean expressions. As a concrete example, the following tern expression

(Not (And (Val 'in1') (Val 'in2')))

describes the output of an NAND gate whose input nodes are 'in1' and 'in2' respectively.

In the definition of tern expression, we carefully excluded symbols which correspond to $\perp$ and $\top$ in $\mathcal{F}$. This choice will be justified when the semantics of tern and Set expressions is presented later.

A Set expression is defined as:

Set ::= Empty | Element str, Driver | Union Set Set | Join Set Set
Driver ::= (tern, tern)

The constant Empty corresponds to an empty finite state machine (which has neither internal state nor output). Element is the constructor that actually introduces a new node which is named by the string str and defines driver functions for the node. A driver function is given in Driver which is a pair of tern expressions. the first expression is a guard and the second expression is the value being driven when the guard is evaluated to 1. For example,

Element 'n' (One, Zero)

creates a circuit node whose value is always 0.

The constructor Union is used to create a collection of Element definitions.

$$\text{Union } S_1 \; S_2$$

is a collection of node drivers which contains all the node drivers in either S_1 or S_2. If there are both S_1 and S_2 contain a driver for the same node, then the Union constructor will use the greatest lower bound of the values being driven at the same time when both guards are evaluated to 1 (See more in Section 3.2.2).

The expression

$$\text{Join } S_1 \; S_2$$

is similar to Union $S_1 \; S_2$, except when both S_1 and S_2 contain a driver for the same node, then the Join operator will use the least upper bound of the values being driven at the same time when both guards are evaluated to 1.

3.2.2   Interpreting Set Expressions to $\mathcal{F}$

The semantics of Set expressions includes evaluation of tern expressions, and an interpretation of Set expressions. This interpretation effectively translates a Set expression to a next-state function (of a finite-state machine).

Given a tern expression t, t is a constant if it does not contain any Val (such as Val 'a') subexpressions. The evaluation of a constant tern expression t, denoted as $\mathcal{E}(t)$, maps t to an element in $\mathcal{F}$:

1. $\mathcal{E}(\text{Zero}) = 0$, $\mathcal{E}(\text{One}) = 1$, and $\mathcal{E}(\text{X}) = X$;
2. $\mathcal{E}(\text{And } t_1 \; t_2) = \text{AND } \mathcal{E}(t_1) \; \mathcal{E}(t_2)$;
3. $\mathcal{E}(\text{Not } t) = \text{NOT } \mathcal{E}(t)$

Given a Set expression S, a *node of* S is a circuit node name (str) such that either

$$\text{Val str}$$

appears in S, or

$$\text{Element str driver}$$

appears in S. A node space of S is a set of nodes such that every node of S belongs to the set. Apparently, there are more than one node space of a given S. Without losing generality, we assume that there exists a universal set of circuit nodes (names), denoted by $\mathcal{U}$, which includes all the circuit nodes we are interested in. We use $\mathcal{U}$ as the node space of any given Set expression, unless explicitly indicated otherwise.

A state is a function $\varphi : \mathcal{U} \to \mathcal{F}$. An example of such a function is $\bar{\bot} : \mathcal{U} \to \{\bot\}$. That is, for every $a \in \mathcal{U}$, $\bar{\bot}(a) = \bot$.

We now extend the evaluation of a constant tern expression to arbitrary tern expressions: given a state φ, the evaluation of a tern expression t in the state φ, denoted as $\mathcal{E}(t, \varphi)$, is defined recursively as:

1. $\mathcal{E}(\text{Zero}, \varphi) = 0$, $\mathcal{E}(\text{One}, \varphi) = 1$, and $\mathcal{E}(\text{X}, \varphi) = X$;
2. $\mathcal{E}(\text{Val } str, \varphi) = \varphi(str)$;
3. $\mathcal{E}(\text{And } t_1 \ t_2, \ \varphi) = \text{AND } \mathcal{E}(t_1, \varphi) \ \mathcal{E}(t_2, \varphi)$;
4. $\mathcal{E}(\text{Not } t, \ \varphi) = \text{NOT } \mathcal{E}(t, \varphi)$

The interpretation (semantics) of a Set expression S, denoted by $[\![S]\!]$, is a function which maps a state to a state: for any state s,

1. $[\![\text{Empty}]\!](s) = \tilde{\perp}$.
2. $[\![\text{Element } n, (g, v)]\!](s)$ is $\psi : \mathcal{U} \to \mathcal{F}$, such that for every $a \in \mathcal{U}$,

$$\psi(a) = \begin{cases} \mathcal{E}(v, s) & \text{if } \mathcal{E}(g, s) = 1 \text{ and } n = a \\ \perp & \text{Otherwise} \end{cases}$$

3. $[\![\text{Union } S_1 \ S_2]\!](s) = \text{UNION } [\![S_1]\!](s) \ [\![S_2]\!](s)$.
4. $[\![\text{Join } S_1 \ S_2]\!](s) = [\![S_1]\!](s) \sqcup [\![S_2]\!](s)$.

where the function UNION (in $\mathcal{F}$) is defined by the following "truth-table":

	$\perp$	0	X	1	$\top$
$\perp$	$\perp$	0	X	1	$\perp$
0	0	0	$\perp$	$\perp$	0
X	X	$\perp$	X	$\perp$	X
1	1	$\perp$	$\perp$	1	1
$\top$	$\perp$	0	X	1	$\top$

The truth-table of the function UNION

4 Finite State Machine Compositions

The purpose of $\mathcal{F}$ is to provide the capability of composing next state functions of finite state machines. In this section, we show that the Join constructor of Set expression realizes circuit composition. The relationship between the Join and Union operators will also be discussed.

4.1 Circuit Model Compositions

In circuit designs, composition of two (or more) physical circuits means connecting (wiring) nodes with the same name (in different circuits) together to form a new circuit. The following "truth table" gives our understanding of "connecting two node values" in $\mathcal{F}$:

	$\perp$	0	X	1	$\top$
$\perp$	$\perp$	0	X	1	$\top$
0	0	0	$\top$	$\top$	$\top$
X	X	$\top$	X	$\top$	$\top$
1	1	$\top$	$\top$	1	$\top$
$\top$	$\top$	$\top$	$\top$	$\top$	$\top$

The truth table of "connecting" values in the 5-element domain. Note that the truth table is identical to that of the $\sqcup$ operator in the same domain.

Let A, C be any trajectory formulae, S_1, S_2 be any Set expressions. Also let $Y = [\text{Join } S_1 \, S_2]$, $Y_1 = [S_1]_\varrho$, $Y_2 = [S_2]_\varrho$. If $\tau_Y(A)$ does not have any $\top$ element, (i.e. let $\tau_Y(A) = \tau^0 \cdots \tau^n \cdots$. For every $i \geq 0$, $\top$ is not in the range of τ^i.) then

Theorem 2. $\delta(C) \sqsubseteq \tau_{Y_1}(A)$ *implies* $\delta(C) \sqsubseteq \tau_Y(A)$, *and* $\delta(C) \sqsubseteq \tau_{Y_2}(A)$ *implies* $\delta(C) \sqsubseteq \tau_Y(A)$.

This theorem shows properties which are held in components are also held in composition, if the composition is modeled by the constructor Join.

4.2 Implementation of Model Compositions

In practice, the function JOIN poses two problems:

1. Potentially, it may create a large number of $\top$ elements. To catch all these $\top$ elements requires extensive computing resource. According to our experience, it is responsible for up to 5% increase of computing time during verifications.
2. The verification may be too pessimistic: it is may be acceptable to have a circuit node have a $\top$ value. In fact, during transient states, occurrences of $\top$ on a circuit node is quite common and does not necessarily mean that the circuit presents undesirable behavior.

To solve these two problems, we use the Union operator in the Set expressions, rather than Join, to compose two Set expressions.

From the truth-table of UNION (Section 3.2.2), we find that it has the following properties:

1. It generates far less $\top$s than $\sqcup$ does;
2. UNION does not generate a $\top$ unless both operands are $\top$.

3. Except the bottom row and the right-most column, the truth-table of UNION corresponds that of $\sqcup$ in the following way: the corresponding entries in the tables are either equal to each other, or the entry in the table of $\sqcup$ is $\top$ and the entry in that of UNION is $\bot$.

The last property is essential: from verification's perspective, if a circuit node has the value $\bot$, then nothing can be concluded about that circuit node's value, except its being $\bot$; If a node value is $\top$, then it implies that the circuit may have had undesirable behavior, such as short-circuits. This implies a possible malfunction of the circuit implementation unless the specification does not explicitly mention the value of this node $i.e.$ the node has value $\bot$. Therefore, it is safe to replace $\top$ values in a trajectory (sequence of circuit states) by $\bot$ as far as verification is concerned. This property suggests that circuit composition might be realizeded by the Union operator as well. The following theorem confirms this.

Let S_1 and S_2 be two Set expressions, $Y = [\text{Join } S_1\ S_2]$, and $Y' = [\text{Union } S_1\ S_2]$. The following theorem shows the relationship between Y and Y':

Theorem 3. *Given trajectory formulae A and C,*

1. *If $\delta(C) \sqsubseteq \tau_{Y'}(A)$, then $\delta(C) \sqsubseteq \tau_Y(A)$.*

2. *If $\delta(C) \sqsubseteq \tau_Y(A)$ and $\tau_Y(A)$ does not contain any $\top$ value, then $\delta(C) \sqsubseteq \tau_{Y'}(A)$.*

This theorem says that, instead of verifying $\delta(C) \sqsubseteq \tau_Y(A)$, we verify $\delta(C) \sqsubseteq \tau_{Y'}(A)$. If $\delta(C) \sqsubseteq \tau_{Y'}(A)$ holds, then $\delta(C) \sqsubseteq \tau_Y(A)$ holds. Otherwise, either $\delta(C) \sqsubseteq \tau_Y(A)$ fails, or $\tau_Y(A)$ contains $\top$ values. By proving $\delta(C) \sqsubseteq \tau_{Y'}(A)$, we can avoid significant amount $\top$-checking, and avoid to be too pessimistic in interpreting verification results.

5 Relationship Between the $\mathcal{Q}$-Model and the $\mathcal{F}$-Model

This section discusses the relationship between the two different circuit models. The purpose of establishing such a relationship is practical: Voss, a circuit verification system which uses the $\mathcal{Q}$-model, has been developed and proven practical in circuit design verifications. In fact, both tern expressions and Set expressions are the integrated components of the Voss system. Establishing such a relationship between $\mathcal{Q}$-model and $\mathcal{F}$-model allows us conduct verification in Voss and then to interpret the result in $\mathcal{F}$-model.

5.1 Interpreting Set Expressions to the Q-Model

In Q, the functions AND, NOT, $\sqcup$, and UNION are as defined in the following "truth-tables":

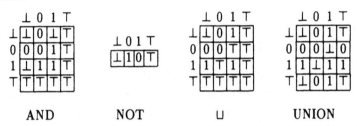

Note that the extensions of AND, and NOT to Q is carefully done so that both functions are monotonic. This is to accommodate the requirement that every next-state function in a Q-model has to be monotonic [1].

Given a tern expression t, it can be evaluated to a value in Q the same way as we evaluate t to $\mathcal{F}$. To distinguish the evaluation of a tern expression to Q and that to $\mathcal{F}$, we add subscripts to the evaluation function $\mathcal{E}$: i.e. $\mathcal{E}_Q$ $(\mathcal{E}_\mathcal{F})$ evaluates a tern expression to Q $(\mathcal{F})$. Similarly, we add subscripts to the semantic function $[\![S]\!]$ i.e. $[\![S]\!]_Q$ and $[\![S]\!]_\mathcal{F}$ respectively, in order to distinguish the different interpretations in Q and $\mathcal{F}$.

We now interpret Set expressions in Q: given a Set expression S,

1. $[\![\mathsf{Empty}]\!]_Q(s) = \vec{\bot}$.
2. $[\![\mathsf{Element}\ n, (g, v)]\!]_Q(s)$ is $\psi : \mathcal{U} \to Q$, such that for every $a \in \mathcal{U}$,

$$\psi(a) = \begin{cases} \mathcal{E}_Q(v, s) & \text{if } \mathcal{E}_Q(g, s) = 1 \text{ and } n = a \text{ and } \mathcal{E}_Q(v, s) \neq X \\ \bot & < \text{Otherwise} \end{cases}$$

3. $[\![\mathsf{Union}\ S_1\ S_2]\!]_Q(s) = \mathsf{UNION}\ [\![S_1]\!]_Q(s)\ [\![S_2]\!]_Q(s)$.
4. $[\![\mathsf{Join}\ S_1\ S_2]\!]_Q(s) = [\![S_1]\!]_Q(s) \sqcup [\![S_2]\!]_Q(s)$.

5.2 The Relationship Between $[\![S]\!]_\mathcal{F}$ and $[\![S]\!]_Q$

The difference between $[\![S]\!]_Q$ and $[\![S]\!]_\mathcal{F}$ is the treatments when $\mathcal{E}(v, s) = X$, $\mathcal{E}(g, s) = 1$, and $n = a$ (in $[\![\mathsf{Element}\ n, (g, v)]\!]_Q(s)$ and $[\![\mathsf{Element}\ n, (g, v)]\!]_\mathcal{F}(s)$). In this situation, $[\![S]\!]_Q(s)(a) = \bot$ whereas $[\![S]\!]_Q(s)(a) = X$. This is because that, in Q, there is no distinction between driven unknown value and undriven unknown value, they both are represented by $\bot$. To further illustrate this subtle difference, let us compare the Set expression Empty with

$$\mathsf{Element}\ 'n'\ (\mathsf{One}, X)$$

Since $[\![Empty]\!]_Q = \bar{\bot}$ and $[\![Element\ 'n'(One, X)]\!]_Q = \bar{\bot}$, the two expressions are equivalent in Q. On the other hand, $[\![Empty]\!]_Q = \bar{\bot}$, $[\![Element\ 'n'(One, X)]\!]_Q$ is a function f such that $f(n) = X$, therefore, $[\![Empty]\!]_{\mathcal{F}} \neq [\![Element\ 'n'(One, X)]\!]_Q$. Although very subtle, this difference is exactly we have been after: Empty represents a finite state machine whose value on the node 'n' is undriven unknown whereas

$$Element\ 'n'\ (One,\ X)$$

represents a finite state machine whose value on the node 'n' is driven unknown.

The following theorem establishes an important relationship between $[\![S]\!]_Q$ and $[\![S]\!]_{\mathcal{F}}$:

Theorem 4. *Let S be any Set expression such that $[\![S]\!]_Q$ is monotonic. Let $Y = [\![S]\!]_{\mathcal{F}}$ and $Y' = [\![S]\!]_Q$. For any trajectory formulae A, and C,*

1. if $\delta(C) \sqsubseteq \tau_{Y'}(A)$, then $\delta(C) \sqsubseteq \tau_Y(A)$.

2. if $\delta(C) \sqsubseteq \tau_Y(A)$, and $\tau_Y(A)$ does not contain any $\top$, then $\delta(C) \sqsubseteq \tau_{Y'}(A)$.

This theorem allows us to conduct verification in the Voss system, which models circuits in Q, and interpret the verification results in $\mathcal{F}$: if $\delta(C) \sqsubseteq \tau_{Y'}(A)$, then $\tau_C \sqsubseteq \tau_Y(A)$. If $\delta(C) \not\sqsubseteq \tau_{Y'}(A)$, then either $\delta(C) \not\sqsubseteq \tau_Y(A)$, or $\tau_Y(A)$ contains $\top$ elements. Neither case is desirable thus $A \Rightarrow C$ is not an assertion of the given circuit. Disallowing $\top$ elements in $\tau_Y(A)$ is a way to obtain pessimistic verification results. The Voss system is capable of checking and reporting whenever a trajectory ($\tau_Y(A)$) contains a $\top$ value.

6 Application

6.1 The Voss System

The Voss system, a formal hardware verification system developed at the University of British Columbia, consists of three major components: a highly efficient implementation of OBDDs; an event driven symbolic simulator with very comprehensive delay and race analysis capabilities [2]. From a user's point of view, the Voss system verifies assertions in the form of

$$FSM\ fsm\ (ante,\ cons)$$

where fsm denotes a finite-state machine and the pair (ante, cons) represents a trajectory assertion to be verified. The finite-state machine fsm is specified by a next-state function.

The following table summaries several verification experiments conducted by the Voss system:

Circuits	# Transistors	Time (sec)	Memory
64 × 32-bit moving data stack	16, 470	35	1.2M
64 × 32-bit stationary data stack	15, 383	200	3.2M
UART	9840	400	9.0M
32-bit Tamarack (synchronous)	7, 214	320	10.1M
32-bit Tamarack (asynchronous)	7, 214	530	10.1M
32-bit RISC core (32 regs)	16, 100	7, 300	35.4M

6.2 Verification by Composition

Originally, our work on verification by composition was motivated by verification of a system of more than one components. It is often the case that the designer of one component does not have finite state machines of other components when his design needs to be verified. In this situation, traditional verification methods may not be applicable due to a lack of a complete description of implementation *e.g.* a finite state machine. Although the designer could make assumptions about the environment (behaviors of other components) in order to valid his own design, such assumptions may disguise design errors even verifications produce positive answers.

To briefly illustrate our approach to the above mentioned problem, let us examine various verification exercises performed on the Tamarack microprocessor [3, 4]. Originally, a memory is an integrated component of the microprocessor's behavioral specification [3]. However, various Tamarack implementations which we are aware of do not contain such a memory module. Instead, memory is treated as part of the environment in which the microprocessor operates. To correlate the verification results to the original specification, assumptions of the behavior of the memory module have to be made.

There are mainly two reasons justifying separation of processor and memory in various verification efforts:

- The complexity of memory would make verification by automated verifiers, such as Voss, computationally intractable.
- Tamarack designs usually do not contain a memory module. Therefore, a complete implementation of Tamarack is usually not available to verification exercises.

However, the correctness of an implementation can only be established in an environment which realizes the memory module faithfully. Conceptually, our approach, *i.e.* verification by composition, proceeds as follows:

1. Obtaining a specification for the processing unit of Tamarack and verify an implementation of the unit against the specification. The work reported in [4] accomplished this in hybrid hardware verification system [5, 6].
2. Obtaining a specification of the memory module and then generating an abstract circuit model automatically:
   Automatically generate a Set expression S_m from the trajectory assertions of memory. In [7], we reported a constructive method which generates a finite-state machine from a given set of trajectory assertions such that $[\![S]\!]_{\mathcal{F}}$ is the next-state function of some circuit model of the given set of assertions. We also proved that such a finite state machine is the smallest machine in terms of the number of internal states [8].
3. Translate the next-state function of Tamarack to an equivalent Set expression S_t, according to the interpretation of Set expressions presented in Section 3.2.2[2].
4. Perform the Union operation on S_m and S_t to obtain a new Set expression $S = Union\ S_1\ S_2..$
5. Finally, verify the original Tamarack specification against S by the Voss system. An important difference between this verification and other verification exercise, such as [4], is that, if succeeds, guarantees the correctness of the processor/ memory composition.

7 Conclusion

In this paper, we presented a 5-element circuit domain which is compositional. The purpose of devising this model is to facilitate verification by composition. The major results can be summarized as follows:

1. We presented a language for finite state machines (SetExpressions). This language is used to describe circuits behaviors. Expressions written in the language can be interpreted to $\mathcal{F}$, as well as to $\mathcal{Q}$. An operator in the language is designed for finite-state machine composition. The semantics of this operator is consistent to our intuitive understanding of "connecting two circuit node together". The major theorem concerning this operator is that it preserves the properties of the finite-state machines which are being composed (Theorem 2).
2. The finite-state machine description can also be interpreted to the model which is used by the Voss system, a circuit verification system. A theorem

---

[2] This is an ongoing work.

(Theorem 4) shows that under certain condition, two interpretations of the same finite-state machine description achieve same verification results. This theorem allows us to perform circuit verification by using the well-developed Voss system, and then interpret the verification result in the model presented in this paper.

References

1. SEGER, C.-J., AND BRYANT, R. Formal verification of digital circuits by symbolic evaluation of partially-ordered trajectories. Tech. Rep. Technical Report 93-8, The Computer Science Department, The University of British Columbia, The Computer Science Department, The University of B.C. Vancouver B.C. V6T 1Z4, 1993. To appear in *Journal of Formal Methods in System Design*.
2. SEGER, C.-J. Voss – a formal hardware verification system, user's guide. Tech. Rep. Technical Report 93-45, The Computer Science Department, The University of British Columbia, The Computer Science Department, The University of B.C. Vancouver B.C. V6T 1Z4, 1993.
3. JOYCE, J. *Multi-level Verification of Microprocessor-Based Systems*. PhD thesis, Computer Laboratory, Cambridge University, 1989.
4. ZHU, Z., JOYCE, J., AND SEGER, C.-J. Verification of the tamarack-3 microprocessor in a hybrid verification environment. In *Proceedings of 1993 international meeting on Higer Order Logic and its Applications, Lecture Notes in Computer Science, Vol 780* (1993), Springer-Verlag.
5. SEGER, C.-J., AND JOYCE, J. A mathematically precise two-level formal verification methodology",. Tech. Rep. Report-92-34, Computer Science Department, The University of British Columbia, 1992.
6. JOYCE, J., AND SEGER, C.-J. Linking bdd-based symbolic evaluation to interactive theorem-proving. In *Proceedings of 30th Design Automation Conference* (1993).
7. ZHU, Z. Construction of circuit models from trajectory specifications. in progress, Janurary 1994.
8. ZHU, Z., AND SEGER, C.-J. Model construction from trajectory assertion and the completeness of a hardware inference system. In *The proceedings of the Sixth International Conference on Computer Aided Verification (CAV94), Lecture Notes in Computer Science, Vol 818* (1994), Springer-Verlag.

Exploiting Structural Similarities in a BDD-based Verification Method

C.A.J. van Eijk and G.L.J.M. Janssen

Eindhoven University of Technology, Department of Electrical Engineering
Design Automation Section, P.O. Box 513, 5600MB Eindhoven, The Netherlands

Abstract. A major challenge in the area of hardware verification is to devise methods that can handle circuits of practical size. This paper intends to show how the applicability of combinational circuit verification tools based on binary decision diagrams (BDDs) can be greatly improved. The introduction of dynamic variable ordering techniques already makes these tools more robust; a designer no longer needs to worry about a good initial variable order. In addition, we present a novel approach combining BDDs with a technique that exploits structural similarities of the circuits under comparison. We explain how these similarities can be detected and put to effective use in the verification process. Benchmark results show that the proposed method significantly extends the range of circuits that can be verified using BDDs.

1 Introduction

The times when researchers in the CAD field could sit in their ivory tower thinking up neat solutions for theoretical problems belong to the past. Nowadays, industry and other funding organisations require projects to come up with useful and practical results. Whereas only 10 years ago, a hardware designer would be surprised when an automatic tool could prove his LSI circuit (a couple of gates and a few flipflops) correct, today, the same designer expects tools that handle his 10,000 gates VLSI circuit. Clearly, the biggest problem faced in the area of hardware verification is that of scale.

In this paper we show how a successful technique for proving equivalence of combinational circuits can be extended to greatly enhance its applicability. Our past work has concentrated on the use of BDDs to represent the circuits' functional behaviour. However, it is a known fact that for many practical circuits no reasonably sized BDDs exist no matter what variable order we choose. Several approaches that try to circumvent this intricate problem have been proposed [1][9][11]. Often, the canonicity requirement for BDDs is dropped; then, of course, the equivalence check becomes harder. It is also possible to introduce extra variables with the intention of obtaining smaller BDDs. We feel that those methods are either not general enough for our purposes, or lack convincingly strong results. The method we propose in this paper is modelled after [6][14], and still uses regular BDDs. It works fully automatically and does not impose any special requirements on the design process.

This paper is organised as follows. In section 2 we momentarily divert into the rather novel technique of dynamic variable ordering. We explain the main idea and show how it can be incorporated in a BDD package. The experimental results presented at the section's end serve both as evidence for the usability of the technique and as a stimulus for further investigations. Section 3 introduces our ideas on using structural information of the circuits at hand to aid in the verification process. We show that with a minimum of extra effort, the available BDD routines can be used in a more clever way to establish equivalence. The results of our first experiments are very encouraging.

2 Dynamic Variable Ordering

In this section, we briefly summarize the issues involved in applying a dynamic variable ordering technique, i.e., changing the order of the variables during BDD construction. We assume that the reader is already familiar with the basic concepts of BDDs [8] and the popular way to implement them [5]. Dynamic variable ordering is a very important addition to a BDD package, because it relieves the user of the burden of specifying a good order a priori, i.e., before the BDDs are constructed. We can define a good order as one that permits the function to be represented by a polynomially sized BDD. For gate level descriptions, several static ordering algorithms have been proposed [10][12] and shown to be successful for many circuits. However, it turns out that dynamic variable ordering often substantially improves on the intermediate and final BDD sizes. In particular when BDDs are used in the area of verification, it is our opinion that dynamic variable ordering becomes a mandatory prerequisite for successfully handling large circuits automatically.

2.1 Basic Principles

It is a well-known fact that the size of a BDD representation for a given boolean function may drastically change when a different variable order is adopted. Therefore, it is very important that a good variable order is used. Because generally it is difficult to predict a good order before the BDDs are actually constructed, it is necessary to use a technique that searches a good order during the construction of BDDs. The problem with changing the order dynamically is that one has to maintain canonicity. In the implementation of a BDD package, canonicity is achieved through a so-called unique table of BDD nodes: a node is identified by its pointer. A new node is only then created when it is not yet present in the unique table; otherwise the pointer stored in the table is returned. It would be very inefficient to construct entirely new BDDs for every different order that is tried. Therefore, dynamic variable ordering is based on a succession of local modifications, each of which can easily be made to preserve canonicity. A natural local modification is the swapping of two consecutive variables, as is illustrated in Fig. 1.

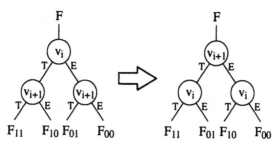

Fig. 1. The effect of a variable swap on a BDD

By repeatedly swapping adjacent variables every variable order can be generated. However, for practical purposes a full exploration of the 'variable orders space' cannot be tolerated; a simple local-search approach with limited hill-climbing is chosen instead. This has become known as the sifting algorithm [16]. In this approach, each variable is tried at all possible positions in the order while the ranks of the other variables remain the same. This search for the best position of that one variable may still lead to the construction of unacceptably large intermediate BDDs, hence the search is aborted as soon as the increase in the total number of nodes exceeds a given limit (we allow an increase of at most 5%). It is easy to see that putting one variable at its best position takes $\mathcal{O}(\#vars \cdot \#nodes)$ time. It makes sense to treat the variables in order of frequency: the variable with the most occurrences in the BDD is sifted first. The rationale is that by changing the position of this variable the largest gain, i.e., decrease in BDD size, may be achieved.

2.2 Implementation Issues

The integration of dynamic variable ordering into a BDD package requires two decisions, namely *where* and *when* the algorithm is applied. Rudell argues that it must be invoked inside the recursive ITE routine, because this is the major source for new nodes. However, dynamic variable ordering violates the invariants of the ITE routine. Since ITE recursively visits the nodes of its three BDD arguments in a depth-first fashion, starting at the top nodes, it is not allowed to swap any variables that have ranks smaller than the node currently under consideration. Dynamic variable ordering may be partly applied to the rest of the variables. This idea has not been persued. Instead, we decided to allow dynamic ordering only to take place outside recursive ITE calls, but potentially after every top-level call.

The question of when to apply dynamic ordering is not an easy one; on the one hand, dynamic variable ordering is a very useful, and for some applications even vital, feature of a BDD package, but on the other hand it takes $\mathcal{O}(\#vars^2 \cdot \#nodes)$ time for a single invocation, and therefore should not be called upon too liberally, especially when many variables are involved. Clearly, there are a number of conflicting interests:

- Dynamic variable ordering should be done as soon as the current order is found to be rather poor.
- The quality of an order can only be assessed relative to the functions that are momentarily represented. There is usually no way to predict the sequence of future BDD operations on the current functions.
- When the initial (or some intermediate) order is good, then the next call to dynamic ordering should be postponed as long as possible.
- When the functions to be represented are such that no good order exists, dynamic ordering should be refrained from completely.

Rudell uses an absolute bound on the total number of BDD nodes to trigger the ordering. After each reordering, it is reset to twice the then existing number of nodes. Our solution is to fix the total time spent in one call to some reasonable constant, say 10 (cpu) seconds, and to use both a relative and an absolute threshold to trigger the variable ordering. The absolute threshold criterion we use is the same as described above. The relative threshold is introduced to be able to anticipate sharp increases in BDD size as a function of the number of top-level ITE calls. Dynamic ordering is triggered when the increase exceeds a factor of 2. This value is chosen because the majority of BDD operations takes two operands, and empirical evidence shows that the size of the result is of the order of the sum of the sizes of the operands; worst-case it would be the product.

The effect of dynamic variable ordering is illustrated in Fig. 2 for a 16-bits rotator circuit with 16-bits data input and output, and a 4-bits control input (20 BDD variables). Mark the difference in scale of the vertical axes: without dynamic variable ordering more than a million BDD nodes are required to compute the rotator's outputs; with dynamic ordering, some 4000 are needed. The ragged behaviour of the graphs is due to the particular way the circuit is described as a table, see Fig. 3. The program that constructs the BDDs handles a table row by row which causes intermediate results that are directly freed again, hence the peaks and plateaus in the figure (careful examination shows that there are precisely 16 of them).

Dynamic variable ordering requires easy access to all BDD nodes with the same variable label. Hence unique tables need to be maintained per variable instead of having one large table shared by all of them. As a consequence and clear disadvantage, memory for those hash tables gets fragmented. On the positive side, it allows for a finer control over the amount of memory allocated for each table. In fact we do this dynamically, i.e., when more nodes for a particular variable are needed the table is extended, and when enough nodes of that variable are freed again the table is shrunk. This approach turns out to be both beneficial with regard to run time and memory usage.

2.3 Experimental Results

We have implemented the algorithm for dynamic variable ordering in the BDD package that is developed by one of the authors. It is based on the work of Karl Brace reported in [5] and papers by Richard Rudell [16][17]. Apart from

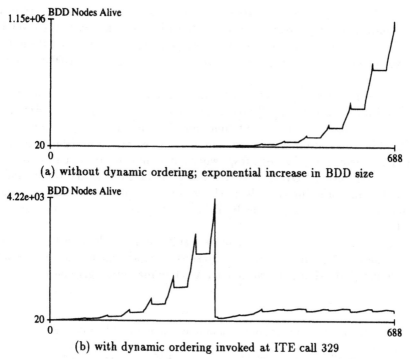

(a) without dynamic ordering; exponential increase in BDD size

(b) with dynamic ordering invoked at ITE call 329

Fig. 2. The number of nodes as function of the number of top-level ITE calls for a 16-bit rotator

```
behaviour rot16 (input In[15:0]; input C[3:0]; output Out[15:0])
{
    table {C ::                     Out;
           0 :: In[15:0]                ;
           1 :: In[14:0] || In[15:15];
           2 :: In[13:0] || In[15:14];
           3 :: In[12:0] || In[15:13];
           4 :: In[11:0] || In[15:12];
           5 :: In[10:0] || In[15:11];
           6 :: In[ 9:0] || In[15:10];
           7 :: In[ 8:0] || In[15: 9];
           8 :: In[ 7:0] || In[15: 8];
           9 :: In[ 6:0] || In[15: 7];
          10 :: In[ 5:0] || In[15: 6];
          11 :: In[ 4:0] || In[15: 5];
          12 :: In[ 3:0] || In[15: 4];
          13 :: In[ 2:0] || In[15: 3];
          14 :: In[ 1:0] || In[15: 2];
          15 :: In[ 0:0] || In[15: 1];
    }
}
```

Fig. 3. Description of the 16-bit rotator circuit

the usual logical operations on BDDs, it includes a rich set of meta routines (e.g. quantification with respect to a set of variables, composition, conversion to sum-of-cubes), routines for statistics (e.g. size, number of minterms), support for development of new operations, and routines to visualize BDDs (e.g. for X-Windows).

Table 1 indicates the effect of dynamic variable ordering on some typical benchmarks, all taken from [13] except for the rotator. #nodes is the size of the final (shared) BDDs for all output functions. The run time is in seconds on a HP9000/735 workstation. The 'good' and 'bad' orders are obtained manually, we don't claim them to be the best, resp. worst; 'dynamic' means dynamic variable ordering is on during BDD construction; 'bad' is taken as initial order, and at the end dynamic ordering is applied exhaustively until no more gain is obtained. The results for min_max include BDDs for both the regular outputs and the next-state functions.

With our implementation, we achieve results comparable with Rudell's [15]. As our experiments point out, there is no such thing as "one medicine cures all". Some tuning of the dynamic variable parameters may often give better results.

Table 1. Experimental results of dynamic variable ordering

Circuit	Good		Bad		Dynamic	
	Nodes	Secs	Nodes	Secs	Nodes	Secs
16-bits rotator	81	<1	1081328	45	81	<1
8-bits adder	36	<1	751	<1	36	<1
16-bits adder	76	<1	196575	12.7	123	<1
32-bits adder	156	<1	>1000000	29	452	1.9
8-bits min_max	893	<1	75377	4.7	892	1.7
16-bits min_max	3305	<1	>1000000	45.2	3303	8.9
32-bits min_max	12545	1.4	>1000000	33	33971	48.5
8-bits multiplier	9084	2.1	16697	2.9	8958	13.3
10-bits multiplier	72916	24.5	159278	32.8	72204	204.8
12-bits multiplier	598463	238.6	1513070	360.5	560216	1012.3

3 Exploiting Structural Similarities

In many design environments, formal verification is applied to establish the equivalence of two circuit descriptions. Typically, one description has been derived from the other by one or more design steps. Many of these steps have little effect on the global structure of a circuit. Therefore, it is likely that there exist structural similarities between the two circuits to be verified, i.e., not only the outputs are functionally equivalent, but also some of the internal signals. For example, when the correctness of a technology mapping step is verified, both circuits have a similar global structure, and when a designer wants to check the correctness of some manual modifications, only small parts of the circuit may have changed. Intuitively, it should be easier to prove the equivalence of two circuits which have a similar structure, than to prove the equivalence of two circuits with a totally different structure. However, this is typically not true for BDD-based verification tools; the efficiency of these tools is almost completely determined by the compactness of the BDD representations for the outputs of the circuits. Therefore, the efficiency depends on the *type* of circuit that is verified and not on the actual differences between the circuits that are compared.

One of the first verification methods to use structural equivalences was presented in [3]. However, this method may lead to so-called *false negatives*, i.e., the method may fail to prove the equivalence of two equivalent circuits. A more advanced method was described in [4]. This method still suffers from false negatives, although the authors suggest how BDDs can be used to avoid this problem. Recently, two verification methods were presented that exploit equivalences between internal signals in combination with a test generator [6][14]. These methods do not suffer from false negatives, and the presented benchmark results demonstrate the efficacy of using structural similarities.

In this section, it is shown how a BDD-based verification method can be extended to exploit structural equivalences of the circuits that are compared. This improves the method's ability to deal with circuits that are structurally similar, even if these circuits cannot be represented efficiently by BDDs. First, it is described how equivalent internal signals can be used to improve the verification method; the ideas used are similar to some of the ideas presented in [4]. Then, it is shown how these equivalences can be detected during the verification of the circuits.

3.1 Using Structural Equivalences

Suppose we want to verify that two outputs f and g are functionally equivalent, i.e., we want to establish the equivalence of the corresponding functions $F : B^n \rightarrow B$ and $G : B^n \rightarrow B$, where n denotes the number of primary inputs. This means we have to prove that for every input vector $\underline{i} \in B^n$,

$$F(\underline{i}) \oplus G(\underline{i}) = 0 . \tag{1}$$

Now assume that the fanin cones of f and g both contain a signal with the function $H : B^n \rightarrow B$. Then, this function can be used to decompose F and G, i.e., there are functions $F_{\mathrm{d}} : B^{n+1} \rightarrow B$ and $G_{\mathrm{d}} : B^{n+1} \rightarrow B$, such that:

$$\begin{aligned} F(\underline{i}) &= F_{\mathrm{d}}(\underline{i}, H(\underline{i})) , \\ G(\underline{i}) &= G_{\mathrm{d}}(\underline{i}, H(\underline{i})) . \end{aligned} \tag{2}$$

In that case, (1) can also be written as follows:

$$(\exists v \in B : (F_{\mathrm{d}}(\underline{i}, v) \oplus G_{\mathrm{d}}(\underline{i}, v)) \wedge (v = H(\underline{i}))) . \tag{3}$$

This formula indicates how a structural equivalence can be used to decompose the calculation of the difference. First, the difference of the two functions is calculated while the common subfunction H is represented by an extra variable. If this difference is zero, the two functions are equivalent. However, if it is not zero, it cannot be decided directly that the functions are not equivalent; this decision can only be made if the difference expressed in the original variables differs from zero. Otherwise, there must apparently exist variable assignments for which the calculated difference does not equal zero, but these assignments may not be valid because the variables representing subfunctions cannot be assigned values independently. To avoid these false negatives, it is necessary to replace the introduced variable by the function it represents.

A verification method can repeatedly use structural equivalences to compare two circuits. The main advantage of this approach is that the BDD representations remain compact if sufficient equivalences exist between the circuits. The functions of the signals are not only expressed in terms of the primary inputs, but also in terms of variables representing common internal signals. Because the functional dependencies between these variables are not taken into account while constructing BDDs, this approach leads to smaller BDDs, even though the number of variables is increased. The main disadvantage is of course that the equivalence check now requires substitutions, possibly resulting in a large BDD representation for the difference of two signals. However, in case the two circuits not only have some internal signals in common, but also use these signals in a similar manner, the difference may be relatively small, or it may even be zero, in which case there is no need for substitutions at all.

The two circuits in Fig. 4 are used to illustrate how the verification method uses equivalent internal signals.

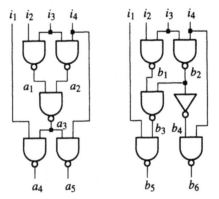

Fig. 4. An example of two similar circuits

First, it is verified whether the outputs a_4 and b_5 are equal. Obviously, this should be easy, because the parts of the circuits which define these outputs, are not only functionally but also structurally identical. During the verification, it is detected that the signals a_1 and b_1, and a_2 and b_2 are equal, and therefore, two new variables v_1 and v_2 are introduced to represent the functions of these signals.

Then, it is detected that a_3 and b_3 are equal, because for both, the function $\neg(v_1 \wedge v_2)$ is computed. The variable v_3 is introduced for this function. Similarly, the function $\neg(i_1 \wedge v_3)$ is calculated for both a_4 and b_5, and the equivalence of the outputs is established. This shows that the proposed method can handle structural equivalence without any problem.

The verification of the outputs a_5 and b_6 is slightly more difficult. If the same equivalences as detected earlier are used, the function $\neg(i_4 \wedge v_3)$ is computed for output a_5, and the function $\neg(i_4 \wedge \neg v_2)$ for output b_6. The difference $a_5 \oplus b_6$ is $i_4 \wedge (v_3 \oplus \neg v_2)$. Because this is not equal to zero and still depends on variables representing internal signals, the function $\neg(v_1 \wedge v_2)$ is substituted for variable v_3. Then, the difference becomes $i_4 \wedge v_2 \wedge \neg v_1$. This is still not equal to zero, and therefore, also v_2 is replaced by the function it represents. This results in the difference $i_4 \wedge \neg(i_3 \wedge i_4) \wedge \neg v_1$, which can also be written as $i_4 \wedge \neg i_3 \wedge \neg v_1$. After the replacement of v_1 by its function, this becomes $i_4 \wedge \neg i_3 \wedge i_3 \wedge i_2$, which equals zero. This proves that the outputs a_5 and b_6 are equivalent.

3.2 Detecting Structural Equivalences

We will now address the problem of detecting structural equivalences of two circuits. A naive approach is to compare every signal calculated in one circuit with all signals calculated in the other circuit. However, this is not very efficient, since every equivalence check may require a series of substitutions; especially the comparison of non-equivalent signals can be quite expensive, both in terms of run time and memory usage. Of course, it is also possible to detect only those equivalences for which no substitutions are required. In that case, it is not guaranteed that all equivalences are found, because it is only tested if two signals have exactly the same BDD representation. However, this is not an effective approach. If an equivalence is not detected because the difference of the BDDs is apparently not equal to zero, it is very likely that all other equivalences in the fanout cone of these signals also remain undetected. In fact, it is very likely that the difference becomes larger while it propagates towards the outputs. Therefore, we introduce an extra preprocessing step to detect which signals are most likely to be equivalent.

An effective preprocessing step is to calculate a signature for every signal by simulating both circuits for a limited number of randomly generated input vectors. Because two signals cannot be equivalent if they have different signatures, such a signature can be used as a coarse check for the equivalence of two signals. This technique does not exclude any equivalences, and therefore does not suffer from the same disadvantage as selecting signals on basis of the calculated BDD representation. However, there are circuits for which random input vectors are not very effective in distinguishing signals. In that case, many inequivalent internal signals will be compared, possibly requiring many substitution operations. The effects of this problem can be diminished by exploiting the result of the equivalence check. This result is namely the set of all input vectors that distinguish the two signals under comparison. Therefore, it is possible to randomly select vectors from this set, and perform extra signature calculations to update the lists of potentially equivalent signals. This way, the negative outcome of an equivalence check is also used to check all other potential equivalences.

Structural equivalences can only be used effectively if they are detected as soon as possible; otherwise, the BDDs for the outputs are constructed before any extra variables are introduced and the method cannot perform better than a conventional BDD-based method. Therefore, it is necessary to calculate signals from both circuits simultaneously and to compare potentially equivalent signals as soon as possible. This means that the calculation order has to group potentially equivalent signals together as much as possible. To do this, we adopt the following strategy. The calculation order is determined by maintaining a so-called 'ready list' of signals. This list contains all signals for which potential equivalences exist and for which all predecessors with potential equivalences have already been calculated; initially, it contains the primary inputs of both circuits. From this list, the signal with the lowest topological level is selected, together with all signals that are potentially equivalent to it. These signals are calculated together with the parts of the corresponding fanin cones that have not yet been calculated. Then, the equivalence of these signals is actually checked, and the ready list is updated.

3.3 Implementation Issues

We have implemented the presented verification method in C++ using the BDD package mentioned in Sect. 2. Dynamic variable ordering is used to order the variables; variables representing internal signals are created at the front of the order. A signature calculation is used to partition the signals of both circuits into groups of signals that are likely to be equivalent. Signatures are calculated by simulating both circuits simultaneously for 32 randomly generated input vectors. These signatures are used to create the initial partition. Then, the partition is refined by calculating new signatures. This is repeated until the partition does not change during five successive runs. Under the assumption of perfect hashing, the calculation of these potential equivalences takes $\mathcal{O}(C \cdot n)$ time and $\mathcal{O}(C)$ space, where C denotes the size of both circuits that are compared, and n is the number of times the partition is refined.

The notion of structural equivalence is slightly extended to allow for signals that are functionally equivalent modulo complementation. Every signal is assigned a 'sign' that is determined by the value of that signal for a randomly selected input vector. If a signal has a negative sign, its function is complemented before we check for equivalences. If we detect structural equivalences and decide to introduce a new variable, the function of a signal with a negative sign is replaced by a negated variable; this has no further consequences for the equivalence check. To limit the number of extra variables introduced by the method, a new variable is created only if the size of the BDD for the corresponding function exceeds a given threshold. In our implementation, we use a threshold of 32 nodes. If it is decided not to introduce a new variable, the BDDs of the equivalent signals are still unified by selecting the smallest BDD as the representation for all these signals.

3.4 Experimental Results

To evaluate the efficiency of the presented verification method, we use it to verify the correctness of some circuits that have been synthesized with the logic synthesis systems EUCLID [2] and SIS, a system developed at University of California, Berkeley. All tests in this section are performed on a HP9000/735 workstation with a memory limit of 100 Mb. As benchmarks, we use instances of the min_max circuit and the unsigned bit multiplier, which are both described in [13]. The notations mM[n] and mult[n] are respectively used to refer to the n-bit instances of these circuits. Because our verification method only deals with combinational circuits, every latch in the min_max circuit is modeled by an extra input and output.

Table 2 shows the performance of a conventional BDD-based verification tool on some selected instances of the benchmarks. As these results illustrate, the min_max circuit is an 'easy' example, which can be represented compactly with BDDs and thus scales well. The multiplier circuit on the other hand is very difficult to verify; the required amounts of memory and run time grow rapidly with increasing bit width. With a memory limit of 100 Mb, the applicability of this tool is restricted to instances with $n \leq 14$.

Table 2. Benchmark results of a conventional BDD-based verification tool

Circuits	CPU time (s)	BDD mem. (kb)
mM[16], mM[16]sim	1.7	340
mM[32], mM[32]sim	13.9	699
mM[48], mM[48]sim	24.0	1301
mM[64], mM[64]sim	29.6	2112
mult[8], mult[8]sim	14.3	741
mult[10], mult[10]sim	168.7	5036
mult[12], mult[12]sim	1725.4	43806

As a first experiment for the presented verification method, we test if all constants in the original circuit descriptions are propagated correctly by EUCLID. The resulting circuits are denoted by the suffix 'sim'. The results of these experiments are given in Table 3. The first two columns show the run times and the amount of memory required for storing the BDDs. The last column shows the number of variables representing equivalent signals. As the results demonstrate, the proposed verification method performs significantly better than the conventional one on these simple examples. This clearly illustrates that the efficiency of the method does not depend solely on the type of circuit that is verified, but also on the actual differences between the two circuits.

In the second set of experiments, we resynthesize the benchmarks with SIS. After the usual optimizations such as constant propagation, the circuit is partially collapsed and factored. This results in circuits that are harder to verify, because large parts of the original structure are modified. The suffix 'res' is used

Table 3. Experimental results for simple benchmarks

Circuits	CPU time (s)	BDD mem. (kb)	Eq. used
mM[16], mM[16]sim	0.6	395	41
mM[32], mM[32]sim	1.1	455	66
mM[48], mM[48]sim	8.9	661	144
mM[64], mM[64]sim	6.4	592	137
mult[8], mult[8]sim	0.8	379	56
mult[16], mult[16]sim	9.7	594	296
mult[24], mult[24]sim	23.6	1422	765
mult[32], mult[32]sim	33.8	2813	1407

to indicate that a circuit is synthesized with this method. The results are shown in Table 4. As these results show, it still pays off to exploit the structural equivalences, even though the verification of the larger instances of the multiplier cannot be completed; for mult[24]res and mult[32]res only 20 and 17 outputs are respectively verified successfully. This is not caused by the absence of structural equivalences, but by the size of the intermediate results during a sequence of substitutions.

Table 4. Experimental results for resynthesized circuits

Circuits	CPU time (s)	BDD mem. (kb)	Eq. used
mM[16], mM[16]res	2.4	385	15
mM[32], mM[32]res	11.2	686	43
mM[48], mM[48]res	11.4	565	32
mM[64], mM[64]res	17.9	1452	72
mult[8], mult[8]res	1.4	353	32
mult[16], mult[16]res	92.6	2134	183
mult[24], mult[24]res	—	—	—
mult[32], mult[32]res	—	—	—

In the third set of experiments, we use EUCLID to map the benchmarks onto another technology, namely a complete library of (3,3)-AOI cells [2]. Because the functions implemented by these cells are larger than the expressions in the original descriptions, the synthesis strategy also involves partial collapsing and factoring of the circuits, which modifies the original structure. The remapped circuits are denoted by the suffix 'map'. The results of these experiments are shown in Table 5. When compared to the results of the previous experiments, it can clearly be observed that the circuits resulting from these synthesis steps are easier to verify, because they have been collapsed less strongly. In this case, 46 and 62 outputs are respectively verified successfully of mult[24]res and mult[32]res.

In the fourth set of experiments, the EUCLID system is used to decrease the maximum delay of the circuits resulting from the previous experiment. This

Table 5. Experimental results for remapped circuits

Circuits	CPU time (s)	BDD mem. (kb)	Eq. used
mM[16], mM[16]map	1.5	387	16
mM[32], mM[32]map	2.9	454	63
mM[48], mM[48]map	7.0	556	54
mM[64], mM[64]map	10.8	638	74
mult[8], mult[8]map	0.8	367	39
mult[16], mult[16]map	17.2	640	201
mult[24], mult[24]map	—	—	—
mult[32], mult[32]map	—	—	—

means that segments of the critical paths in the circuits are restructured in order to improve the delay of these paths. For the min_max circuits, a speedup of 30% is obtained, and for the multipliers a speedup of 15%. The circuits that are synthesized with this approach are given the suffix 'fast'. The results are given in Table 6. Mark that in this case, the results are not verified against the original descriptions, but against the results of the previous experiments.

Table 6. Experimental results for circuits resynthesized for speed

Circuits	CPU time (s)	BDD mem. (kb)	Eq. used
mM[16]map, mM[16]fast	0.4	404	27
mM[32]map, mM[32]fast	0.9	436	46
mM[48]map, mM[48]fast	5.9	489	99
mM[64]map, mM[64]fast	6.8	540	133
mult[8]map, mult[8]fast	0.5	373	41
mult[16]map, mult[16]fast	7.0	498	203
mult[24]map, mult[24]fast	21.2	1288	505
mult[32]map, mult[32]fast	28.1	2575	926

In order to compare our results with [14], we verify some of the more difficult ISCAS benchmarks [7] against the same circuits after redundancy removal [18]. We also verify mult[16] against the ISCAS benchmark c6288, which is a 16-multiplier with a similar structure. The results are shown in table 7. Although it is generally difficult to compare run times measured on different machines, the results indicate that our verification method is at least an order of magnitude faster than the method presented in [14] on the selected benchmarks. For the other ISCAS benchmarks, the differences are relatively smaller. The verification of the 16-bit instance of the multiplier against the ISCAS benchmark c6288 is performed very efficiently, because both circuits have the same architecture, and therefore, many structural equivalences exist.

To measure the effect of dynamic variable ordering on the results, we have also performed some tests with dynamic variable ordering turned off. The re-

Table 7. Experimental results for some ISCAS benchmarks

Circuits	CPU time (s)	BDD mem. (kb)	Eq. used
c3540, c3540nr	4.5	429	94
c5315, c5315nr	13.5	505	115
c7552, c7552nr	46.7	1104	145
mult[16], c6288	24.7	703	201
mult[16], c6288nr	15.2	621	200

sults of these experiments are given in Table 8. As these results show, dynamic variable ordering has a significant impact on the memory usage for the more difficult tests. It succeeds in finding good variable orders to compactly represent intermediate results of a series of substitutions. For the easier examples, the introduction of new variables for intermediate signals ensures that the number of BDD nodes does not grow very fast in case of many structural equivalences. Therefore, dynamic variable ordering cannot improve much in these cases.

Table 8. Influence of dynamic variable ordering

Circuits	Dyn. order.		No dyn. order.	
	Time (s)	Mem. (kb)	Time (s)	Mem. (kb)
mM[64], mM[64]sim	6.4	592	2.6	653
mult[32], mult[32]sim	33.8	2813	20.4	3092
mM[64], mM[64]res	17.9	1452	8.6	1520
mult[16], mult[16]res	92.6	2134	42.5	5802

4   Conclusions and Future Work

In this paper, we have discussed two techniques that enhance the ability of BDD-based verification methods to handle larger circuits. The technique of dynamic variable ordering has been briefly discussed, because it virtually removes the need to worry about a reasonable initial variable order. Therefore, it makes the verification method more robust. We have shown how this technique can be incorporated in a BDD package in a way that is fully transparant to an application of the package. The penalty is an increase in run time, estimated at a factor 4 on average.

We have also shown how a BDD-based verification method can be extended with a technique to exploit structural similarities of the circuits under comparison. These similarities are detected fully automatically. Experimental results demonstrate that the new technique significantly extends the ability of BDD-based verification methods to deal with circuits that are structurally similar. The successful verification of various multipliers shows that it is possible to handle circuits that cannot be represented compactly with BDDs. Our current research focuses on finding a more effective technique to perform the required substitutions, because we believe that the size of the intermediate results during a sequence of substitutions is the bottleneck in our current method. We also intend to incorporate our method in an industrial design system.

Acknowledgements

We kindly acknowledge IBM Corporation, Yorktown, for making their BSN design system available to us. We would also like to thank the anonymous reviewers for their valuable comments, and Harm Arts for his help with synthesizing the benchmarks.

References

1. Ashar, P., Ghosh, A., Devadas, S.: Boolean Satisfiability and Equivalence Checking using General Binary Decision Diagrams. Proc. ICCD-91, pp. 259–264, 1991
2. Berkelaar, M.R.C.M., Theeuwen, J.F.M.: Logic Synthesis with Emphasis on Area-Power-Delay Trade-Off. Journal of Semicustom ICs, September 1991, pp. 37–42
3. Berman, C.L.: On Logic Comparison. Proc. DAC-81, pp. 854–861, 1981
4. Berman, C.L., Trevillyan, L.H.: Functional Comparison of Logic Designs for VLSI Circuits. Proc. ICCAD-89, pp. 456–459, 1989
5. Brace, K.S., Rudell, R.L., Bryant, R.E.: Efficient Implementation of a BDD package. Proc. DAC-90, pp. 40–45, 1990
6. Brand, D.: Verification of Large Synthesized Designs. Proc. ICCAD-93, pp. 534–537, 1993
7. Brglez, F., Fujiwara, H.: A Neutral Netlist of 10 Combinational Benchmark Circuits and a Target Translator in FORTRAN. Special session on the 1985 IEEE Int. Symposium on Circuits and Systems, 1985
8. Bryant, R.E.: Graph Based Algorithms for Boolean Function Representation. IEEE Transactions on Computers, vol. C-35 no. 8, pp. 677–691, August 1986
9. Burch, J.R.: Using BDDs to Verify Multipliers. Proc. DAC-91, pp. 408–412, 1991
10. Fujita, M., Fujisawa, H., Matsunaga, Y.: Variable Ordering Algorithms for Ordered Binary Decision Diagrams and Their Evaluation. IEEE Transactions on CAD, vol. 12 no. 1, pp. 6–12, January 1993
11. Jain, J., et al.: IBDDs: an Efficient Functional Representation for Digital Circuits. Proc. EDAC-92, pp. 440–446, 1992
12. Jeong, S.-W., et al.: Variable Ordering and Selection for FSM Traversal. Proc. ICCAD-91, pp. 476–479, 1991
13. Kropf, T.: Benchmark-Circuits for Hardware-Verification. February 1994
14. Kunz, W.: HANNIBAL: An Efficient Tool for Logic Verification Based on Recursive Learning. Proc. ICCAD-93, pp. 538–543, 1993
15. Mets, A.A.: Dynamic Variable Ordering for BDD Minimization. Student report Eindhoven University, Department of Electrical Engineering, January 1994
16. Rudell, R.: Dynamic Variable Ordering for Ordered Binary Decision Diagrams. Workshop Notes Int. Workshop on Logic Synthesis, 1993

17. Rudell, R.: Dynamic Variable Ordering for Ordered Binary Decision Diagrams. Proc. ICCAD-93, pp. 42–47, 1993
18. Tromp, G.J., van de Goor, A.J.: Logic Synthesis of 100-percent Testable Logic Networks. Proc. ICCD-91, pp. 428–431, 1991

Studies of the Single Pulser in Various Reasoning Systems

Steven D. Johnson,[1] Paul S. Miner,[2] and Albert Camilleri[3]

[1] Indiana University, Bloomington, IN 47405-4101 USA
[2] NASA Langley Research Center, Hampton, VA 23681-0001 USA
[3] Hewlett-Packard Company, Roseville, CA 95747-5596 USA

Abstract. The *single pulser* is a clocked sequential device which generates a unit-time pulse on its output for every pulse on its input. This paper explores how a single-pulser implementation is verified by various formal reasoning tools, including the PVS theorem prover for higher-order logic, the SMV model checker for computation tree logic, the DDD design derivation system, and the Oct Tools design environment. By fixing a single, simple example, the study attempts to contrast how the underlying formalisms influence one's perspective on design and verification.

KEYWORDS AND PHRASES: Formal methods, hardware verification, formal verification, theorem prover, higher order logic, model checker, design derivation, logic synthesis, PVS, SMV, CTL, DDD, Oct Tools.

1 Introduction

One of the problems confronting the transfer to practice of formal verification methodology is the confounding variety of formal systems available. This paper is an initial attempt to contrast several mechanized formalisms as applied to a simple common problem, a circuit called a *single pulser*. The single pulser is arguably the simplest sequential circuit that does anything interesting. Even so, its verification exposes interesting issues in each of the studies we have undertaken.

We look at four systems in this paper. The first and most general is the PVS theorem prover [11], which operates on higher-order logic expressions over inductive types. Next, we look at SMV, a symbolic model checker [10]. The propositional temporal logic on which SMV is founded is far less expressive than the higher-order logic of PVS; and consequently, the proofs are far more automatic. Third, we try DDD, a design derivation system based the algebra of first-order functional expressions [9]. Finally, we apply the Oct Tools suite of design synthesis tools [13] to the single-pulser problem.

As discussed in the following section, there are many ways of looking at a single pulser. The four studies we have done are not directly comparable because they address different facets of the verification task. However, there are common impressions resulting from these studies. In each case, there are conceptual, methodological, and representational questions to understand before the system can be applied. In our view, there is also evidence to support the need for more integrated reasoning environments, in which tools such as these can be used in consort to solve design problems.

Our goal in reporting these studies is to gain perspective on the variety of approaches that have been taken to formalize reasoning about hardware. We hope eventually to expand this project to include a greater number of systems. This work illustrates how design aspects are represented in different formal systems. It also conveys some of the intuitive feel of working within these systems. We feel that tutorial material such as this will help researchers and practitioners understand how reasoning tools relate to one another. More systematic and rigorous comparisons would certainly be valuable, but are well beyond the scope of these studies. Even if one were to have criteria for comparisons, a more diverse set of examples would be needed to make them meaningful.

More detailed accounts of some of these studies (e.g. transcripts of interactions, script files, etc.) will be made available through [1]. Sections 3 through 6 contain references to more detailed descriptions of the individual systems under study. For a broader view of hardware verification systems and methods, Gupta's survey [7] contains an extensive bibliography and a partial taxonomy; it is a good starting point for further reading although it specifically mentions only one of the systems used in this work, SMV.

2 Informal Description of the Single Pulser

The *single pulser* comes from a textbook by Winkel and Prosser on clocked-synchronous design [12]. Their original English specification reads:

> **Problem Statement.** We have a debounced pushbutton, on (true) in the down position, off (false) in the up position. Devise a circuit to sense the depression of the button and assert an output signal for one clock pulse. The system should not allow additional assertions of the output until after the operator has released the button.

In all of the studies that follow, there is no attempt to account for debouncing an analog input signal; in fact, we shall also assume that input is synchronous. The single pulser device, *SP* has a one-bit input and a one-bit output:

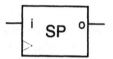

Its observable behavior might stated as follows:

"SP emits a single unit-time pulse on o for each pulse received on i."

Actually, this sentence is an understatement of the specification, if formulated literally, although the necessary details would no doubt be inferred by a designer. The Winkel-Prosser problem statement indirectly says that there is exactly one output pulse for every input pulse. Furthermore, simplicity demands that the output pulse occur in the neighborhood of the input pulse (rather than occuring in independent periodic bursts, for example). Let us take the timing diagram below as a somewhat more rigorous informal specification.

Ellipses indicate time intervals of undetermined duration. The left region of the diagram is intended to say that there are no extra output pulses and the right region that there is just one output pulse some time during every input pulse.

Any reasonable hardware implementation would pick either the beginning or the end of the interval to generate the output pulse. However, the diagram also admits some impossible implementations, such as one in which the output pulse occurs at the midpoint of the input pulse.

The *SP* timing diagram is like a *requirements specification*, describing the expected observable properties of the device. Similarly, could call one of the finite-state diagrams below a *design specification*. Each of them describes a synchronous process with, it is claimed, the required behavior.

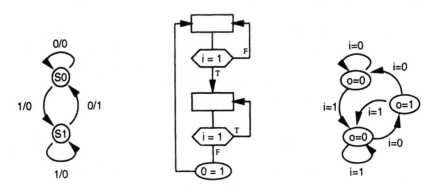

We show several kinds of state-machine diagrams because we are going to adopt perhaps the least familiar one, in the middle. It is the Algorithmic State Machine (ASM) diagram used by Winkel and Prosser. In ASM-diagrams, rectangles represent states, diamonds represent control-flow decisions, and ovals represent conditional (or "Mealy") outputs. There is also a convention that conditions occurring in the diagram are denied unless they are explicitly asserted; thus, "$o = 1$" is false (i.e., $o = 0$) on all but one of the paths. ASMs are relatively expressive in practice but more difficult to formalize. A timing diagram for the state machine is:

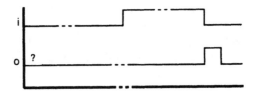

This diagram satisfies (or implements) the *SP* timing diagram, assuming that we don't care what happens at "time zero," and further, that the output pulse is sufficiently "within the neighborhood" of the input pulse.[4] An *implementation description* of the single pulser is represented by the circuit diagram:

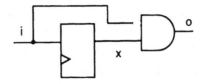

---

[4] The Reader is invited to jot down an appropriate definition of "neighborhood" at this point, as this is a topic we shall return to later.

It is easy to work out that this circuit has the timing diagram:

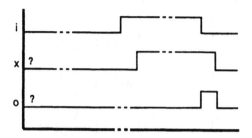

Hence, it is an implementation of both the design and requirements specifications.

Winkel and Prosser make two additional observations about the single pulser which are relevant to its role as a tutorial example. The first of these remarks has to do with the duality of control and architecture in hardware descriptions. The *SP* circuit shown earlier is systematically derivable from the ASM specification:

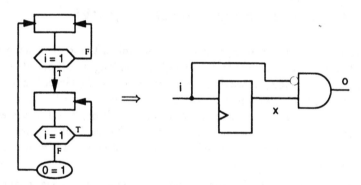

However, now that we have the circuit we can look at it (specifically the delay element) as a piece of architecture under the control of the one-state (i.e., vacuous) ASM:

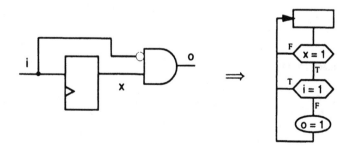

Similarly, the isolated view we have taken of the single pulser as a separate device overshadows another, equally valid, abstraction of *SP* as a synchronization protocol. "Being a single pulser" might be taken to mean that a larger circuit contains the same handshake as the abstracted single pulser:

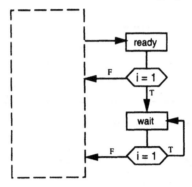

We could also think of SP as being abstract with respect to the kinds of event sequences it recognizes as a "pulse." None of these generalizations arise in the studies we have carried out so far, but it might be good to keep them in mind.

3 Verification of a Single Pulser using PVS

PVS (Prototype Verification System) is a mechanical theorem proving system developed at SRI International [11]. Its specification language is based on simply-typed higher-order logic; it provides an interactive proof checker employing sequent calculus proof rules and it has decision procedures for linear arithmetic. One interacts with PVS through an ASCII text editor, but the system also provides formatting facilities for printed reports. Those facilities were used to generate the logic formulas in this section. The proof presented in this section is also discussed in [8].

Since the single pulser is a simple circuit, we bypass the state machine level and directly prove a candidate implementation with respect to a high-level specification. A plausible specification of a single pulser is the following (i and o' are functions from time to values, where time ranges over the integers and values are taken from $\{0, 1\}$. Variables m, n, j, and k are of type time):

$$
\begin{aligned}
&\mathrm{spec1}(i, o) : \mathrm{bool} \\
&= \\
&(\forall\, n, m : \\
&\quad \mathrm{Pulse}(i, n, m) \\
&\qquad \supset \\
&\quad (\exists\, k : \\
&\qquad n \leq k \\
&\qquad \land\ k \leq m \\
&\qquad \land\ o(k) = 1 \\
&\qquad \land\ (\forall\, j : (n \leq j \land j \leq m \land o(j) = 1 \\
&\qquad\qquad \supset j = k))))
\end{aligned}
$$

An English reading of this formula is:

> Whenever there is a pulse on the input signal i, say from time n to time m, there is a unique time k in the vicinity of the input pulse when the output signal is asserted.

A graphical representation of Pulse(f, n, m) is given by:

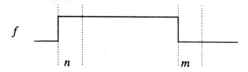

The definition of predicate Pulse in PVS is:

$$\text{Pulse}(f, n, m) : \text{bool}$$
$$= (n < m \land f(n - 1) = 0 \land f(m) = 0$$
$$\land (\forall t : (n \le t \land t < m \supset f(t) = 1)))$$

Our candidate single pulser circuit is

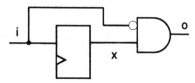

Using the style advocated by Gordon [6] we can represent this circuit as follows:

$$\text{imp}(i, o) : \text{bool} = (\exists x : (\text{delay}(i, x) \land \text{and}_\bullet^0(i, x, o)))$$
$$\text{delay}(i, o) : \text{bool} = (\forall t : (o(t + 1) = i(t)))$$
$$\text{and}_\bullet^0(a, b, y) : \text{bool} = (\forall t : (y(t) = (- a(t)) \times b(t)))$$

It is fairly simple to verify that this implementation satisfies spec1. A summary of the PVS proof is given in Figure 1. Each of the subgoal sequents displayed in Figure 1 can be easily verified using appropriate instantiations of quantified variables in the assumptions.

Unfortunately, spec1 is not sufficient. It only specifies the behavior of the circuit in the neighborhood of an input pulse. There are no constraints on the behavior of imp between pulses. As an example, a simple inverter satisfies our proposed single pulser specification. This is illustrated by

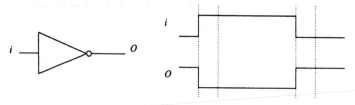

where the inverter is defined by

Verbose proof for `single_pulse1`.

`single_pulse1`:

$[1]$ $(\forall\,(o, i : \text{signal}) : \text{imp}(i, o) \supset \text{spec1}(i, o))$

Expanding the definitions of spec1, imp, delay, and $^{\circ}_{\bullet}$, $-$, Pulse
Applying (SKOSIMP*),
Instantiating the top quantifier in $+$ with the terms: m!1
Applying propositional simplification and decision procedures,
which yields 2 subgoals:

`single_pulse1.1`:

$\{-1\}$ $(\forall\,(t : \text{time}) : (x'(t + 1) = i'(t)))$
$\{-2\}$ $(\forall\,(t : \text{time}) : (o'(t) = (1 - i'(t)) \times x'(t)))$
$\{-3\}$ $n' < m'$
$\{-4\}$ $i'(n' - 1) = 0$
$\{-5\}$ $i'(m') = 0$
$\{-6\}$ $(\forall\,(t : \text{time}) : (n' \le t \wedge t < m' \supset i'(t) = 1))$
$\overline{\{1\}\quad o'(m') = 1}$

`single_pulse1.2`:

$\{-1\}$ $n' \le j'$
$\{-2\}$ $j' \le m'$
$\{-3\}$ $o'(j') = 1$
$[-4]$ $(\forall\,(t : \text{time}) : (x'(t + 1) = i'(t)))$
$[-5]$ $(\forall\,(t : \text{time}) : (o'(t) = (1 - i'(t)) \times x'(t)))$
$[-6]$ $n' < m'$
$[-7]$ $i'(n' - 1) = 0$
$[-8]$ $i'(m') = 0$
$[-9]$ $(\forall\,(t : \text{time}) : (n' \le t \wedge t < m' \supset i'(t) = 1))$
$\overline{\{1\}\quad j' = m'}$

Fig. 1. PVS proof trace for `single_pulse1`

$$\text{inv}(i, o) : \text{bool} = (\forall\, t : (o(t) = 1 - i(t)))$$

There are a number of ways to remedy this situation. One possibility is to modify our definition of Pulse to extend our notion of neighborhood until the beginning of the next pulse.[5] However, spec1 still does not constrain the behavior for all possible input streams. In the cases where the input has been high forever or remains high forever, the implementation is unconstrained. In addition to spec1, we need to show that the implementation also satisfies the following:

$$\text{spec2}(i, o) : \text{bool}$$
$$=$$
$$(\forall\, k :$$
$$o(k) = 1$$

---

[5] See Footnote 4, in Section 2.

$\supset$ SinglePulse(o, k)

$\wedge$ $(\exists\, n, m : n \leq k \wedge k \leq m \wedge \text{Pulse}(i, n, m)))$

This formula states that whenever the output is asserted, it is a single pulse and is in the vicinity of an input pulse. It is fairly simple to show that the output is a single pulse, if it is ever asserted. We can also easily show that the output is asserted after the input transitions from high to low. However, we have no means to show that the input pulse has been low previously. In other words, this circuit actually implements a synchronous falling-edge detector. It can only be a single pulser under the assumption that at some time in the past, the input was low. This is obviously a reasonable assumption; thus, we modify our verification condition by adding a clause about the environment in which the circuit is intended to operate, giving the following additional verification condition:

env1(f): bool $= (\forall\, t : (\exists\, t_1 : t_1 < t \wedge f(t_1) = 0))$
single_pulse2 : LEMMA env1(i) $\supset$ (imp(i, o) $\supset$ spec2(i, o))

The additional assumption about the environment requires some additional reasoning unrelated to the correct operation of the circuit. In particular, it was necessary to establish from env1 that a greatest such time exists.

This completes our verification of the single pulser using PVS. Our example lacks some characteristics of larger verification efforts. In our proofs we expand the definitions of the circuit elements to their representation and invoke the built-in decision procedures of PVS. In a larger verification, we would prove properties about our representation and simplify accordingly, rather than employing the brute force approach shown here.

4   Verification of the Single Pulser Using Temporal Logic

We used the SMV (*Symbolic Model Verifier*) system [10] to specify and verify the single pulser circuit. An implementation is represented by a finite-state machine and the specification is represented by a formula in *computation-tree logic* (CTL). The system automatically verifies whether the given state-machine satisfies that formula, that is, provides a model which makes the formula true.

A CTL formula specifies a possibly infinite computation tree describing the intended behavior of a correct design [4, 3]. Universal and existential quantifiers, A and E, refer to paths in the tree. The modalities F (some future state), G (all states), X (next state) and U ("(strong) until," or an interval between states) refer to the totally ordered set of states along a path. The simple variables of a CTL formula are propositional with '!' ('&', '|', '->') standing for logical negation (conjunction, disjunction, implication). A state is specified by the propositions that hold in it.

An SMV description of the single pulser design is given by

```
VAR state: {ready, wait};
    i : boolean;

ASSIGN
 init(state) := ready;
 next(state) :=
   case
   state = ready & !i : ready;
   state = ready &  i : wait;
   state = wait  &  i : wait;
   1                  : ready;
   esac;

DEFINE
   o :=
   case
   state = ready & !i : 0;
   state = ready &  i : 1;
   state = wait  &  i : 0;
   1                  : 0;
   esac;
```

Arbitrarily, this version of the state machine issues its output pulse at the rising edge of an input pulse rather than the falling edge as before. The state-machine description could be simplified to

```
ASSIGN next(state) := case i: wait, 1: ready; esac;
DEFINE o := (state = ready) & i;
```

(and even further once a boolean assignment is made for ready and wait) but we made a direct translation from the diagram, thinking that a more automatic system should tolerate a mechanistic approach. A representation of the implementation circuit is equally straightforward if we associate a state transition with a unit-time interval, so that a clocked register can be represented by the SMV next operator.

```
VAR
  x: boolean;

ASSIGN
  init(i) := 0;
  next(x) := i;

DEFINE
  rising_edge := i & !x;
```

The need to preset i corresponds to the initial-state assignment in the state machine. However, as was also the case with PVS, this reset assumption obscures some possibly significant anomalies that often occur during power-up intervals in physical circuits. We have immediately that the circuit implements the state machine. The keyword SPEC indicates an assertion to be verified.

```
SPEC  o = rising_edge
```

SMV confirms this to be true as it does with all of the SPEC statements that follow.

In order to express the more general single-pulser specification in SMV we must as before settle on an acceptable notion of *neighborhood* within which to look for an output pulse. In this case, let us define *neighborhood* to be the interval between two successive rising edges on i (Notice above that rising_edge refers not to the transition event on i but to the first unit-length period thereafter).[6] The single pulser specification decomposes into three overlapping properties as follows.

(a) SPEC AG (rising_edge -> (AF (o)))
  Whenever there is a rising edge o is 1 some time later.

(b) SPEC AG (o -> AX (A[!o U rising_edge]))
  Whenever o is 1 it becomes 0 in the next state (ie, one unit later) and it remains 0 at least until the next rising edge on i

(c) SPEC AG (rising_edge -> (!o -> (AX A[!rising_edge U o])))
  Whenever there is a rising edge, and assuming that the output pulse doesn't happen immediately, there are no more rising edges until that pulse happens. In other words, there can't be two rising edges on i without a pulse on o between them.

Property (b) is actually not valid because the semantics of the *until* operator would require rising_edge to become true on all paths—the so-called *strong until*. However, SMV has a provision called FAIRNESS which asserts that a condition holds infinitely often on all paths. If we assert FAIRNESS rising_edge then Property (b) becomes valid for the single pulser state machine.

We should comment on the difficulties and uncertainties we experienced in reaching this form of the specification. For the novice users, it took a long period to adjust to the expressive limitations of CTL, and it is still very much a matter of debate when a particular logical representation is "well put." Beginners quite often fell into logical traps, for example, writing formulas like AG(!i) -> AG(!o) to express, "In the case that i is never true, neither is o." This formula is true, but only because AG(!i) is a false premise. Coming up with the definition of

---

[6] This choice of the interval from rising edge to rising edge was based on the experience of the PVS study. See Footnote 5.

`rising_edge` seemed to be a key. Without this definition the CTL would be much more complex. Similarly, one should not try to say too much with a single sentence. Our specifications improved significantly once we decomposed the specification into a set of simpler properties.

Once the logical representations are determined, the proofs are immediate and automatic. Failed proofs produce scenarios that are useful in refining the formulations. However, in this study, invalid theorems always revealed inadequacies in the specification, rather than mistakes in the design.

5 Derivation of a Single Pulser using DDD

DDD (*Digital Design Derivation*) is a specialized transformation system for digital system design [9, 2]. It operates on two dialects of first-order functional expressions concretely represented by Scheme (Lisp) s-expressions. These dialects correspond to behavioral and structural forms of hardware description.

The goal of transformation is to reduce a higher-level algorithmic specification into a hierarchical network of processes. A *derivation* is a sequence of transformations and constructions applied to an initial expression, which is typically a design specification. Most DDD transformations preserve functional equivalence, although some are contingent on assumptions about the context (input-output timing, for example); and others add detail to the design (data representations, for example). As a formal object, a derivation together with any side conditions it synthesizes constitutes a proof of "implementation correctness," that is, equivalence between the design specification and its circuit implementation.

Typically, there are three phases in a derivation. The first is to manipulate the behavioral form to achieve certain architectural goals. The second is to construct and refine the structural expression describing that architecture. The third is to project the design to a concrete data representation, ultimately, a boolean system to which logic synthesis tools are applied.

In the single pulser study, we began with the specification SP shown in Fig. 2. SP is a proper Scheme program in which the states of the single pulser ASM represented by a system of tail-recursive functions definitions, READY and WAIT. This is a standard way to model finite-state control with function expessions. Within each function definition, the *let*-bindings for X and I provide a simple input/output interface: SP executes as a Scheme program to animate the specification. Thus, the DDD formulation differs from predicate formulations in that the expressions on which it operates can be directly executed to explore the design behavior. However, the I/O supporting animation contaminates the expressions when considered as a hardware description. The algebra used to transform these expressions into hardware can also be used to isolate and factor out the modeling interface, as is demonstrated in this study.

```
(define SP
(lambda (IN)
  (letrec
    ((READY (lambda (0)
        (let ((X (out 0))
              (I (= (inp IN) 1)))
             (if I
                 (WAIT 0)
                 (READY 0)))))

     (WAIT  (lambda (0)
        (let ((X (out 0))
              (I (= (inp IN) 1)))
             (if I
                 (WAIT 0)
                 (READY 1))))))

     (READY 0))))
```

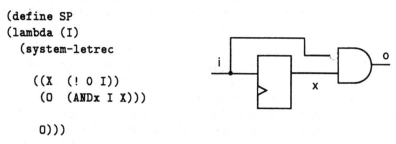

Fig. 2. DDD design specification of a single pulser

A second dialect of functional modeling expression is used to describe circuit structure. The target implementation would be:

```
(define SP
(lambda (I)
  (system-letrec

    ((X  (! 0 I))
     (0  (ANDx I X)))

    0)))
```

The system-letrec expression defines a network of non-terminating *streams* representing infinite sequences of values over time. The defining expression for X uses DDD's delay operator, '!'' to (arbitrarily) initialize sequence X with a zero. Sequence 0 is obtained by extending the binary operation ANDx element-wise to the streams I and X. Recursion in system-letrec expressions corresponds to feedback in the stream network, but there is no feedback in the single pulser example. Streams are not standard Scheme constructs but are added as a syntactic extension [2].

The first step in the derivation constructs an initial structural description. The two components of this description are (a) a *selection combination*, which represents the control structure of the specification, and (b) an initial sequential system.

```
(a) (define select
        (lambda* ((s p0) v0 v1 v2)
          (case s
            [READY (if p0 v0 v1)]
            [WAIT (if p0 v0 v2)])))

(b) (define SP.1
      (lambda (IN)
        (system-letrec
          ((STATUS (XPS STATE I))
           (I       (SELECT
                      STATUS
                      (= (INP IN) 1)
                      (= (INP IN) 1)
                      (= (INP IN) 1)))
           (STATE   (! READY (SELECT STATUS WAIT READY READY)))
           (O       (! 0 (SELECT STATUS 0 0 1)))
           (X       (SELECT STATUS (OUT 0) (OUT 0) (OUT 0))))
          (list STATUS I STATE O X))))
```

Essentially, the streams in SP.1 are execution traces of the formal parameters in specification SP. The STATE and STATUS streams, together, implement a control automaton. The select combinator contains one other extension of standard Scheme syntax: nested lambda parameters such as ((s p0) v0 v1 v2) make it easier to manipulate multiple-output functions.

The next step is to isolate the single pulser circuit from the artifacts of the modeling interface, namely, the signal X, the subexpression (= (INP IN) 1), and the "register" represented by the '!' in the defining expression for O. In order to isolate the O register, we had to expand '0' by its defining expression in the equation for X, and then identify the common subexpression (SELECT STATUS 0 0 1). Once this was done, DDD was instructed to partition SP into two subsystems, one of which encapsulated the modeling interface. After a term simplification, the residual subsystem is the pure single pulser:

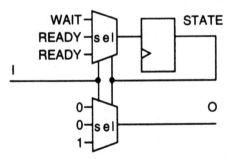

```
(c) (define SP.1.1
      (lambda (I)
```

```
(system-letrec
  ((STATUS (XPS STATE I))
   (STATE (! READY (SELECT STATUS WAIT READY READY)))
   (0 (SELECT STATUS 0 0 1)))
   (XPS STATUS STATE 0))))
```

Once binary values are assigned to WAIT and READY, this system can be projected to logic synthesis. DDD expands and simplifies the occurrence of SELECT in the defining expression for STATE to get:

```
(define select-state
  (lambda* (p0) (if p0 wait ready)))
```

Thus, if we choose WAIT = 0 and READY = 1 then STATE simplifies to I. Under this state assignment, the Oct Tools logic minimizer also finds a single-gate realization of (SELECT STATUS 0 0 1), and we end up with the desired circuit:

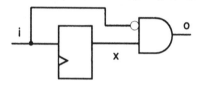

In all, six DDD commands were involved in the derivation, plus a rather cumbersome script of logic-synthesis commands.

6 Synthesis of a single-pulser using Oct Tools

A standard CAD system—we use Oct Tools [13] as a readily available example— can automatically synthesize efficient implementations of the single-pulser. We shall not pursue whether a set of CAD tools should be classified as a formal system, but it is software that supports a specific kind of reasoning. This point is discussed further at the end of this section.

Within the Oct Tools environment are two hardware description languages, a behavioral HDL called *Bdsyn* and a structural HDL called *Bdnet*. Figure 4 shows the control-specification fragment of the single-pulser in *Bdsyn*. An affiliated *Bdnet* structural description, not shown, specifies the kind of storage device used to hold pres_state, among other things.

From this description, an intermediate boolean system description (blif file) was generated (using the bdsyn/bdnet translation tools). A minimizer and technology mapper were applied to the intermediate file (using misII). Physical placement (octflatten) and routing (wolfe) tools were then used to create the VLSI layout shown in Fig. 3. When targeted to standard-cell technology, this process resulted in an implementation containing just one flipflop and one gate, as we would hope. There was no manual intervention beyond writing the specification and choosing the appropriate synthesis libraries and tactics. A *Bdsyn/Bdnet* description of the stateless version of the single-pulser (discussed in Sec. 2) resulted in an apparently identical VLSI layout, as did Oct Tools synthesis of the implementation derived in DDD.

The *Bdsyn* specification in Figure 4 includes an explicit assignment of boolean values to the state labels READY and WAIT. This assignment did affect the standard-cell layout. The opposite assignment READY = 1, WAIT = 0 introduced an inverter, which did not change the net number of transistors in the realization but did enlarge the layout area by 12%. The inverter might have been eliminated by a re-timing tool, but this avenue was not explored.

We believe it would be hard to differentiate between the kind of user-tool negotiation employed in the previous paragraph, and that seen when using (say) a general purpose theorem prover. For this reason, we are inclined to include a CAD environment among the other reasoning systems we are exploring, even if it does broaden the sense of the term "reasoning."

7 Conclusion

The four studies presented in this paper address distinct aspects of the verification task, as illustrated in Fig. 5. They involved four modes of description, three of which are formal. We likened the single-pulser timing diagram to a true specification because it details the observable properties of interest. The state-machine representation should be called a design description; among other things, it determines the particular wave form used to satisfy the specification. The circuit diagram, then, is an implementation description, detailing how this wave from is generated. Finally, there is a realization of the implementation in the form of a VLSI layout.

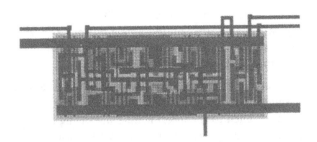

Fig. 3. Oct Tools synthesized layout of the single pulser as displayed by vem

```
MODEL SP

! OUTPUTS next_state<0>, out<0>
=
! INPUTS pres_state<0>, in<0>;

CONSTANT
          READY = 0, WAIT = 1,
          FALSE = 0, TRUE = 1;

ROUTINE single_pulser_control;
  next_state = READY;
  out = FALSE;
  SELECT pres_state FROM
    [READY]: BEGIN
      IF in EQL TRUE THEN
      BEGIN
         next_state = WAIT;
      END ELSE BEGIN
         next_state = READY;
      END;
    END;
    [WAIT]: BEGIN
      IF in EQL TRUE THEN
      BEGIN
         next_state = WAIT;
      END ELSE BEGIN
         next_state = READY;
         out = TRUE;
      END;
    END;
  ENDSELECT;
ENDROUTINE;
ENDMODEL;
```

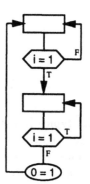

Fig. 4. SP in *Bdsyn*

Although diagrams are used informally in this paper, efforts are underway to make their meaning precise, with the goal of extending formal-reasoning systems to be more visually oriented [8, 5].

In PVS, the most general of the formal systems used, a proof directly establishes that the implementation satisfies the requirements of the specification. The intermediate representation of the state machine was not needed, although it, too, could have been formalized in PVS. A more complicated proof exercise

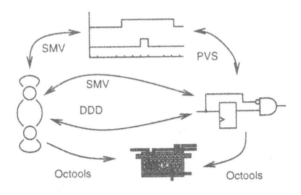

Fig. 5. Relating the studies

in PVS would typically use the design description. We had difficulties getting the specification right. It was never difficult to prove that a correct circuit met the specification, but the proof process exposed pathologies that would make unacceptable circuits correct. In particular, our notion of *neighborhood* had to be strengthened before we could reject unacceptable implementations, and we had to explicitly exclude the pathological case of an infinite pulse.

In the SMV study, temporal logic formulas were found to represent the timing-diagram specification, and SMV definitions and assignments were used to represent the state machine and circuit descriptions. We made many unsuccessful attempts at specifying a single pulser's behavior. As was the case in the PVS study, a bad specification usually resulted in a valid but inadequate theorem. In other words, SMV would affirm that the implementation was correct, but cross examination showed that the specification was too weak or simply wrong. Typically, our failures abused implication in some way, and the model checker would find a path that falsified the premise. The limited expressiveness of CTL made it more challenging to produce an acceptable form of the specification, but the proofs were always automatic and immediate.

DDD applies to the problem of deriving an implementation from a design. In the DDD environment, the boolean optimization is done by logic synthesis tools. Since the DDD algebra is specialized for manipulating sequential system descriptions, we did not encounter the kinds of logical pathologies found in PVS and SVM. The derivation study actually focused on isolating the design description from its interface to a modeling environment. This was a good exercise of the algebra, and also showed how other modeling tasks can interfere with formal reasoning processes. Nevertheless, We had to use some very round-about tactics to get the DDD algebra to do exactly what we wanted.

Synthesis of the single pulser was straightforward for an experienced Oct Tools user. Experimenting with minor variations, such as changing the state assignment, led to a pattern of reasoning that we found quite similar to the interaction with the other systems.

In each of the studies, application of the tool is complicated by some aspect of problem representation, but it is a different aspect in each case. In PVS a refinement of environmental constraints—no infinite pulses—was needed to carry the proofs through. In the SMV study, there were similar problems with pathalogical cases, and while the proofs were automatic, we found it hard to represent the specification in the less expressive CTL. In DDD, the modeling context interferred, as discussed earlier. Finally, the complexity in using a design synthesis environment lies in understanding what tools to apply and when to be satisfied with the outcome.

We plan to continue collecting and comparing studies of the single pulser, and will maintain the results in [1].

8 Acknowledgments

We are especially grateful to Shyamsundar Pullela for his contributions to earlier versions of this paper, and to both Shyam and Kathi Fisler for their help in the SMV study. Kamlesh Rath has participated in many of the discussions leading to this paper and provided expertise in the use of Oct Tools. Esen Tuna participated in the development of the DDD derivation of the single-pulser. The original PVS study was done during discussions of visual inference with Jon Barwise, Gerry Allwein, and Kathi Fisler. Zheng Zhu has made several valuable comments about these studies and earlier drafts of this paper.

This research was supported, in part, by the National Science Foundation under grant number MIP92-08745. We are also grateful to the National Aeronautics and Space Administration for granting leave to Paul Miner to pursue graduate study at Indiana University.

References

1. Further details about these studies can be obtained through the World Wide Web via URL www.cs.indiana.edu. Access the *Single Pulser Study* through the *Hardware Methods Group* thread in the list of departmental research projects. Individuals wishing to contribute to this collection should contact sjohnson@cs.indiana.edu or write Hardware Methods Laboratory, Indiana University Computer Science Department, Bloomington Indiana, USA.

2. Bhaskar Bose. DDD - A Transformation system for Digital Design Derivation. Technical Report 331, Computer Science Dept. Indiana University, May 1991.

3. J.R. Burch, E.M. Clarke, D.L. Dill, and K. L. McMillan. Sequential circuit verification using symbolic model checking. In *Proceedings of the 27th ACM/IEEE Design Automation Conference*, June 1990.

4. E.M. Clarke, E.A. Emerson, and A.P. Sistla. Automatic verification of finite-state concurrent systems using temporal logic specification. *ACM Transactions on Programming Languages and Systems*, 8(2), April 1986.

5. Kathi Fisler. Extending formal reasoning with support for hardware diagrams, 1994. This volume.

6. M.J.C. Gordon. Why higher order logic is a good formalism for specifying and verifying hardware. In G.J. Milne and P.A. Subrahmanyam, editors, *Formal Aspects of VLSI Design: Proceedings of the 1985 Edinburgh Conference on VLSI*, pages 153–177. North Holland, 1986.

7. Aarti Gupta. Formal hardware verification methods: A survey. *Formal Methods in System Design*, 1:151–238, 1992.

8. Steven D. Johnson, Gerard Allwein, and Jon Barwise. Toward the rigorous use of diagrams in reasoning about hardware. In Gerard Allwein and Jon Barwise, editors, *Working Papers on Diagrams and Logic*. Indiana University Logic Group Preprint IULG-93-24, May 1993.

9. Steven D. Johnson and Bhaskar Bose. A system for mechanized digital design derivation. In P.A. Subramanyahm (ed.), Participants' procedings of the *ACM/SIGDA International Workshop on Formal Methods in VLSI Design*, Miami, Florida, USA, January 1991., December 1990. With appendix as Indiana University Computer Science Department Technical Report No. 323.

10. K. L. McMillan. *Symbolic Model-Checking: An Approach to the State Explosion Problem*. PhD thesis, Carnegie Mellon University, 1992.

11. S. Owre, J.M. Rushby, and N. Shankar. PVS: A prototype verification system. In Deepak Kapur, editor, *11th International Conference on Automated Deduction (CADE)*, volume 607 of *Lecture Notes in Artificial Intelligence*, pages 748–752, Saratoga, NY, June 1992. Springer Verlag.

12. Franklin Prosser and David Winkel. *The Art of Digital Design*. Prentice-Hall, Englewood Cliffs NJ, second edition, 1987.

13. Rick L. Spickelmier. *Release Notes for Oct Tools Distribution 5.1*. Electronics Research Laboratory, University of California, Berkeley, August 1991.

Mechanized Verification of Speed-independence

Michael Kishinevsky and Jørgen Staunstrup

Department of Computer Science, Tech. Univ. of Denmark,
DK–2800 Lyngby, Denmark.
e-mail: {mik,jst}@id.dth.dk

Abstract. Speed-independence is a property of a circuit ensuring correct operating regardless of the magnitude of delays in all its gates. In this paper, circuits are modeled by formal transition systems, and speed-independence is characterized by state predicates expressing constraints on the transition system. This makes it possible to define a formal condition corresponding to speed-independence, and to mechanically verify that a given transition system satisfies the condition. The condition is formulated in such a way that the transition system, and hence also the circuit design, can be checked in a modular way, i.e., by checking the circuit design module by module. This means that large designs can be checked in smaller pieces and without providing an explicit circuit realization of the environment.

A number of designs have been verified using the approach described here, including a speed-independent RAM cell, a complex switch of a data-path, and a number of standard components such as counters, FIFO registers, and various Muller C-elements.

1   Introduction

The correct operation of a speed-independent circuit does not depend on the delays of its components (gates). Such circuits are very robust to data and parameter variations. This may have significant practical advantages [11, 14], for example, a potential reduction of power dissipation [18]. However, to realize a design by a speed-independent circuit, the design must meet some constraints excluding behavior that depends on timing details of the components. Hence, a designer must not violate these constraints. There are several ways to achieve this, one would be to follow a "correct by construction" approach [11]; in this paper another alternative is explored, using mechanical tools to check that a high-level description of the designs behavior meets certain conditions (ensuring speed-independence). The following standard example is used throughout the paper to introduce and motivate the approach.

Example: Modulo-N Counter. The modulo-N counter with constant response time is a simple, yet interesting, example of a speed-independent design [5]. To save space, it is assumed that N is a power of two, and therefore the counter is called a modulo-2^n counter. The counter has one input, a, and two outputs p and q. Every signal change on the input a is acknowledged by a signal change of either p or q. The first $2^n - 1$ up-going changes on a are acknowledged by up-going changes on p and the last, 2^n-th, by an up-going change on q. The same with down-going changes. The counter cell is composed of a toggle element, a pipeline latch, and an OR-gate. The design used in this paper uses a four-phase protocol and was done by Christian D. Nielsen [15], it has many similarities with the two-phase design described in [5].

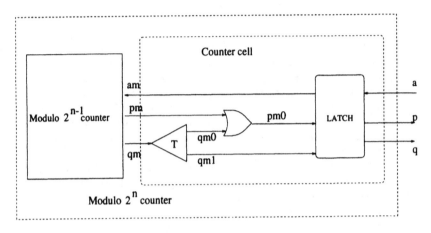

Fig. 1. A Diagram of the Modulo-2^n Counter

End of example

This paper introduces a technique for checking that a high-level description of a design, such as the modulo-N counter, allows for a speed-independent circuit realization. By using a high-level description, it becomes possible to check the design early in the design process. The behavior of a circuit defined by a high-level description is modeled as a transition system, and the conditions ensuring speed-independence are expressed as predicates on the state space of the transition system. The conditions are formulated in such a way that the transition system, and hence also the circuit design, can be checked in a modular way, i.e., by checking the circuit design module by module. The modulo-2^n counter consists of a toggle module (called T in Fig. 1), a few gates, and a modulo-2^{n-1} counter module (which itself may have further submodules). The basis of the recursive specification is a modulo-1 counter which simply connects the input a to the output q by a wire, and assigns the constant value false to the other output p.

The hierarchical nature of a design is exploited in the speed-independence check by treating modules as black boxes where the internal details are hidden. As a consequence, it is also possible to check a particular module, such as the toggle, without providing an explicit circuit realization of the environment, i.e., to check non-autonomous designs. This means that large designs can be checked in smaller pieces, and that designs can be checked without providing an explicit circuit model of the environment.

There are already several verification tools for checking speed-independence based on model-checking. For example, in [1, 8] different methods have been presented based on a direct construction of the circuits state space or its essential subspaces. In [3, 4] speed-independence is verified as an absence of choking (or computational interference) in a trace-based specification of a circuit. These approaches use special purpose tools aimed at verifying a limited set of properties. The work described in this paper differs in several respects. Most importantly with respect to the level of the circuit specification. It is possible to check designs with composite data-types, e.g., n-bit words, hierarchy (which is maintained in the verification), and non-autonomous parameterized designs. Another interesting difference is that the tools used to check for speed-independence are general purpose and not constructed specifically for speed-independence. Exactly the same tools (the Larch Prover [7] and a translator [12]) are used for verifying other safety properties of design descriptions. The price for the generalization, is an increased computational complexity of the verification algorithm. However, the hierarchical verification technique compensates for this. A number of designs have been verified, including a speed-independent RAM cell, a complex switch of a data-path, and a number of standard components such as various Muller C-elements, FIFO registers, and counters.

This paper is organized as follows. Section 2 describes the design language SYNCHRONIZED TRANSITIONS used for modeling circuits. In Sect. 3 it is shown how to define and verify constraints, called invariants and protocols, on a design and its environment. Section 4 presents the definition of speed-independence, and informally describes a condition called persistency that guarantees speed-independent behavior of a design. Section 5 describes how to check the persistency condition with a combination of a theorem prover and a simple model-checking tool. Section 6 demonstrates the application of the method to the mechanical checking of a recursively described design. The appendix contains a number of definitions of concepts that are introduced informally in the main text of the paper.

2 Modeling Circuits

Speed-independence is a property of a physical circuit ensuring that the circuit operates correctly regardless of the magnitude of delays in all gates of the circuit. To make formal analysis of speed-independence possible, a model of the physical circuit is required. In this paper formal transition systems described in the design language SYNCHRONIZED TRANSITIONS are used to model physical circuits. As an example, consider a circuit component for a Muller C-element, this is described as follows:

$$\ll a = b \; -> \; y := a \gg.$$

In this example, $a, b,$ and y are boolean state variables, and whenever $a = b$, it is possible to assign the value of a to y. If $a \neq b$, then y keeps its current value. This construct is called a *transition*, and it models a single independent component of a circuit. A circuit with many components (operating in parallel) is described by composing a number of such transitions (one for each component).

$$\ll a \neq b \;\text{->}\; y := a \gg \;||\; \ll a := NOT\; y \gg \;||\; \ll b := y \gg$$

This describes a simple oscillator, if initialized in any state then the oscillator describes a computation where the three state variables a, b, and y alternate between *TRUE* and *FALSE* indefinitely. The boolean expression appearing before -> in a transition is called the precondition, when this is the constant *TRUE*, it can be omitted as exemplified by the last two transitions. State variables are introduced by a variable declaration.

STATE y : BOOLEAN

The name of the state variable (y) denotes the value of the variable. The type of the state variable (given after the ":") specifies its domain, i.e., the set of possible values. The state variable y can, for example, take the boolean values *TRUE* and *FALSE*. The value of a state variable is changed by executing a transition where the name of the state variable (e.g., y) appears on the left-hand side of an assignment ($:=$).

SYNCHRONIZED TRANSITIONS has a number of additional constructs that are not explained here, see [16] for a comprehensive introduction. The appendix defines the concepts used in this paper.

2.1 Operational Model

A design specifies a set of transitions (fixed throughout the computation) each of which may execute whenever it is enabled. Although a design description in SYNCHRONIZED TRANSITIONS has some similarity with a program in a high-level programming language, the interpretation is very different. An assignment statement in a high-level program is only executed when the control of the program is at the point of the statement. There is no similar global control flow determining the computations of a design in SYNCHRONIZED TRANSITIONS, which just specifies a fixed set of transitions. Unlike a Pascal or C program, the order in which transition descriptions are written does not influence the computation. SYNCHRONIZED TRANSITIONS is similar to UNITY [2] which describes a computation as a collection of conditional data-flow actions without any explicit control-flow. Operationally, the computation can be modeled as repeated nondeterministic selection and execution of an enabled transitions. In this model, transitions are executed:

- *one at a time*, i.e., only one active transition is executed in any state [1],
- *repeatedly*, each time it has been executed, it is immediately ready to be selected again,
- *independently*, of the order it appears in the design description.

---

[1] This corresponds to an interleaving semantics of parallel processes.

It is not required that a transition is executed immediately after it becomes enabled, because other enabled transitions may be selected. In fact, there is no upper bound on when a transition is selected. This corresponds to the unbounded gate delay in logic circuits [10]. For example, the transition $\ll y := a\ OR\ b \gg$ describes an OR-gate. The implicit precondition, $TRUE$, specifies that it is always allowed to set the output, y, to the logical OR of the inputs, a and b; however, an arbitrary delay may elapse between a change of the inputs and the changing of the output.

SYNCHRONIZED TRANSITIONS can be used to model circuits at different level of abstraction. The examples given above show modeling at the gate level. However, the same language is also used at higher levels, for example, a multiplier can be described as follows:

$$\ll z := s * t \gg.$$

Here s, t, and z are state variables of type integer and the transition describes a state change where the product of s and t is assigned to z.

By modeling a circuit as a design in SYNCHRONIZED TRANSITIONS, it is possible to formally verify properties of the design, for example functional properties or refinement. In this paper, the emphasis is on verifying speed-independence, and the next sections describe how to capture this property as a condition on the transition system.

3   Invariants and Protocols

To verify a certain property formally, it is necessary to formulate it rigorously in such a way that it can be determined unambiguously whether a given design has the property or not. In this paper, the focus is on formally expressing the property that a design is speed-independent, however, this is only a special case, and one can envision many other properties that a designer may wish to verify. In general, the verification tools for SYNCHRONIZED TRANSITIONS support two ways of rigorously specifying properties that are to be verified.

Invariants: are predicates over the state variables. They define a restriction on the allowable *subset of the state space* (the states for which the predicate holds).

Protocols: are predicates on pairs of states, *pre, post*, defining a restriction on the allowable *transitions between states* (to ones where the pre- and post-state satisfy the predicate).

The language has constructs for writing invariants and protocols. As an illustration, consider an invariant stating that two state variables x and y are never $TRUE$ simultaneously (mutual exclusion for x and y).

$INVARIANT\ NOT\ (x\ AND\ y)$

The following is an example of a protocol which states that whenever x changes, it gets the value of either y or z.

PROTOCOL x.pre $\neq$ x.post $\Rightarrow$ x.post=y.pre OR x.post=z.pre

x.pre denotes the value of *x* immediately before the transition (the pre-state), and similarly *x.post* is the value of *x* immediately afterwards (the post-state). The same notation is used to write the value of an expression *E* in the pre-state as *E.pre* and in the post-state as *E.post*.

Invariants and protocols are typically used for verifying safety properties of a design, for example, mutual exclusion, or that variables in the interface follow a convention such as four-phase signaling (hence the name protocol). An in depth treatment is given in [16]. Establishing a safety property involves the following steps:

- finding a way to express a property as an invariant or/and a protocol;
- verifying that the invariant holds in any initial state of the design;
- checking that the invariant holds in all states reachable from the initial states;
- verifying that the protocol holds for any possible transition between reachable states.

Notice that invariants and protocols are stated by the designer, and that they express important safety properties that should hold for any computation of the design. It is the aim of the verification to check whether they really hold, i.e., whether they hold for any reachable state of the design and for any possible transition between reachable states. In section 4 it is is shown how speed-independence can be expressed as a protocol which may then be verified as any other protocol.

3.1 Environment and Non-determinism

In general, the computation of a design depends on the behavior of the environment. Therefore, to verify the design one needs a way of specifying the environment. In [3, 4] the behavior of the environment is expressed by the set of possible signal traces, i.e., using the same model used for specifying the internal behavior. In [1, 8] the verification is done on an autonomous circuit that is constructed by composing the original circuit with an environment making the composition autonomous. This makes it difficult to express a non-deterministic behavior of the environment. Here we describe behavior of the environment implicitly by defining protocols and invariants constraining the state space and possible transitions of *external* state variables.

By making invariants and protocols express information about permissible environments, it becomes feasible to verify a component in isolation, i.e., without checking the global behavior of the entire design. When combining several components, it is of course necessary to check that their invariants and protocols are consistent, i.e., that they have a consistent view of their interface. In [16, 17] it is shown how such a verification is done in a localized manner using the same techniques and tools that are used in this paper for checking speed-independence.

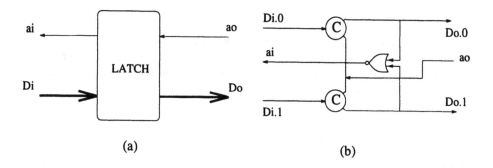

Fig. 2. The interface to the latch (a), its gate-level realization (b).

Example: A Pipeline Latch. A simple speed-independent pipeline latch is used to illustrate how to specify a non-deterministic environment. The interface to the latch consists of four state variables: two boolean acknowledgments, ai, ao, and two duals, Di, Do, modeling a one bit data-path, see Fig. 2.a. The domain for the data-path variables contains three possible values of a data bit: $\{E, T, F\}$ ("empty", "true", and "false"). The variables ao (the output acknowledgment) and Di (input data) are changed by the environment. Fig. 2.b shows a possible gate-level realization of a latch based on two C-elements and one NOR-gate [14]. The latch is described in SYNCHRONIZED TRANSITIONS as follows:

> $CELL\ latch(ai, ao:\ BOOLEAN;\ Di, Do:\ dual)$
> $BEGIN$
> $\ll\ ai:=\ empty(Do)\ \gg\ \|\ \ll\ ao \neq empty(Di)\ ->\ Do:=\ Di\ \gg$
> $END\ latch$

The boolean function *empty* returns the value $TRUE$ when the value of its dual parameter is equal to E.

The environment of the latch supplies the dual input, Di, and one can imagine different behaviors of the environment, it may for example, follow the return-to-empty-convention where Di has the value E between any two valid values. This external behavior is described by the following protocol (the predicate $same(E)$ is a shorthand for $E.pre = E.post$):

> $PE_1 \equiv (NOT\ same(Di)) \Rightarrow$
> $\quad ((ai.post \neq empty(Di.post))\ AND\ (ai.post = empty(Di.pre)))$

This expresses that whenever Di changes ($NOT\ same(Di)$) then its empty status must change to the opposite value of the acknowledge ai. Even though this is a very restrictive protocol, it allows the environment to non-deterministically choose between the two valid values T and F.

Below a less restrictive protocol is specified where Di is allowed to change directly from one of the valid values to the other.

> $PE_2 \equiv (NOT\ same(Di)) \Rightarrow (ai.post \neq empty(Di.post))$

In both cases, ao, must behave as follows:

$(NOT\ same(ao)) \Rightarrow (ao.post = empty(Do.post))$

To describe a pipeline latch for a wider data-path, e.g., a 16 bit word, the only change in the design description is replacing the type dual with another type. However, the gate level realization of a wider data-path would require a significant change (to compute *empty*).

4 Characterizing Speed-independence

This section gives an informal description of the persistence condition which is a protocol that must be met by a design if is going to be realized as a speed-independent circuit. In David Muller's original work, speed-independence was defined through the notion of "final classes" of behavior [13]. This paper follows the more recent trend defining a circuit to be speed-independent if its correct operation is independent of gate delays. A practically useful check for speed-independence cannot be based directly on this definition, because that would require checking a possibly infinite number of different combinations of gate delays. Instead, we have found a condition, called the *persistency condition*, which is both mechanically checkable and sufficient to ensure speed-independence.

A transition t is persistent, if once it becomes active, it remains active, providing the same post-value for the write variable, while other transitions occur. A design meets the persistency condition, if it can be shown that all transitions are persistent both with respect to changes made by other transitions and with respect to transitions made in the environment. In the appendix it is shown how to formulate the persistency condition as a protocol which makes it possible to check that a design meets the protocol mechanically.

As an illustration, consider again the simple oscillator presented in Sect. 2.

$$\ll a \neq b\ ->\ y := a \gg\ ||\ \ll a := NOT\ y \gg\ ||\ \ll b := y \gg$$

The first transition is persistent because once it is active, it must be the case that $y \neq a$ and $y = b$. The second and third transitions are not active when this is the case, hence neither a nor b can change value. Similarly, it can be argued that the other two transitions are persistent.

To illustrate a non-persistent design consider the following modified oscillator:

$$\ll y := a\ OR\ NOT\ b \gg\ ||\ \ll a := NOT\ y \gg\ ||\ \ll b := y \gg$$

Consider a state where $a, b, y = FALSE, FALSE, FALSE$. The second transition is active, but so is the first. If the first transition changes y to $TRUE$, then the second is no longer active, and hence it is not persistent.

Example: The Pipeline Latch (continued). To illustrate the use of the persistency condition on a non-autonomous design, consider again the pipeline latch from Fig. 2. It turns out that the speed-independence of the latch depends on how strong assumptions can be made about the environment. To show that the latch design satisfies the implementation condition persistency, it is necessary to verify the persistency protocol for each of the two transitions; for brevity only the protocol for last one is shown below:

$$(ao.pre \neq empty(Di.pre)) \wedge (Do.pre \neq Di.pre) \wedge same(Do)$$
$$\Rightarrow (ao.post \neq empty(Di.post)) \wedge (Do.post \neq Di.post) \wedge same(Di)$$

The left side of the implication indicates that the transition is active in the pre-state. The expression $same(Do)$ makes the condition hold trivially for the transition itself (as stated in the appendix, this is just a way of encoding the condition $t_1 \neq t_2$). The right side of the implication requires that after the state change from pre to $post$ the transition is still active (in the state $post$), and provides the same value for the variable Do (clause $same(Di)$).

It must be shown that for any pair of states (pre, $post$) satisfying the protocol, PE, of the pipeline latch, the persistency protocol holds. In Sect. 3.1 two different protocols are considered. For the weakest of the two protocols:

$$PE_2 \equiv (NOT\ same(Di)) \Rightarrow (ai.post \neq empty(Di.post))$$

it is *not* possible to show the persistency protocol, and hence, it cannot be expected that the pipeline latch is speed-independent when placed in an environment where only PE_2 can be assumed. Consider instead, the stronger assumption about the environment defined by PE_1:

$$PE_1 \equiv (NOT\ same(Di)) \Rightarrow$$
$$((ai.post \neq empty(Di.post))\ AND\ (ai.post = empty(Di.pre)))$$

It turns out that with this assumption, it is possible to show that the persistency protocol is met (see Sect. 5.1), and also that the other parts of the definition are met.

5 Mechanizing the Check

This section describes how to mechanize the check of the persistency condition using a combination of a theorem prover and a simple model-checker. One of the possibilities offered by these tools is the ability to check the pipeline latch separately without giving an explicit circuit realization of the environment.

The tools make it possible to take a design description like the one shown in Sect. 3.1 and automatically generate verification conditions corresponding to the persistency condition. These may then be given to the theorem prover for verification. In some cases, like the two oscillators, no additional information is needed, however, in more interesting cases, it is necessary to provide an invariant excluding some or all of the states that the design will never enter. For the latch, it is for example the case that:

$$(NOT\ empty(Do))\ AND\ (NOT\ empty(Di)) \Rightarrow Di=Do$$

This information may be added as an explicit invariant, and for the simple latch example it is possible to manually generate the few extra invariants needed (part of it is shown above), but in larger and more complicated examples, it can be a significant help to use a model-checking tool to characterize the reachable part of the state space; this is explained further in the next section.

5.1 Reachability Invariant

The *reachable state space* of a design is usually a small subset of the state space defined by the cartesian product of the domains of all state variables. A characterization of this reachable state space is useful for almost any kind of formal verification, and also for verifying speed-independence which is the topic of this paper. One way to characterize the reachable state space is as an invariant, called the *reachability invariant*. Below, it is discussed how to generate this invariant.

Let $\mathcal{D}$ be a design in SYNCHRONIZED TRANSITIONS with the set of transitions $t_1, ..., t_n$, the external invariant I_E, and the external protocol P_E. Each transition, t_i, defines a state transition predicate $t_i(pre, post)$. In other words, each transition t_i defines a state transition relation R^{t_i}. The union of these relations define a state transition relation for the *internal* state variables of the design. This relation, R, is defined as follows:

$$(S_1, S_2) \in R \Leftrightarrow I_E(S_1) \wedge I_E(S_2) \wedge (P_E(S_1, S_2) \vee \exists t_i \, t_i(S_1, S_2))$$

Two states S_1 and S_2 are in relation R if they differ either by the value of an external variable or by the value of the write variable of transition t_i. This variable can change its value from state S_1 to state S_2 either as defined by $P_E(S_1, S_2)$ or by $t_i(S_1, S_2)$.

The transitive closure of R, denoted by R^*, is called the *reachability relation* of the design. If the initial predicate, U_0, is given, then the set of states reachable from the initial states can be characterized by a predicate, called the *reachability invariant*, I_{U_0}, that is defined as follows:

$$I_{U_0}(S) \Leftrightarrow \exists u_0 \in U_0 \, (u_0, S) \in R^*$$

For simple designs, derivation of this invariant can be done manually, but for more challenging designs this is too laborious. However, the reachability invariant can be derived automatically using a model-checking tool. For our experiments we have used the state generation kernel of the TRANAL system, included in the FORCAGE system [8]. The algorithm is based on a breadth-first search of the state transition graph, this is illustrated in Fig. 3. The states of a design are represented as boolean vectors and transitions as boolean vector representations of boolean functions. The algorithm proceeds in stages. At the kth stage a kth layer of states is derived. All states in this layer are reachable from the initial set of states through k transitions. For the next iteration, the states reachable from kth layer are found, thus giving the $(k + 1)$-th layer of states that are reachable in $k + 1$ transitions. The algorithm ends when no more new states can be found. To simplify the check for the fixed point of the generation process, a heuristic is described in [8] allowing one to compare two layers of the reachability set instead of performing comparison of the whole set of states in the current layer.

State explosion is a potential danger, of model-checking, however when used as here to generate invariants of a modular design, we hope to avoid the problem. The individual cells of a modular design contain relatively few state variables and model-checking is only used on such limited modules, whereas the composition of modules is verified using localized verification. However, our experience with the approach is still limited.

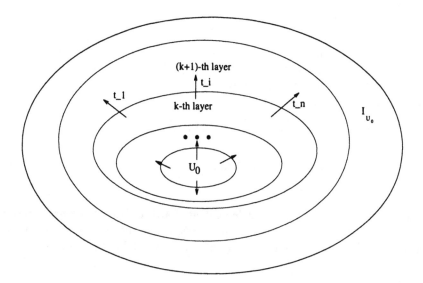

Fig. 3. Breadth-first search for reachability invariant

Example: The Pipeline Latch (continued). The following expression characterizes the invariant derived automatically for the pipeline latch:

INVARIANT
   (Di = Do) OR (ai AND empty(Do)) OR ((NOT ai) AND empty(Di))

This invariant, called I below, can be used to verify that the pipeline latch meets the persistency condition. If the invariant is inserted in the design description for the pipeline latch and if this is given to the automatic translator, the verification conditions shown below are generated (they have been simplified a little to make this presentation more clear). The restrictions on the interface constrains the environment, but it must also be met by the latch itself, in Sect. 4, this restriction was defined by the following protocol (PE).

PROTOCOL
   ((NOT same(ao))⇒ (ao.POST = empty(Do.POST))) ∧
   ((NOT same(Di)) ⇒   (ai.POST ≠ empty(Di.POST)) ∧
                 (ai.POST = empty(Di.PRE)))

For each (of the two) transitions of the design, it must be shown that it satisfies the persistency protocol of the other transitions (in this case there is only one), and the protocol PE describing the environment.

   $I(pre) \wedge t_i(pre, post) \Rightarrow Persistent^{t_j}(pre, post)$
   $I(pre) \wedge PE(pre, post) \Rightarrow Persistent^{t_j}(pre, post)$

Where $Persistent^{t_j}(pre, post)$ is a protocol describing the persistency condition for transition t_j, $i, j \in 1, 2$.   **End of example**

The verification technique described above has been used to check a number of designs, for example, a complicated switch of a data-path, various realizations of

```
(* latch *)
 ≪ am:= NOT (p OR q) ≫ ||
 ≪ a = pm0 -> p:= a ≫ ||
 ≪ a = qm1 -> q:= a ≫ ||
(* OR-gate *) ≪ pm0:= pm OR qm0 ≫ ||
toggle(qm, qm0, qm1) ||
{ N = 1 | count_one(am, pm, qm) } ||
{ N > 1 | count_even(am, pm, qm, N DIV 2) }
```

Fig. 4. The modulo-2^n counter, $N = 2^n$

a Muller-C element, and a speed-independent RAM design. In the next section, it is shown how a parameterized recursive design description can be checked for speed-independence without unfolding the recursion or binding the parameters to particular values.

6 Verification of Modular Designs

In this section it is shown how localized verification [17] is used to check a modular, parameterized, and recursively defined design. The modulo-N counter is an example of such a design; the value of N is left unspecified in the description which is a recursive composition of a modulo-$N/2$ counter, a toggle, and a few simple gates. A 2^n counter can be viewed as a composition of a 2^{n-1} counter and a counter cell, as shown in Fig. 1. The counter cell is composed of a toggle element, a pipeline latch (from Fig. 2.b), and an OR-gate. A SYNCHRONIZED TRANSITIONS description of a modulo-2^n counter is given in Fig. 4. The complete counter that works for arbitrary numbers has been designed and verified, it is omitted in this paper, because it does not show anything substantially new compared to the modulo-2^n counter.

The design shown in Fig. 4 is the same as the one shown in Fig. 1. The first three transitions correspond to the latch (shown in Fig. 2) and the fourth transition is the OR-gate. The last three lines describe the sub-cells. The first is a toggle element. Then follows two conditional instantiations, if $N > 1$ a new instance of the count-even cell is made (the modulo-$N/2$ counter) and if $N = 1$ the recursion stops by instantiating a simple modulo-1 counter.

Using the approach described in this paper, the verification of the speed-independence of the modulo-2^n consists of the following steps:

1. describe the cell interfaces by protocols (and invariants),
2. generate the persistency protocol for each cell,
3. generate the reachability (or another sufficiently strong) invariant of each cell,
4. generate the verification conditions,
5. verify the verification conditions.

The last four steps can be done mechanically whereas the first step requires some insight in the design in order to capture all essential aspects of the environment of a cell. Each step is now illustrated using the modulo-2^n counter as an example.

Protocols for the Cell Interfaces. When the interface variables of the counter cell change, it is always known what the new value is, e.g., whenever p (or q) change it becomes equal to a. This is described by the following protocol:

> PROTOCOL
> $NOT\ same(p) \Rightarrow p.post = a.pre$ AND
> $NOT\ same(q) \Rightarrow q.post = a.pre$ AND
> $NOT\ same(a) \Rightarrow a.post \neq (p.pre\ OR\ q.pre)$

Similar protocols are needed to describe the interfaces to the toggle cell and the modulo-1 counter. It requires some experience and insight in the design to come up with these protocols. If they are too weak (does not contain enough information) it will not be possible to verify the desired properties and the entire verification cycle must be repeated.

The Persistency Protocol. It can be verified that the counter is speed-independent by showing that it meets definition 4. This is done by including persistency protocols in the design description, for example, in the cell *count_even*, the persistency protocol for the four transitions is:

> $(\ (am.pre \neq NOT(p.pre\ OR\ q.pre)\ AND\ SAME(am)\ ) \Rightarrow$
> $\quad (am.post \neq NOT(p.post\ OR\ q.post))\ )$ AND
> $(\ ((a.pre = pm0.pre)\ AND\ (p.pre \neq a.pre)\ AND\ SAME(p)\ ) \Rightarrow$
> $\quad ((a.post = pm0.post)\ AND\ (p.post \neq a.post)))$ AND
> $(\ ((a.pre = qm1.pre)\ AND\ (q.pre \neq a.pre)\ AND\ SAME(q)\ ) \Rightarrow$
> $\quad ((a.post = qm1.post)\ AND\ (q.post \neq a.post))\ )$ AND
> $(\ (pm0.pre \neq (pm.pre\ OR\ qm0.pre)\ AND\ SAME(pm0)\ ) \Rightarrow$
> $\quad (pm0.post \neq (pm.post\ OR\ qm0.post))\ )$

Similar protocols are needed in the other cells (toggle and count-one).

The Reachability Invariant. To verify that the design meets the consistency protocols (and the other protocols formulated by the designer) it is necessary to find an invariant for each cell. The toggle and the count-one cells are simple enough that a suitable invariant can be formulated by the designer. However, for the count-even cell, it is possible to generate the following invariant automatically as described in section 5.1:

INVARIANT
 ((NOT pm) AND (NOT qm0) AND (NOT p) AND (NOT pm0) AND
  ((am AND qm AND qm1) OR
 ((NOT am) AND ((NOT qm) AND (NOT qm1) OR (q AND qm1))))))
 OR ((NOT q) AND (NOT qm1) AND (am AND (NOT pm) AND qm AND
  ((pm0 AND qm0) OR
 ((NOT p) AND (NOT pm0)))
 OR (pm0 AND qm0 AND p AND (NOT am) AND (NOT pm)) OR
 ((NOT qm) AND (NOT qm0) AND (pm AND pm0 AND (am OR p) OR
 (am AND (NOT p) AND (NOT pm0)) OR
(p AND (NOT am) AND (NOT pm))))))))

This is by construction an invariant, and hence, it need not be verified again, it is however included in the design description and utilized to verify the persistency and other protocols.

The Verification Conditions. To verify that the modulo-N counter meets the persistency protocols the design description is translated into a number of verification conditions for the theorem prover. The localized verification technique embedded in the translator utilizes the modular structure of the design; therefore, the number of verification conditions that are generated to verify speed-independence (or any other safety property that can be described as an invariant or protocol) is linear in the size of the design description. In case of the modulo-N counter, the number of verification conditions is independent of N (which determines the depth of recursion).

First, all transitions of each cell are verified locally by showing that the (local) invariants and protocols of that cell (including persistency) will hold after executing the transition, assuming that the invariant held in the pre-state. This is a standard way of verifying invariance [6].

The second step is to verify that no cell instantiation results in a design where the invariants and protocols for the instantiated cell are violated by the environment and vice versa. This is done using the localized verification technique [17] without considering the individual transitions of the cell or the environment. Instead the invariants and protocol of the instantiating cells are assumed to express sufficient constraints on the state changes of the environment to show an implication ensuring that the protocols and invariants of the instantiated cell hold. For example, to verify the instantiation of the cell *toggle* the following two implications must be shown:

$$I_{ce}(pre) \wedge I_t(pre) \wedge I_t(post) \wedge P_t(pre, post) \Rightarrow I_{ce}(post) \wedge P_{ce}(pre, post)$$

$$I_{ce}(pre) \wedge I_t(pre) \wedge I_{ce}(post) \wedge P_{ce}(pre, post) \Rightarrow I_t(post) \wedge P_t(pre, post)$$

I and P are the invariants and protocols in the toggle cell (subscript t) and the count-even cell (subscript ce). Note that the two conditions avoid considering the individual transitions of the two cells. Two similar conditions are generated for each cell instantiation in the design. Note also that the protocol, for example

P_t, includes the persistency protocols. Hence, these proof obligations cover part 2 of definition 4.

A more detailed description of the localized verification technique is given in [17] which also contains a proof of the soundness of the technique.

The Verification. The verification of the counter with the mechanical tools shows that the persistency condition is met; therefore, the counter design is speed-independent. The complete verification of the counter with the LARCH PROVER on a DEC Alpha station took approximately 22 min, and the run time is independent of the actual size of the counter (parameter N).

A drawback of theorem provers is that they often require intensive human interaction during the verification. Our translator generates a script that controls the verification and usually only a little additional manual interaction is needed. To verify speed-independence of the modulo-N counter, it was necessary to manually help the theorem prover approximately once, for each verification condition.

7 Conclusion

This paper has described a set of mechanical tools that can be used to assist a designer in verifying that a design is speed-independent. There are already model-checking tools available for this. However, as demonstrated above, the use of general purpose tools gives some new possibilities like checking high-level and incomplete designs. Furthermore, the use of general purpose theorem provers avoids the need for constructing specialized tools. Although the run-time for the method described in this paper is worse than for checking speed-independence by known model-checking tools, we find that it is still reasonable for practical applications. For example, all components of a vector multiplier design were verified within a few hours of run-time on a DEC Alpha station. Several mistakes in the design were found. The inherent universality of theorem provers allows one to verify any properties expressible in the logic theory underlying the particular theorem prover. The persistency condition is an example of a safety property, and any other safety property can be mechanically verified using the same combination of a theorem prover and model-checker.

Appendix. Notation and Terminology

A design consists of a set of state variables, $s_1, s_2, \ldots, s_n$, and a number of transitions, $t_1, t_2, \ldots, t_m$. For a given design, the set of state variables and transitions are fixed. In this paper, we focus on a subset of SYNCHRONIZED TRANSITIONS in which all variables are of finite types, i.e., each variable has a finite set of possible values from a fixed finite domain (determined by the type).

Each transitions, t, has a form $\ll C_t \;-> z_t := E_t \gg$, where C_t – is a predicate called the *precondition*, z_t is a state variable, and E_t is an expression, that has a unique value in any state. The transition t is *enabled* (in a state s) if and only if C_t is satisfied (in s). The transition t is *active* (in a state s) if and only if it is enabled and $z_t \neq E_t$ (in s). For each transition, t, the predicate $active^t$ is defined as follows:

$$active^t \;\equiv\; C_t \wedge (z_t \neq E_t)$$

When necessary, this predicate is explicitly applied in a particular state, s, and then written as $active^t(s)$.

A transition defines a set of ordered pairs of states, for each such pair, the first element is called the pre-state and the second the post-state. More formally, each transition t defines a predicate $t(pre, post)$ on the pairs of states such that: $active^t(pre)$ and state $post$ differs from the state pre only by the value of the variable z_t, i.e., $z_t.post = E_t.pre$, where $z_t.post$ denotes the value of z_t in the post-state, and similarly $E_t.pre$ is the value of an expression E in the pre-state. The predefined predicate $same(x)$ is a shorthand for $x.pre = x.post$, where x can be any state expression.

A design defines a set of computations that are sequences of states, $S_0, S_1, \ldots$, where S_0 is an initial state, and for each pair, S_i, S_{i+1}, there is a transition, t, such that $t(S_i, S_{i+1})$, i.e., S_i is a pre-state of t, and S_{i+1} is a post-state of t. Note, that there is only a single transition accounting for one step of the computation (going from S_i to S_{i+1}). Alternatively it might be external variables of the design that change value between S_i and S_{i+1} according to the external protocol.

The *initial state* (or set of states) is specified by defining the initial value of some or all of the state variables.

$$INITIALLY\; a = FALSE\; b = FALSE$$

State variables get their initial value before any transitions are executed, and they retain this value until a different value is assigned to them. It is not required that all state variables are given an initial value.

A1. Well-behaved Designs

To ensure a one to one correspondence between a design description in SYN-CHRONIZED TRANSITIONS and its circuit realization, it is necessary to exclude certain designs, for example, those that contain contradictory assignments to the same state variables.

The *write set*, W^t, of a transition, t, is the set of state variables appearing on the left-hand side of assignments. In this paper multi-assignments are not considered, therefore the write set has a single element: $W^t = \{z_t\}$. Similarly, the *read set*, R^t, of a transition is the set of state variables that appear in the precondition and on the right-hand side of the assignment.

Definition 1. The transitions $t_1, t_2, \ldots t_n$ meet the *exclusive write* condition if and only if:

$$\forall i, j \in [1..n] : W^{t_i} \cap W^{t_j} \neq \emptyset \Rightarrow \neg(C_{t_i} \wedge C_{t_j})$$

This condition ensures that there are no states where different enabled transitions can assign to the same state variable.

Definition 2. The transitions $t_1, t_2, \ldots t_n$ meet the *unique write* condition if and only if for each state variable z and for each value, v, in the domain of state variable, z, there is a unique transition that can assign the value v to the state variable z.

This condition ensures that for each value that a state variable may get, it is possible to identify a unique transition assigning that value.

Definition 3. A SYNCHRONIZED TRANSITIONS design is called *well-behaved* if it obeys the exclusive write and unique write conditions.

It can be shown that the expressiveness of well-behaved SYNCHRONIZED TRANSITIONS designs is enough to code any asynchronous logic circuit.

A2. Characterizing Speed-independence

Let t be a transition of a design. The protocol $Persistent^t(pre, post)$ is defined as follows:

$$Persistent^t(pre, post) \equiv Active^t(pre) \Rightarrow (Active^t(post) \wedge same(E_t))$$

Intuitively, $Persistent^t$ defines the constraint that transition t stays active, providing the same post-value for the write variable, while other transitions occur. If the write variable of the transition t is of type boolean, then the latter conjunct, $same(E_t)$, is redundant.

Definition 4. Let D be a design with the invariant I_U and the protocol $P_E(pre, post)$. Then D satisfies the *persistency* condition, if the following can be shown:

1. for all pairs of transitions t_1, t_2 in D, $t_1 \neq t_2$:
   $t_1(pre, post) \wedge I_U(pre) \Rightarrow Persistent^{t_2}(pre, post)$
2. for any transition t in D:
   $P_E(pre, post) \wedge I_U(pre) \Rightarrow Persistent^t(pre, post)$.

When a design meets the persistency condition, it is ensured that no active variable is disabled by the state changes of other transitions or by the state changes of the external variables. The justification of the persistency condition is given in [9], where it is argued that *a well-behaved* SYNCHRONIZED TRANSITIONS *design is speed-independent, if and only if it satisfies the persistency condition.*

The persistency protocol generalizes the notion of a *conflict state* [8], that is used for analysis of semi-modular binary circuits.

In the mechanical tools described in this paper the following formulation of the protocol $Persistent^t(pre, post)$ is used:

$$Persistent^t(pre, post) \equiv$$
$$Active^t(pre) \wedge same(W^t) \Rightarrow (Active^t(post) \wedge same(E_t))$$

An additional clause $same(W^t)$ guarantees that transition t has not occurred between states *pre* and *post*. This clause ensures that the $t_1 \neq t_2$ condition holds as required by definition 4.

References

1. P.A. Beerel and T.H.-Y. Meng. Semi-modularity and testability of speed-independent circuits. *Integration, the VLSI journal*, 13(3):301–322, September 1992.

2. K. Mani Chandy and Jajadev Misra. *Parallel Program Design: A Foundation*. Addison-Wesley, 1988.

3. D.L. Dill. *Trace Theory for Automatic Hierarchical Verification of Speed-Independent Circuits*. The MIT Press, Cambridge, Mass., 1988. An ACM Distinguished Dissertation 1988.

4. Jo C. Ebergen and S. Gingras. A verifier for network decompositions of command-based specifications. In *Proc. Hawaii International Conf. System Sciences*, pages 310–318. IEEE Computer Society Press, 1993.

5. Jo C. Ebergen and Ad M. G. Peeters. Design and analysis of delay-insensitive modulo-N counters. *Formal Methods in System Design*, 3(3), December 1993.

6. R.W. Floyd. Assigning meanings to programs. In J.T. Schwartz, editor, *Proceedings of the Symposium in Applied Mathematics*, volume 19, pages 19–32. American Mathematical Society, 1967.

7. John V. Guttag, James J. Horning with S.J. Garland, K.D. Jones, A. Modet, and J.M. Wing. *Larch: Languages and Tools for Formal Specification*. Springer-Verlag Texts and Monographs in Computer Science, 1993. ISBN 0-387-94006-5, ISBN 3-540-94006-5.

8. M. A. Kishinevsky, A. Y. Kondratyev, A. R. Taubin, and V. I. Varshavsky. *Concurrent Hardware. The Theory and Practice of Self-Timed Design*. John Wiley and Sons Ltd., 1994.

9. Michael Kishinevsky and Jørgen Staunstrup. Checking speed-independence of high-level designs. In *Proceedings of the Symposium on Advanced Reserch in Asynchronous Cirsuits and Systems*, Utah, USA, November 1994. to appear.

10. L. Lavagno and A. Sangiovanni-Vincentelli. *Algorithms for synthesis and testing of asynchronous circuits*. Kluwer Academic Publishers, 1993.

11. Alain J. Martin, Steven M. Burns, T. K. Lee, Drazen Borkovic, and Pieter J. Hazewindus. The first asynchronous microprocessor: the test results. *Computer Architecture News*, 17(4):95–110, June 1989.

12. Niels Mellergaard. *Mechanized Design Verification*. PhD thesis, Department of Computer Science, Technical University of Denmark, 1994.

13. D. E. Muller and W. C. Bartky. A theory of asynchronous circuits. In *Annals of Computing Laboratory of Harvard University*, pages 204–243, 1959.
14. David E. Muller. Asynchronous logics and application to information processing. In H. Aiken and W. F. Main, editors, *Proc. Symp. on Application of Switching Theory in Space Technology*, pages 289–297. Stanford University Press, 1963.
15. Christian D. Nielsen. *Performance Aspects of Delay-Insensitive Design*. PhD thesis, Technical University of Denmark, 1994.
16. Jørgen Staunstrup. *A Formal Approach to Hardware Design*. Kluwer Academic Publishers, 1994.
17. Jørgen Staunstrup and Niels Mellergaard. Localized verification of modular designs. *Formal Methods in System Design*, 1994. accepted for publication.
18. Kees van Berkel, Ronan Burgess, Joep Kessels, Ad Peeters, Marly Roncken, and Frits Schalij. A Fully-Asynchronous Low-Power Error Corrector for the DCC Player. In *ISSCC 1994 Digest of Technical Papers*, volume 37, pages 88–89, San Francisco, 1994.

Automatic Correctness Proof of the Implementation of Synchronous Sequential Circuits Using an Algebraic Approach

Junji Kitamichi, Sumio Morioka, Teruo Higashino and Kenichi Taniguchi

Department of Information and Computer Sciences, Osaka University
Machikaneyama 1-3, Toyonaka, Osaka 560, Japan
Tel: +81-6-850-6607   Fax: +81-6-850-6609
E-mail: {kitamiti,morioka,higashino,taniguchi}@ics.es.osaka-u.ac.jp
WWW: http://sunfish.ics.es.osaka-u.ac.jp/

Abstract. In this paper, we propose a technique for proving the correctness of the implementations of synchronous sequential circuits automatically, where the specifications of synchronous sequential circuits are described in an algebraic language ASL, which we have designed, and the specifications are described in a restricted style. For a given abstract level's specification, we refine the specification into a synchronous sequential circuit step by step in our framework, and prove the correctness of the refinement at each design step. Using our hardware design support system, to prove the correctness of a design step, we have only to give the system some invariant assertions and theorems for primitive functions. Once they are given, the system automatically generates the logical expressions from the invariant assertions and so on, whose truth guarantees the correctness of the design step, and tries to prove those truth using a decision procedure for the prenex normal form Presburger sentences bounded by only universal quantifiers. Using the system, we have proved the correctness of the implementation of a GCD circuit, the Tamarack microprocessor, a sorting circuit and so on, in a few days. The system has determined the truth of each logical expression within a minute.

1. Introduction

Recently, many hardware description languages such as VHDL [8], Verilog HDL [17] and SFL [13] have been proposed. In order to develop reliable circuits, the specifications must be described formally, and the semantics of the specification languages and the correctness of the refinements should also be defined formally. Formal description techniques (FDT), which have those features, are studied widely [1-3,5,9,14]. In the fields related with FSM, for example, the verification using Larch Prover for proving the correctness of the design of pipelined CPU's has been described in [15]. Some properties of sequential circuits have been verified in [16]. We describe the functions of a sequential circuit as a requirement specification. For example, as the requirement specification of a GCD circuit, we describe that the output of circuit must be the greatest common divisor of two inputs. Here we want to prove that synchronous sequential circuits satisfy their functional requirements, and we don't treat the verification for timing, temporal properties such as liveness property and so on. To specify and verify such properties, the higher order logic approaches or the temporal logic approaches are suitable.

In this paper, we propose a technique for proving the correctness of the implementations of synchronous sequential circuits automatically, where the specifications of synchronous sequential circuits are described in an algebraic language ASL, which we have designed [10] [1]. We have developed a hardware design support system using ASL. For a given abstract level's specification, we refine the specification into a synchronous sequential circuit step by step in our framework, and prove the correctness of the refinement at each design step using our hardware design support system automatically.

In our approach, the specifications and implementations of synchronous sequential circuits are described as sequential machine style specifications that correspond to FSMs with registers. The control flows of the circuits may depend on not only the current control state but also the current values of the registers.

We introduce a type *state* representing the abstract state of the system. Each transition corresponds to the action for changing the values of the registers and is treated as the *state transition function* which returns the next abstract state from the current abstract state. The content of each register is expressed by the *state component function* which returns the value of the register at the current abstract state. In the abstract levels' specifications, a complicated action may be specified as a transition.

At each level (denoted as level k), the relations between the current registers' values and those values after each transition is executed are described as the axioms (D_k). The order and execution conditions of the transitions are also described as the axioms (C_k) [2].

At the next level (denoted as level $k+1$), such a complicated action is refined as the execution of a sequence of some more concrete actions and its repetitions. The correspondence from the functions (state components and state transitions) in level $k+1$ to the functions in level k is also described as the axioms. The correctness of the refinement is proved by showing that each axiom in level k holds as the theorem on level $k+1$ description $(\langle D_{k+1}, C_{k+1} \rangle)$, the correspondence (M_k) and the theorems for primitive functions/predicates (PRM). We repeat those refinements until we can get a synchronous sequential circuit [3].

In this paper, we adopt a restriction for describing each level's specification. The restriction is as follows: we only use a variable s of sort *state* in the axioms for describing the relations between the current registers' values and those values

---

[1] The name "ASL" is also used in [18]. Of course, these two languages are different.

[2] On the concrete level's specification, the definitions of the registers (including the state register in the controller) and the combinatorial circuits are described. The connection between these components are also described.

[3] After some repetitions of the refinements, the redundant transitions may be generated in the concrete level's circuit. However, such redundant transitions are deleted using our system [11].

after the transition is executed. We don't use other variables. For example, let $MEM(T(s))$ denote the contents of a memory after the transition T is executed, and let $I1(s)$ and $I2(s)$ denote the pointers for the memory before the transition T is executed. If we want to describe the property that the contents of the memory $MEM(T(s))$ are arranged in ascending order from position $I1(s)$ to position $I2(s)$, then we describe the axiom as follows.

$$0 \leq I1(s) \leq I2(s) \leq N$$
$$imply\ ordered(MEM(T(s)), I1(T(s)), I2(T(s))) == TRUE$$

where $ordered(a, i, j)$ is a primitive predicate which represents that the contents of memory a are arranged in ascending order from position i to position j. We don't use, for example, the following description style because it includes a variable i other than **s**.

$$0 \leq I1(s) \leq i < I2(s) \leq N$$
$$imply\ MEM(T(s))[i] \leq MEM(T(s))[i+1] == TRUE$$

Under the restriction of description style, the verification of a refinement can be done as follows.

(1) If a transition of level k is refined using some repeated executions of the transitions at level $k+1$, we use a *Nötherian induction* on transitions to prove the correctness of the implementation. In the induction, first, we assign the *invariant assertions* for some intermediate states. We don't use any variable except the variable **s** of sort *state* for describing the assertions.

(2) At the each step of the induction, we construct a logical expression P representing that the assertion (or the property to be proved) holds after a transition t of the lower level $k+1$ is executed if the assertion holds before the transition is executed. The constant S is substituted for the variable **s** in the assertions and so on, therefore P doesn't contain any variables.

(3) Consider the proof of the expression P. First, we make a precondition R corresponding to both the axioms AX representing the content of the transition t and the theorems for primitive functions/predicates PRM. The following is an example of theorems for the primitive predicate *ordered*:

$$0 \leq i < j \leq N \wedge a[i] \leq a[i+1] \wedge ordered(a, i+1, j)$$
$$imply\ ordered(a, i, j) == TRUE$$

The constant S is substituted for the variable **s** in the expressions corresponding to AX. The terms such as $f(S)$ or $f(t(S))$ representing the values of the state components at state S or state $t(S)$ are substituted for the variables in the expressions corresponding to PRM. Therefore, the expression R doesn't contain variables. Then, we construct the expression $R\ imply\ P$.

(4) Assume that $R\ imply\ P$ consists of Boolean operators, integer operators and some terms whose sorts are integers or Booleans. If $R\ imply\ P$ is true regardless of the values of those terms, that is, if $R\ imply\ P$ is true for any integer values of those integer terms and any Boolean values of those Boolean terms, then we conclude that P is true.

(5) If the operators among those terms are restricted to "$\wedge$, $\vee$, $\neg$, $+$, $-$, $=$, $>$", then the condition that the expression $R\ imply\ P$ is true regardless

of the values of the terms can be expressed in a prenex normal form *Presburger sentence* bounded by only universal quantifiers. (The form is like $\forall v_1 \forall v_2 \cdots \forall v_n \ EXP(v_1, v_2, \cdots, v_n)$.) It is decidable whether the Presburger sentence is true [7] [4]. For example, Presburger sentence which corresponds to an expression

$$(r1(S) + r2(S) = r1(S) \times r2(S) \vee pred(r1(S)) \wedge \cdots) \ imply \ (r1(S) > 0 \vee \cdots)$$

is as follows.

$$\forall v_1 \forall v_2 \forall v_3 \forall v_4 \cdots (\ (v_1 + v_2 = v_3 \vee v_4 \wedge \cdots) \ imply \ (v_1 > 0 \vee \cdots))$$

The variables v_1, v_2, v_3 and v_4 correspond to the terms $r1(S)$, $r2(S)$, $r1(S) \times r2(S)$ and $pred(r1(S))$, respectively. Here, v_1, v_2 and v_3 are integer variables, and v_4 is a Boolean variable.

We have developed a verification support system (verifier) where the above verification method is used. The verifier has a routine to decide efficiently whether a given prenex normal form Presburger sentence bounded by only universal quantifiers is true [5].

Under this verification method, we have proved the correctness of the implementations for the GCD circuit given as the TPCD94 benchmark, the Tamarack microprocessor used as a verification example in [9] and a maxsort circuit we have designed. The design and verification of the GCD circuit have been carried out within two days which contain the time used for trial and error in the verification.

As we mentioned above, when we prove the correctness of a refinement using this verifier, one have only to give the system some invariant assertions, theorems for the primitive functions and the substitutions for the variables of the theorems. Once they are given, the verifier automatically tries to prove, and one need not send the system many commands interactively.

The paper is structured as follows. The description style of synchronous sequential circuits is explained in Section 2. In Section 3, we describe a stepwise refinement. In Section 4, we define the correctness of the refinement in our framework formally and explain basic techniques for verifying the correctness of the refinements. In Sections 2, 3 and 4, we use the GCD circuit as an example for explanation. The experimental results of the verification of the GCD circuit are given in Section 4, and those of the CPU and maxsort circuit are also given in Section 5. In Section 6, we give the concluding remarks.

2. Algebraic Language ASL and Descriptions of Synchronous Sequential Circuits

In general, the synchronous sequential circuits can be modeled as the finite state machines with registers. The specification S of a synchronous sequential

---

[4] Usually the Presburger sentence doesn't contain any Boolean variable. However, the truth of the sentences which contain some Boolean variables is decidable. In this paper, we also call such sentences "Presburger sentences".

[5] The routine can decide the truth of the sentences which have some Boolean variables. We have implemented the routine by integrating the tautology checking algorithm for the propositional logic into the Cooper's algorithm given in [4].

circuit consists of a pair $\langle D, C \rangle$ of a description D of the contents of transitions and a state diagram C. In D, the relations between the current registers' (memories') values and those values after the transition is executed are described. In C, the condition to execute each transition at each finite state is described, and the next finite state after the transition is executed is also specified.

Here, we describe the specifications of synchronous sequential circuits in our algebraic language ASL [10]. A specification in ASL is a tuple $t = (G, AX)$, where G is a context free grammar without a starting symbol, and AX is a set of axioms. In ASL, we assume that the (infinite) axioms for primitive functions/predicates are given as the definition tables which represent the values of functions/predicates for all input values. An ASL text (specification) can include such definition tables.

Here, we explain a requirement description (level 1's description) of the GCD calculator shown in Table 1. (The descriptions of the grammar are omitted. They are also omitted in the other tables of this paper.) We introduce **R1**, **R2** and **R4** as the registers in the level 1's circuit. These registers correspond to the registers Max, Min and $X2$ given in Fig. 14-1 of the TPCD94 benchmarks, respectively. Then we introduce the abstract transitions **calcgcd** and **nop**. For example, by executing the transition **calcgcd**, the value of the GCD of **R1** and **R2** is calculated and transferred to **R4**. We use the following primitive functions.

- $MaxMember(si)$: A function which represents the maximum value in a set si of integers. This function is defined only when si isn't empty.

- $DivisorSet(i)$: A function which represents the set of divisors of an integer i. This function is defined only when i isn't zero.

- $Intersection(si, sj)$: A function which represents the set of intersection of two sets si and sj of integers.

Table 1. Description of GCD level 1 (D_1 and C_1)

```
init1: R1(init(A,B)) == Max(A,B);
init2: R2(init(A,B)) == Min(A,B);
gcd1:
(1 <= R2(s) and R2(s) <= R1(s) and R1(s) <= N) imply
 R4(calcgcd(s))
   = MaxMember(Intersection(DivisorSet(R2(s)),DivisorSet(R1(s))))== TRUE;
nop1:  R4(nop(s))    == R4(s);
valid1:   VALID(init(A,B))    == TRUE;
valid2:   VALID(calcgcd(s))   == VALID(s) and CONTROL(s)=INIT;
valid3:   VALID(nop(s))       == VALID(s) and CONTROL(s)=END;
control1: CONTROL(init(A,B))  == INIT;
control2: CONTROL(calcgcd(s)) == if CONTROL(s) = INIT then END;
control3: CONTROL(nop(s))     == if CONTROL(s) = END   then END;
```

Using the primitives above, we describe the contents of transitions (D_1). For example, the axiom **gcd1** means that **R4(calcgcd(s))** should be the maximum number of the intersection of the divisor sets of **R1(s)** and **R2(s)** under the assumption **R1(s)** is greater than or equal to **R2(s)** [6]. The values of

---

[6] In this example, we use a formal parameter **N** as the maximum value of the registers. The results of our verification are valid for any positive integer value of **N**.

R1(calcgcd(s)) and R2(calcgcd(s)) aren't specified and hence any values of these terms are permitted. The axioms don't have any variable other than s.

We also describe the state diagram (C_1) as the axioms using both the predicate VALID and the function CONTROL. The predicate VALID represents the execution condition of each transition. The function CONTROL represents the state name. (At the concrete level, the function represents the value of the state register.) For example, the axiom valid2 means that the execution condition of calcgcd is CONTROL(s) = INIT[7]. The axiom control2 means that the value of controller after the execution of calcgcd at state INIT is END.

3. Stepwise Refinements

In this section we explain how to refine a specification.

Let $S_k = \langle D_k, C_k \rangle$ denote a level k's specification. In our framework, the specification $S_{k+1} = \langle D_{k+1}, C_{k+1} \rangle$ is obtained as follows.

(1) We introduce the primitive functions at level $k + 1$.

(2) We introduce some state component functions at level $k + 1$.

(3) We introduce some state transition functions at level $k + 1$ and describe D_{k+1} using the primitive/transition functions. (In the concrete level (circuit level), we describe both the definition of the components in the circuit and the connections between the circuit's components as D_{k+1}. See Section **3.3**.)

(4) We give the correspondence M_k (mapping functions) as follows.

  (4-1) We give the correspondence from the state components functions at level $k + 1$ to those at level k.

  (4-2) We give the correspondence from the state transitions functions at level $k + 1$ to those at level k. Each transition at level k is implemented as the execution of a sequence of some transitions at level $k + 1$ and its repetitions [8].

  We don't give the mapping explicitly if a function F in the level k corresponds to the same function F in the level $k + 1$.

(5) The system can automatically synthesize the level $k+1$'s state diagram C_{k+1} from both the level k's state diagram C_k and the correspondence M_k. (In the concrete level, we describe the circuit's controller as C_{k+1}. See Section **3.3**.)

(6) For the derived level $k+1$'s specification S_{k+1}, we prove that S_{k+1} is a correct implementation of S_k under the correspondence M_k. The proof method is summarized in Section 4.

(7) Some redundant state transitions may be included in S_{k+1}, even if S_{k+1} is a correct implementation of S_k. For example, we may be able to merge two consecutive transitions into one transition. Such an optimization is carried out using our hardware design support system.

---

[7] In general, the execution condition of a transition is written as a sum of products (current state and selection condition).

[8] The mapping is written by mutual recursive expressions with only tail recursions, so that the controller can be implemented by a finite state control.

The process described above is continued until a concrete circuit is derived. In the concrete level, the state component functions (including **CONTROL**) correspond to the hardware components such as the registers, memories, flip-flops and so on. The state transition functions correspond to the data calculations and transfers among those hardware components caused by one machine clock. The details are explained in Section **3.3**.

3.1   Example of refinement to level 2 of GCD circuit

In this section, we explain how we have refined the level 1's specification.

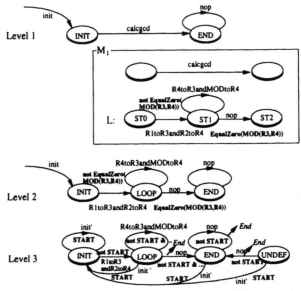

Fig. 1   State diagram of each level of GCD circuit

At first, we have decided to use the Euclid's algorithm to calculate the value of GCD. Then, we have introduced the following primitive function and predicate in the level 2.

- $MOD(i, j)$: A function which represents the value of $i \bmod j$ where i and j are both integers. This function is defined only when the integer j isn't zero.
- $EqualZero(i)$: A predicate whose value is *true* if and only if an integer i is equal to zero.

We have introduced R3 as a new register in the level 2. This register corresponds to the register $X1$ in Fig.14-1 of the TPCD94 benchmarks. The circuit also has three registers R1, R2 and R4 which are the same as the those in the level 1.

We have introduced the following transitions (Table 2). We call the description of the contents of these transitions as D_2.

- **R1toR3andR2toR4** : A transition to execute parallel data transfers; the transfer from R1 to R3, and that from R2 to R4.
- **R4toR3andMODtoR4** : A transition to execute parallel data transfers and calculation; the transfer from R4 to R3, the calculation of the value of **MOD(R3,R4)** and its transfer to R4.

We have described both the order of the execution of transitions and the execution conditions for their executions (Table 3). This is also a description of the correspondence from the transitions in the level 2 to the transition `calcgcd` in the level 1. We call this description M_1. All of the axioms in M_1 and D_2 don't have other variables than **s**.

Then the system automatically derived the state diagram C_2 (Fig. 1) from C_1 and M_1. Here, let L be the state diagram for `calcgcd`. The L has three states $ST0$, $ST1$ and $ST2$. $ST0$ is the initial state of L and $ST2$ is the end state of L. The states $INIT$, $LOOP$ and END in Fig. 1 correspond to $ST0$, $ST1$ and $ST2$, respectively.

Table 2. Description of the contents of the transitions in GCD level 2 (D_2)

```
R3(R1toR3andR2toR4(s)) == R1(s);
R4(R1toR3andR2toR4(s)) == R2(s);
R3(R4toR3andMODtoR4(s)) == R4(s);
R4(R4toR3andMODtoR4(s)) == MOD(R3(s),R4(s));
```

Table 3. Description of the correspondence (M_1)

```
calcgcd(s) ==   S1(R1toR3andR2toR4(s));
S1(s)      ==   if   (EqualZero(MOD(R3(s),R4(s))))
                then  nop(s)
                else  S1(R4toR3andMODtoR4(s));
```

3.2   Refinement to level 3 of GCD circuit

We have designed level 3's specification as follows.

(a) We have introduced an additional new state $UNDEF$ so that the total number of states becomes four, since we assume that the controller is implemented by using two D Flip Flops.

(b) We introduced the input signal $START$. Whenever $START$ signal becomes *true*, the transition *init'* will be executed and the circuit enters $INIT$ [9].

(c) We added the output signal *End* as follows.

$End(s, START, A, B) ==$
$\quad(\neg START \wedge CONTROL(s) = LOOP \wedge EqualZero(MOD(R3(s), R4(s))))$
$\vee (\neg START \wedge CONTROL(s) = END)$
$\vee (\neg START \wedge CONTROL(s) = UNDEF);$

The state diagram of the level 3's specification is shown in Fig. 1.

3.3   Concrete implementation of the GCD circuit

The concrete implementation (level 4's specification) is the same as the implementation of the TPCD94 benchmark [10].

We have defined the circuit's components such as registers (including the state register of the controller) and combinatorial logic circuits (Table 4). For

---

[9] Since the GCD circuit in [12] uses this $START$ signal, we have also introduced it at the level 3. In our circuit, the circuit enters state END and outputs End signal, if $START$ signal isn't given at state $UNDEF$. These are based on [12].

[10] We have found some errors in the Tables 15-2,15-3 and 15-4 of TPCD94 benchmarks v1.0.0 during the verification using our verifier, and we have corrected the errors.

example, the axiom **reg** is the definition of the state component function **REG** (register). The state transition **CK_r** of the register corresponds to the transition caused by a clock signal.

Table 4. Definition of logical components of GCD circuit

```
reg: REG(CK_r(r_s,ctl,data)) == if ctl then data else REG(r_s);
q:   Q(CK_q(d_s,d)) == d;
mul: Mul(n,m,u)      == if u then n else m;
                 :
```

The circuit has four registers and two D-FFs. (These D-FFs compose a state register of the controller.)

The state of the total circuit is described as the tuple of the state of each component. The state of the i-th component can be refered by the primitive function **proj_i**.

The data path between the registers is shown in Fig. 2. In the Table 5, the content of the transition **CK** under the data path shown in Fig. 2 are described. The transition **CK** is caused by a machine clock. The content of **CK** are determined by the current values of registers, the values of the control signals for the circuit's each component and the values of the circuit's input signals. The description D_4 of the content of transition in level 4 consists of the descriptions shown in Table 4 and Table 5.

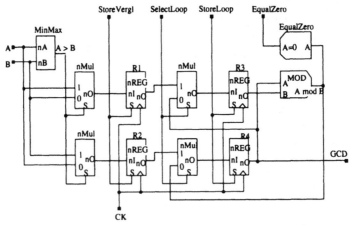

Fig. 2 Data path of GCD circuit

Table 5. Description of data path of GCD circuit

```
define 'outMod' := 'Mod(REG(proj_3(s)),REG(proj_4(s)))';
CK(s,A,B,START,StoreVergl,StoreLoop,SelectLoop,Q1,Q2)
== [ CK_r(proj_1(s),StoreVergl,Mul(A,B,GT(A,B)))
     CK_r(proj_2(s),StoreVergl,Mul(B,A,GT(A,B)))
     CK_r(proj_3(s),StoreLoop,
                     Mul(REG(proj_4(s)),REG(proj_1(s)),SelectLoop))
     CK_r(proj_4(s),StoreLoop, Mul(outMod,REG(proj_2(s)),SelectLoop))
     (* -- note -- the following line is controller *)
     CK_q(proj_5(s),Q1) CK_q(proj_6(s),Q2) ];
```

Then we gave the correspondence M_3 (Table 7). The correspondence consists of the followings: (1) correspondence from the state components in level 4 to those in level 3 (including state assignment), and (2) correspondence from state transition function in the level 4 to those in level 3.

The controller is shown in Fig. 3. The connections of the logical gates in the controller is described in Tbl 6. The description corresponds to C_4.

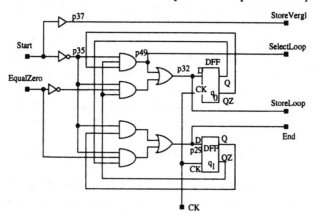

Fig. 3 Implementation of controller of GCD circuit

Table 6. Description of controller of GCD circuit

```
VALID(CK(s,A,B,START,StoreVergl,StoreLoop,SelectLoop,Q1,Q2))
== VALID(s) and StoreVergl = BUF(START)
          and StoreLoop = p32
                  :
          and Q1 = p32
          and Q2 = p29 ;
define 'p37' := 'BUF(START)'
define 'p32' := 'OR(p49,AND3(p35,NOT(eq0),QZ(proj_5(s))))';
define 'p29' := 'OR(AND(p35,Q(proj_5(s))),AND3(p35,eq0,Q(proj_6(s))))';
define 'p35' := 'NOT(START)';
```

Table 7. Correspondence from level 4 to level 3

```
R1(s) == REG(proj_1(s));
R2(s) == REG(proj_2(s));
     :
CONTROL(s) ==[ Q(proj_5(s)) , Q(proj_6(s))];
INIT   == [ FALSE , FALSE ];
LOOP   == [ FALSE , TRUE ];
     :
```

(* Note: "--" means " don't care" *)
    (*   A, B,START,StoreVergl,StoreLoop,SelectLoop, Q1, Q2 *)

R1toR3andR2toR4(s,START)							
== CK(s,--,--,START,			FALSE,	TRUE,	TRUE,FALSE,TRUE);		
R4toR3andMODtoR4(s,START)							
== CK(s,--,--,START,			--,	TRUE,	FALSE,FALSE,TRUE);		

```
nop(s,START)        == .....;
init'(s,A,B,START) == .....;
```

4. Correctness Proof of Implementations

In this section, we define the correctness of the refinements formally, and explain how to prove the correctness algebraically.

4.1 Verification techniques

Let t' denote a specification which contains all sorts and functions in a specification t. We say that t' is a correct refinement of t if and only if $\alpha \equiv_t \beta$ implies $\alpha \equiv_{t'} \beta$ where $\equiv_\tau$ denotes the congruence relation defined by specification τ. Let $\sigma(\xi)$ denote a substitution of ground terms for the variables in the term ξ. If $\sigma(\alpha)$ and $\sigma(\beta)$ are congruent for any substitution σ, then we describe as $\alpha \approx_t \beta$, and "$\alpha \approx_t \beta$" is called a *theorem*. If $l \approx_{t'} r$ holds for any axiom $l == r$ in t, then t' is a correct refinement of t [6,10]. If we can prove that all of the axioms in S_k hold as theorems on S_{k+1}, M_k and theorems for primitive functions/predicates (we call the set of these theorems PRM), then we conclude that S_{k+1} is a correct implementation of S_k.

Under our frame work, the use of the axioms in C_{k+1} is unnecessary for the proof of D_k.

The proof of C_k is unnecessary if C_{k+1} is derived from C_k and M_k automatically. So if the proof of D_k succeed, then S_{k+1} is a correct implementation of S_k. If the designer gives C_{k+1} directly, then the proof of C_k is necessary.

The followings are used to prove a theorem [6].

(A) term reduction by regarding each axiom as a rewrite rule

(B) case analysis of conditional branches

(C) the decision procedure for Presburger sentences :

Let $\xi(x_1, \cdots, x_n)$ denote a term consisting of only integers, integer (or Boolean) variables $x_1, \cdots, x_n$ and operators "$\wedge, \vee, \neg, +, -, =, >$". An algorithm to determine the truth of the following formula is given in [4].

(C-1) $\forall x_1, \cdots, x_n[\xi(x_1, \cdots, x_n)]$

If the formula (C-1) is *true* and all of the values (constructor terms) of terms $t_1, \cdots, t_n$ are defined in t, then we can conclude that $\xi(t_1, \cdots, t_n) \approx_t true$.

(D) the use of the theorems for primitive functions/predicates

(E) Nötherian induction

4.2 Verification of refinement from level 1 to level 2

Now, we'll explain how to prove that the axiom **gcd1** hold as a theorem on D_2, M_1 and PRM. We should prove the followings.

- **Partial correctness** : After the execution of the state diagram L, the value of the register **R4** should be equal to the value of the GCD of **R1** and **R2** before its execution.

- **Termination** : The execution of the state diagram L should terminate.

4.2.1 Proof of partial correctness

In order to prove the partial correctness, we use the *Nötherian induction*. In the induction, first, we assign the invariant assertion IA_{ST1} (Table 8) to the intermediate state $ST1$ in L. Let *Pre* and *Post* (Table 8) be the precondition and the postcondition from the axiom **gcd1**, respectively. Then we prove that the following conditions hold under the precondition *Pre*.

- **first:** IA_{ST1} should hold immediately after reaching the state $ST1$ from the state $ST0$ by executing the transition **R1toR3andR2toR4**.
- **induc:** IA_{ST1} should hold immediately after reaching the state $ST1$ from the same state $ST1$ by executing the transition **R4toR3andMODtoR4**.
- **last:** *Post* should hold immediately after reaching the state $ST2$ from the same state $ST1$ by executing the transition **nop**.

Table 8. Assertions used for proof

```
< Precondition of gcd1: Pre(S0) >
   1 <= R1(S0) <= R2(S0) <= N
< Invariant assertion for ST1: IAST1(s,S0) >
   1 <= R3(s) <= R1(S0) and
   1 <= R4(s) <= R2(S0) and
   R4(s) <= R3(s) and
   SameSet(Intersection(DivisorSet(R1(S0)),DivisorSet(R2(S0))),
           Intersection(DivisorSet(R3(s)),DivisorSet(R4(s))))
< Postcondition of gcd1: Post(s,S0) >
   R4(s) = MaxMember(Intersection(DivisorSet(R1(S0)),
                                  DivisorSet(R2(S0))))
```

Using our verifier, the proof work is carried out as follows.

Step 1: The system displays a figure of L graphically on the X Window (see Fig. 4) automatically by analyzing the description M_1. This figure provides a facility of the user interface where the user can point the state to assign the invariant assertions by clicking the state in the figure directly.

Step 2: The system automatically generates $Pre(S0)$ and $Post(s, S0)$, which correspond to the precondition part and the postcondition (conclusion) part of the axiom **gcd1**, respectively.

Step 3: We assigned the invariant assertion $IA_{ST1}(s, S0)$ to the state $ST1$. (The proof may succeed even if another assertion is given.) This assertion expresses that the set of the common divisors of **R3** and **R4** at $ST1$ should be equal to that of the common divisors of **R1** and **R2** at $ST0$. The relation between the values of **R3** and **R4** which should hold at $ST1$ is also expressed in IA_{ST1}.

We don't use other variables except **s** in the invariant assertions.

Step 4: The system automatically generates three expressions (conditions) P_{first}, P_{induc} and P_{last} corresponding to **first**, **induc** and **last**, respectively. The general form for these expressions is as follows. (Here, we consider a path PT from the state ST_i to ST_j through a transition T where $cond_{PT}$ denotes a condition for executing PT, IA_i denotes an assertion assigned to ST_i and IA_j denotes an assertion assigned to ST_j.)

$Pre(S0)$ *imply*
$$(IA_i(S, S0) \land cond_{PT}(S) \text{ imply } IA_j(T(S), S0))$$

For example, the expression P_{induc} is as follows.

$Pre(S0)$ *imply*
$$(IA_{ST1}(S, S0) \land \neg EqualZero(MOD(R3(S), R4(S))))$$
$$\text{imply } IA_{ST1}(R4toR3andMODtoR4(S), S0))$$

We should prove that the conditions P_{first}, P_{induc} and P_{last} hold under D_2, M_1 and PRM. Now, we'll explain how to prove one of these conditions, using P_{induc} as an example.

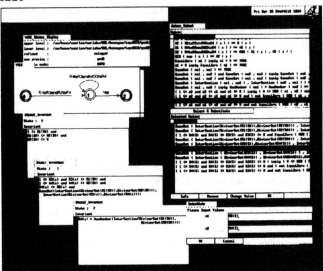

Fig. 4 Display of our verifier

Step 5: The system automatically rewrites each expression by treating the axioms of D_2 as the rewrite rules. Let P'_{induc} be an expression obtained from P_{induc} by the term rewriting.

Step 6: The system automatically selects from D_2 the axioms which describe the content of the transition **R4toR3andMODtoR4** but were not used in the term rewriting at Step 5. Then the system substitutes the constant **S** for the variable **s** in these axioms, replace "==" in the axioms with "=" and make their logical products. Let AX be an expression obtained in this way.

Step 7: We write some statements for primitive functions/predicates which are thought to be included in PRM. We call the set of the statements TH (see Table 9). Each statement is written in the same form as the axiom.

We substituted the values of the registers such as **R3(S)** and **R4(R4toR3andModtoR4(S))** for the variables of each statement in TH [11]. Then the system replace "==" in the axioms of TH with "=" and make their logical products. Let TH' be an expression obtained in this way.

Step 8: The system automatically constructs the logical expression $TH' \wedge AX$ $\text{imply } P'_{\text{induc}}$. Let Q_{induc} be this expression.

Step 9: The system automatically determines the truth of each logical expression (now Q_{induc}) as follows.

At first, the system gets, from Q_{induc}, an expression Q'_{induc} consisting of "$\wedge, \vee, \neg, +, -, =, >$" and integer (Boolean) variables which satisfies the following conditions.

---

[11] The system can find the candidates of the substitution, that is, most general unifiers of the logical expression AX $\text{imply } P'_{\text{induc}}$ and the statement in TH.

Table 9. Statements for primitive functions/predicates (TH)

```
prim1:  EqualZero(n1) imply n1 = 0 == TRUE;
prim2:  n1 = 0 imply EqualZero(n1) == TRUE;
prim3:  SameSet(ns1,ns1) == TRUE;
prim4:  SameSet(ns1,ns2) and SameSet(ns2,ns3)
        imply SameSet(ns1,ns3) == TRUE;
prim5:  SameSet(ns1,ns2) and SameSet(ns1,ns3)
        imply SameSet(ns2,ns3) == TRUE;
prim6:  SameSet(Intersection(ns1,ns2),Intersection(ns2,ns1))== TRUE;
prim7:  SameSet(ns1,ns2) imply MaxMember(ns1) = MaxMember(ns2) == TRUE;
prim8:  (1 <= n2 and n2 <= n1 and n1 <= N)
        imply (0 <= MOD(n1,n2) and MOD(n1,n2) <= n2 - 1) == TRUE;
prim9:  (1 <= n2 and n2 <= n1 and n1 <= N and (not EqualZero(MOD(n1,n2))))
        imply 1 <= MOD(n1,n2) == TRUE;
prim10: (1 <= n1 and n1 <= n2 and n2 <= N and EqualZero(MOD(n2,n1)))
        imply MaxMember(Intersection(DivisorSet(n1),DivisorSet(n2)))
              = n1 == TRUE;
prim11: (1 <= n1 and n1 <= n2 and n2 <= N and not EqualZero(MOD(n2,n1)))
        imply SameSet(Intersection(DivisorSet(n2),DivisorSet(n1)),
                      Intersection(DivisorSet(n1),DivisorSet(MOD(n2,n1))))
        == TRUE;
```

- Q'_{induc} must be obtained by replacing the integer (Boolean) sub-terms in Q_{induc} with the integer (Boolean) variables. The same terms must be replaced with the same variable, and the different terms must be replaced with the different variables.
- The outermost operator (or function symbol) of the subterm replaced must be other than "$\wedge, \vee, \neg, +, -, =, >$".

The system gets, from Q'_{induc}, a prenex normal form Presburger sentence Q''_{induc} by bounding all of the variables by the universal quantifiers.

Then the system applies the decision procedure for the prenex normal form Presburger sentences bounded by only universal quantifiers to Q''_{induc}. If the result of the decision is *"true"*, then P_{induc} holds on D_2, M_1 and TH[12]. Therefore, if the result is *"true"* then we conclude that P_{induc} holds on D_2, M_1 and PRM under the assumption TH is a subset of PRM. (Statements in TH are "correct" theorems for primitive function/predicates.)

We have implemented a decision procedure for the prenex normal form Presburger sentences bounded by only universal quantifiers. The procedure uses the transformation rule called "quantifier elimination" which is used in Cooper's algorithm [4]. For the speed-up of the algorithm, we have devised a way to decide the ordering for deleting variables depending on the form of a given expression[13].

---

[12] It's necessary to show that the terms replaced have their own values under the assumption that the values of the state component functions at states S0 and S are determined.

[13] For example, in the syntax tree of a given expression, a variable close to the root will be deleted first.

The truth of Q''_{induc} has been determined about 1 second by Sun Classic (see Table 10). In the table, the column "length of Presburger sentences" describes the total number of occurrences of the variables and operators in each Presburger sentence.

Step 10: Trial and error will be needed until the proof for all of the steps of the induction succeeds. In many cases, the proof failures have been caused by the lack of the relations of the state components in the invariant assertions, the lack of the statements for the primitive functions, and so on.

The system automatically manages which steps of the induction are already proved, and which steps aren't proved yet. And the system automatically changes the attributes of the proper steps from "Already proved" to "Not yet proved" if the user modifies an invariant assertion.

Table 10. CPU time used for deciding the truth of Presburger sentences

path	length of Presburger sentence	CPU time
ST0 → ST1	186	0.45 sec
ST1 → ST1	207	1.18 sec
ST1 → ST2	186	0.18 sec

Sun Classic

4.2.2 Proof of termination

In order to prove the termination for the axiom **gcd1**, we proved the following two conditions.

– At the state $ST1$, the value of **R4** should be always positive.
– After an execution of the transition **R4toR3andMODtoR4**, the value of **R4** must be less than the value before the execution.

To show these two conditions hold, we proved the following condition hold under D_2, M_1 and PRM.

$$Pre(S0)\ imply$$
$$(IA_{ST1}(S, S0) \wedge execution\ condition\ of\ R4toR3andMODtoR4$$
$$imply\ 1 \le R4(S) \wedge R4(R4toR3andMODtoR4(S)) < R4(S))$$

Here, the condition $IA_{ST1}(S, S0)$ can be used in the precondition part, because it has been already proved that $IA_{ST1}(s, S0)$ is the invariant at $ST1$ under D_2, M_1, PRM and the precondition of **gcd1**.

The proof was carried out using the same verification method written in Section **4.2.1**. (We applied Step 5 ∼ Step 9 to the condition.) The CPU time used for the verification was 1 second (Sun Classic).

4.3 Verification of refinement from level 2 to concrete implementation

The level 3's specification satisfies the level 2's specification, because S_3 is obtained by adding some axioms to S_2. But the implementation of transition *init* doesn't obey our framework.

At the proof of the correctness of the refinement from level 3 to level 4, we proved that each axiom of S_3 holds as theorem on the S_4, M_3 and PRM [14]. The proof was carried out automatically using the term rewriting facility and decision procedure for the prenex normal form Presburger sentences bounded by only universal quantifiers in our verifier. The CPU time used for the verification was 11 seconds (Sun Classic).

5. Other Examples of Verification

5.1 The Tamarack Microprocessor

We have designed the Tamarack microprocessor in [9] and proved the correctness of the implementation. In the level 1, we described the same requirement (the same instruction set as that of [9]). In the level 3, we have used the same architecture and microprograms as those given in Fig.3 and Appendix of [9].

Our verifier has proved the correctness of each refinement automatically without trial and error, using the term rewriting and the decision procedure for the prenex normal form Presburger sentences bounded only by universal quantifiers. The CPU time used for the verification from level 1 to level 2 was 93 seconds (Sun Classic), and that from level 2 to level 3 was 55 seconds (Sun Classic).

The major reasons why the proofs were carried out without trial and error are as follows.

(1) We had no need to use any invariant assertion at the proofs, because there are no loops in the state diagram.

(2) We had no need to use the theorems for primitive functions/predicates at the proofs.

For such a case, the verification could also be carried out automatically by the symbolic simulation such as [9]. However, our verifier can treat even the case that the relations between the values of the current state component functions and those values at the next state are described as the predicates. This is a merit of our approach.

5.2 Maxsort Circuit

We have implemented a sorting circuit and proved the correctness of the implementation.

The requirement of the sorting circuit is described in Table 11. The predicate $seteq(a, i, j, b, k, l)$ represents that the set of elements between the positions i and j of array a is equal to the set of elements between the positions k and l of array b as the multi set. The predicate $arrayeq(a, i, j, b)$ represents that the integer elements between the positions i and j of array a are the same as that of array b in order.

The axiom **sort1** represents that the elements between the positions $I1(INIT)$ and $I2(INIT)$ (both $I1$ and $I2$ are registers) of the memory $MEM(sort(INIT))$ after the transition **sort** are sorted where $INIT$ is the initial state.

---

[14] At this proof we had to prove that each axiom of C_3 holds as theorem on the S_4, M_3 and PRM, since we gave C_4 (the definition of the components of the state register and the connection of the components in the circuit) directly.

The state diagram of each level is shown in Fig. 5.

We have implemented the transition **sort** using max-sort algorithm. In the level 2, the new registers **max**, **maxpt** and **bound** are introduced. The register **bound** denotes the next position of the lowest position in the sorted area.

Table 11. Requirement of sort circuit

```
sort1: 0<=I1(INIT)<=I2(INIT)<=N imply
       ordered(MEM(sort(INIT)),I1(INIT),I2(INIT)) == TRUE
sort2: 0<=I1(INIT)<=I2(INIT)<=N imply
       seteq(MEM(INIT),      I1(INIT),I2(INIT),
           MEM(sort(INIT)),I1(INIT),I2(INIT)) == TRUE
sort3:0<=I1(INIT)<=I2(INIT)<=N imply
     ( (0<I1(INIT) imply
           arrayeq(MEM(INIT),0,I1(INIT)-1,MEM(sort(INIT),0,I1(INIT)-1)))
       and (I2(INIT)<N imply
           arrayeq(MEM(INIT),I2(INIT)+1,N,MEM(sort(INIT),I2(INIT)+1,N)))
     ) == TRUE
```

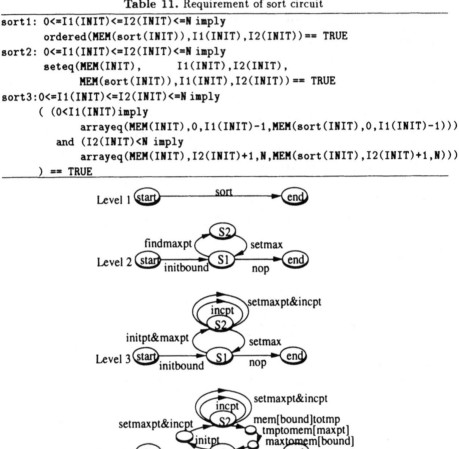

Conditions to execute each transition are omitted

Fig. 5 State diagram of each level of maxsort circuit

The new transitions **initbound**, **findmaxpt**, **setmax** and **nop** are also introduced. By executing the transition **findmaxpt**, the maximum value in the unsorted area is stored into **max** and the position of the maximum value is stored into **maxpt**. By executing the transition **setmax**, the content at the position pointed by **maxpt** in **MEM** is replaced by the content at the position pointed by **bound**, the content at the position pointed by **bound** becomes equal to the content of **max**, and the value of **bound** is decreased by one.

In the verification of the correctness of the refinement from level 1 to level 2, we have used the Nötherian induction. To prove that the axiom **SORT1** holds, we have assigned the following invariant assertion to the state **S1**.

$$I1(s) = I1(S0) \land I2(s) = I2(S0) \land$$
$$I1(s) \leq bound(s) \leq I2(s) \land$$
$$(bound(s) < I2(s) \; imply$$
$$ordered(MEM(s), bound(s) + 1, I2(S0)) \land$$
$$isMaxPos(MEM(s), I1(S0), bound(s) + 1, bound(s) + 1))$$

Predicate $isMaxPos(a, i, j, k)$ represents that $a[k]$ is greater than or equal to any of $a[i]$, $a[i + 1]$, ..., $a[j]$.

In the level 5, we have described both the data path (Fig. 6) and the controller (Fig. 7) of the circuit.

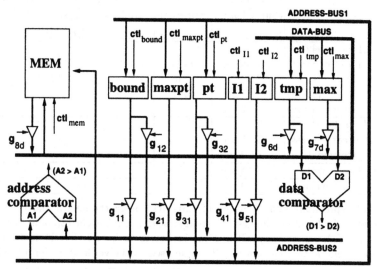

Fig. 6 Data path of maxsort circuit

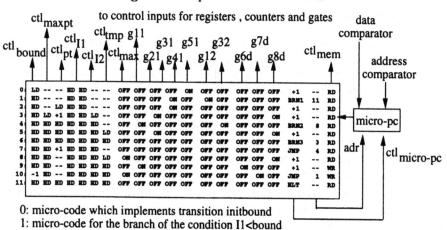

0: micro-code which implements transition initbound
1: micro-code for the branch of the condition I1<bound
:

Fig. 7 Implementation of controller of maxsort circuit

The experimental results for the verification are described in Table 12. In the verifications of the correctness of the refinement from the level 1 to level 2, totally 20 theorems for the primitive functions were used. Therefore the sizes of the prenex normal form Presburger sentences bounded by only universal quantifiers to be checked became rather large. (There were some sentences whose length exceeds 1000 with more than 40 different variables.) However, our verifier could decide their truth within about 1 minute. If the value of a given Presburger sentence is false, then the result '*false*' will be obtained very quickly (within a second in most case). The design and verification were carried out in three days including trial and error.

Table 12. Data of the verification of the sorting circuit

	CPU Time (Sun Classic)	Length of expression	Number of variables	Number of theorems (substitutions) of primitives
Level1 to Level2				
(proof of sort1)	64.85 sec	1093	42	16 (27)
(sort2)	28.30 sec	645	31	10 (16)
(sort3)	15.67 sec	511	29	4 (8)
(termination)	2.2 sec			
Level2 to Level3				
(partial correctness)	19.17 sec			
(termination)	1.9 sec			
Level3 to Level4	17.45 sec			
Level4 to Level5	8.80 sec			

"Length of expression" and "Number of variables" mean the length and the number of different variables of the longest Presburger sentence (that corresponds to the path $S_1 \rightarrow S_2 \rightarrow S_1$), respectively. CPU Time in Level2 to Level3, Level3 to Level4 and Level4 to Level5 means the sum of CPU time used at the verification for all axioms.

6. Conclusion

In this paper, we have proposed a method for refining a given specification to a synchronous sequential circuit and verifying the correctness of the refinements automatically. We have developed a hardware design support system based on algebraic methods. We also gave three examples of implementation verifications.

In the method proposed here, the specifications and invariant assertions must be written in a restricted style. (As the variables, only the use of a variable of sort *state* is permitted.) In the verification, it is proved that rather strong sufficient conditions hold. However, for the GCD circuit, CPU and the maxsort circuit used in this paper as examples, we succeeded in proof.

After giving the assertions and theorems for primitive functions/predicates and the substitutions for the variables of the theorems, the verification is carried out automatically although it may fail under those assertions and theorems. The design and verification of those circuits were carried out in a few days, respectively, including trial and error. This depends on the features that our verifier has many user friendly verification facilities and an efficient decision procedure for the prenex normal form Presburger sentences bounded by only universal quantifiers.

References

1. S. Bose and A. Fisher : "Automatic Verification of Synchronous Circuits Using Symbolic Logic Simulation and Temporal Logic", Proc. IMEC-IFIP Int. Workshop on Applied Formal Methods for Correct VLSI Design, pp.759-764, 1989.

2. M. Browne, E. M. Clarke, D. Dill and B. Mishra : "Automatic Verification of Sequential Circuits using Temporal Logic", IEEE Trans. on Computer, Vol. C-35, No. 12, pp.1035-1044, 1986.

3. H. Busch : "Transformational Design in a Theorem Prover", IFIP Conf. on Theorem Provers in Circuit Design, North-Holland, pp.175-196, 1992.

4. D.C.Cooper : "Theorem Proving in Arithmetic without Multiplication", Machine Intelligence, No.7, pp. 91-99, 1972.

5. M. J. C. Gordon : "HOL : A Proof Generating System for Higher Order Logic", in VLSI Specification, Verification and Synthesis, G. Birtwistle and P. A. Subrahmanyam ed-s., Kluwer Academic Publishers, pp.73-128, 1988.

6. T. Higashino and K. Taniguchi : "A System for the Refinements of Algebraic Specifications and their Efficient Executions", Proc. of the IEEE 24-th Hawaii Int. Conf. on System Sciences (HICSS-24), Vol. II, pp.186-195 (Jan. 1991).

7. J.E. Hopcroft and J.D. Ullman : "Introduction to Automata, Theory, Languages, and Computation", Addison-Wesley, 1979.

8. IEEE : "IEEE Standard VHDL Language Reference Manual", IEEE,1988.

9. J. J. Joyce : "Formal Verification and Implementation of a Microprocessor", VLSI Specification, Verification and Synthesis, Kluwer Academic Publishers, pp.129-157, 1988.

10. T. Kasami, K. Taniguchi, Y. Sugiyama and H.Seki : "Principles of Algebraic Language ASL/*", Trans. of IECE Japan, Vol.69-D, No.7, pp.1066-1074, July 1986 (in Japanese).

11. J. Kitamiti, T. Higashino, K. Taniguchi and Y. Sugiyama : "Top-Down Design Method for Synchronous Sequential Logic Circuits Based on Algebraic Technique", Trans. of IEICE Japan, Vol.77-A, No.3, March 1994 (in Japanese).

12. T. Kropf : "Benchmark-Circuits for Hardware - Verification : v.1.0.0", the Benchmark-Circuits for 2nd Conf. on Theorem Proving in Circuits Design, FTP from Univ. of Karlsruhe (129.13.18.22), Germany, 1994.

13. Y. Nakamura : "An Integrated Logic Design Environment Based on Behavioral Description", IEEE Trans. on Computer-Aided Design Integrated Circuits & Systems, Vol. 6, No. 3, pp.322-336, May 1987.

14. V. Stavridou, J. A. Goguen, A. Stevens, S. M. Eker, S. N.Aloneftis and K. M. Hobley : "FUNNEL and 2OBJ : Towards as Integrated Hardware Design Environment", IFIP Conf. on Theorem Provers in Circuit Design, North-Holland, pp.197-223, 1992.

15. J.B.Saxe S.J.Garland, J.V.Guttag and J.J.Horning, "Using Transformations and Verification in Circuits Design", Designing Correct Circuits, North-Holland, pp.1-25, 1992.

16. G. Thuau ,B.Berkane, "Using the Language LUSTRE for Sequential Circuit Verification", Designing Correct Circuits, North-Holland, pp.81-96, 1992.

17. Open Verilog International : "Verilog Hardware Description Language Reference Manual", 1991.

18. M. Wirsing : "Structured Algebraic Specifications : A Kernel Language", Tech. Report, TU Mchen, 1984.

Mechanized Verification of Refinement

Niels Maretti

Department of Computer Science, Technical University of Denmark,
DK-2800 Lyngby, Denmark. e-mail: nbm@id.dtu.dk

Abstract. This paper describes a mechanized approach to verifying that one concrete design is a refinement of another abstract design. A widely used notion of refinement is trace inclusion, which implies that each externally visible behavior of the concrete design can also be caused by the abstract design. In some cases this is too restrictive and the verification technique proposed here is based on a more liberal notion where information about the environment is exploited. A verification technique is presented for designs written in the design language SYNCHRONIZED TRANSITIONS. The verification technique is supported by a prototype tool for mechanizing 1) the axiomatization of the design descriptions in the logic of an existing theorem prover, and 2) the generation of proof obligations. Based on the axiomatization of the design descriptions, the proof obligations can be discharged using the theorem prover.

1 Introduction

This paper describes a mechanized approach to verifying that one concrete design is a refinement of another abstract design. For mechanized verification to be practical it is important to find verification techniques which break the proof into a number of independent steps of modest complexity. The main contribution of this work is a notion of refinement which is both powerful enough to allow interesting designs to be verified and yet simple enough to make mechanization feasible.

Several approaches exists for formal verification of refinement, for example [1, 5, 7, 8]. [5, 8] use forward and backward simulation, whereas [1, 7] use refinement mappings and prophecy variables. In the field of hardware verification, two very different approaches are [2] and [4]. In [4] the hardware is described in higher order logic and within this logic refinement corresponds to equivalence or implication. Bryant [2] uses a simulator to prove refinement. Common for all of the approaches is that they use trace inclusion as the fundamental notion of equivalence.

Trace inclusion means that each externally visible behavior of the concrete design can also be caused by the abstract design. However, in some cases trace inclusion is too restrictive. Consider, for example, an abstract description of a multiplier that performs the multiplication of two positive integers in a single operation, i.e., if two inputs x, y are provided, the result s ($= x * y$) is available immediately afterwards. Following an example in [4], the multiplier can be realized by accumulating y in s x times. In this case a number of intermediate results $(y, 2 * y, \ldots, (x - 1) * y)$ are observed, before the right value of s $(x * y)$ is observed. Consequently, the notion of trace inclusion does not apply to this example. In case of the multiplier, it would be useful to be able to disregard s during the computation and only to focus on the final value.

This paper proposes a notion of refinement that allows for temporarily leaving s out of consideration by taking the environment into account. Based on information about the environment, the notion of refinement ensures that the concrete design can correctly replace the abstract in the environment, even though the concrete design does not refine the abstract according to trace inclusion. The notion of refinement is based on trace inclusion, but extended to take the environment into account. A prototype tool is developed for verifying that one design refines another according to this notion of refinement.

Section 2 defines the model of a design. Section 3 explains how the environment is taken into account. Section 4 defines the notion of refinement in terms of the model. To verify that one design is a refinement of another, section 5 provides a verification technique for designs written in the design language SYN-CHRONIZED TRANSITIONS. The verification technique is developed in preparation for using a theorem prover. In section 6, the verification technique is applied to part of the Tamarack microprocessor [6].

2   Computational Model

Above, the terms design and environment have been used informally. Below, designs and environments are described as *cells*. The computational model of a cell, C, is a transition system identified by (1) the state space, $\mathcal{S}_C$, spanned by state variables in $\mathcal{V}_C$, (2) the set of initial states, $\mathcal{I}_C$, and (3) the set of state transitions, $\mathcal{T}_C$. These components are described below.

A cell, C, operates on a set of typed state variables, $\mathcal{V}_C$. Some of the state variables are hidden from the environment, they are called *local* variables, $\mathcal{L}_C$. The rest of the state variables are accessible for the environment, they are called *interface* variables, $\mathcal{E}_C$. Communication between cells takes place by means of shared interface variables. The state space of C, determined by the state variables and their corresponding types is denoted $\mathcal{S}_C$. For $v \in \mathcal{V}_C$ and $\sigma \in \mathcal{S}_C$, $v.\sigma$ denotes the value of v in σ. A state in which x has the value 5 and y has the value 2 is written $\{x = 5, y = 2\}$. For $x \in \mathcal{V}_C$, let $val(x)$ denote the set of values that x can have according to its type. For $v \in val(x)$, $\sigma[v/x]$ denotes the state σ where the value of x is replaced by the value v, and the values of the rest of the variables are unchanged.

In order to compare parts of states, projection is defined. Let $\sigma \in \mathcal{S}_C$, $m \subseteq \mathcal{V}_C$. $\sigma \downarrow m$ is the state containing variables in m only, where the values of the variables in m are the same in σ and $\sigma \downarrow m$. For example

$$\{x = 5, y = 2, s = 0\} \downarrow \{x, y\} = \{x = 5, y = 2\}$$

$(\mathcal{S}_C)^\omega$ denotes the set of infinite sequences of states of C. Let t denote the sequence $< \sigma_0, \sigma_1, \ldots, \sigma_i, \ldots >$. Element i in t, is denoted $t[i]$, i.e. $t[i] = \sigma_i$.

Projection is extended to apply to sequences and sets of states, i.e.

$$< \sigma_0, \sigma_1, \ldots > \downarrow m \ = \ < \sigma_0 \downarrow m, \sigma_1 \downarrow m, \ldots >$$

$$\{\sigma_0, \sigma_1, \ldots\} \downarrow m \ = \ \{\sigma_0 \downarrow m, \sigma_1 \downarrow m, \ldots\}$$

A cell, C, defines a set of initial states called $\mathcal{I}_C$, and a set of state transitions, called $\mathcal{T}_C$. Each state transition is a pair (σ, σ') meaning that C can perform a state change from σ to σ'. Returning to the multiplier, the pair $(\{x = 3, y = 2, s = 0\}, \{x = 3, y = 2, s = 6\})$ belongs to $\mathcal{T}$.

A cell determines a computation, represented by a trace $< \sigma_0, \sigma_1, \sigma_2, \ldots >$, i.e., an infinite sequence of states. Each state belongs to $\mathcal{S}_C$. σ_0 belongs to $\mathcal{I}_C$, and for each pair of succeeding states (σ_i, σ_{i+1}),

$$(\sigma_i, \sigma_{i+1}) \in \mathcal{T}_C \ \vee \ \sigma_i = \sigma_{i+1}$$

In section 4.1 it is explained why repetitions $(\sigma_i = \sigma_{i+1})$ are allowed. The set of traces, $\mathcal{B}_C$, that a computation of a cell, C, can determine is defined by

$$\mathcal{B}_C = \{s \in (\mathcal{S}_C)^\omega \,|\, s[0] \in \mathcal{I}_C \wedge \forall i \geq 0 : (s[i], s[i+1]) \in \mathcal{T}_C \ \vee \ s[i] = s[i+1]\}$$

The states present in any trace of $\mathcal{B}_C$ are referred to as the reachable states.

2.1 Composition

Above, the set of traces of one cell is defined. Often designs are composed of several cells. Below, composition of cells is defined.

For two cells, C and D, composition is denoted $C\|D$. In order to avoid renaming, it is assumed that interface variables of C and D are identically named and typed, and that local variable names of C and D are not overlapping. This means that $\mathcal{S}_{C\|D}$ is identified by $\mathcal{V}_C \cup \mathcal{V}_D$ as explained above. $\mathcal{I}_{C\|D}$ is the set of states from $\mathcal{S}_{C\|D}$ such that the variables of C and D are initialized according to the states in $\mathcal{I}_C$ and $\mathcal{I}_D$, respectively. $\mathcal{T}_{C\|D}$ is the set of pairs of states from $\mathcal{S}_{C\|D} \times \mathcal{S}_{C\|D}$ such that for each pair (σ, σ'), the variables of $\mathcal{V}_C$ are changed according to $\mathcal{T}_C$ and the variables of $\mathcal{L}_D$ are unchanged, or vice versa.

3 Interface Protocols

The interface protocol is an important part of the interface description (together with the types of the interface variables, etc.). The interface protocol documents the communication pattern of the cell and the environment. If the communication pattern is formalized, it can be used for reasoning about cells, for example, for refinement purposes.

Informally, the information about the communication is stated as a condition for each interface variable. The condition indicates whether the value of the variable can be used by the environment in a given state. For each state, this defines a subset of the interface variables containing variables that the environment may use. This subset is referred to as the *observable* part of the interface.

Instead of focusing on the whole interface, the notion of refinement defined below is only concerned with the observable part of the interface. Recall, for example, the multiplier where the aim is to disregard s until s has reached the value of $x * y$. Following the outline given here, s is kept out of the observable part of the interface until the correct value ($x * y$) has been reached, whereupon s is included in the observable part of the interface.

Below, interface protocols are defined, and it is described what it means for an environment to respect an interface protocol.

3.1 Definition

A condition, $V(x)$, is associated with each interface variable x. The condition is an assertion on the interface variables. The collection of interface conditions constitute the interface protocol. The variable x is defined to be in the observable part of the interface in a state σ iff $V(x)$ evaluates to true in σ, written $V(x).\sigma$.

Example. Consider again the multiplier. The interface of the multiplier contains the variables x, y, s and rdy. In the multiplier, rdy is set true to indicate that the computation of s has finished. Therefore the environment should not use the value of s unless rdy is true. This is formalized using the interface protocol

$$V(s) = rdy$$

Consider the state $\sigma = \{x = 4, \ y = 2, \ s = 4, \ rdy = F\}$. The observable part of σ is $\{x = 4, \ y = 2, \ rdy = F\}$. This is also the case for $\sigma' = \{x = 4, \ y = 2, \ s = 0, \ rdy = F\}$. Consequently, σ and σ' can be treated identically when arguing about the states from the environments point of view. This is exploited in the definition of refinement given below.
End of example

For an interface variable x, $V(x)$ is expressed by means of operators and other interface variables. The variable y *appears in* $V(x)$ if y is among the interface variables that are used to express $V(x)$. Since $V(x).\sigma$ is supposed to determine whether the environment is allowed to observe x all the variables that appear in the predicate $V(x)$ should be observable. To avoid circular dependencies, for example

$$V(x) = y \qquad V(y) = x$$

it is sufficient to require that if y appears in $V(x)$, then $V(y)$ must be the constant true. $V(x)$ defaults to true, i.e. leaving out $V(x)$ implies that x always belongs to the observable part of the interface.

The possibility of exploiting the interface protocol relies on the cell and the environment to have the same understanding of which part of the interface is observable in a given state. This is dealt with next.

3.2 Respecting an Interface Protocol

Let $ns(\sigma)$ denote the set of next states that the environment can cause by performing one state transition from $\mathcal{T}_E$ started in σ.

$$ns(\sigma) = \{\sigma' \in \mathcal{S}_E \mid (\sigma, \sigma') \in \mathcal{T}_E \}$$

Informally, the environment *depends on* a variable, y, in a state, σ, if the value of y influences the changes of other variables caused by the environment. This is denoted $dep(y, \sigma)$. Let $\tilde{y}$ denote the set $\mathcal{V}_E \setminus \{y\}$.

$$dep(y, \sigma) \;=\; \exists v \in val(y) : ns(\sigma) \downarrow \tilde{y} \neq ns(\sigma[v/y]) \downarrow \tilde{y}$$

Consider any cell, C, with an interface protocol, V, and environment, E. Let (σ, σ') denote any state transition that E can perform. The environment is not allowed to depend on a variable y in σ when $V(y).\sigma$ is false. This leads to the following condition, **ND** (no dependency)

$$\textbf{ND} \qquad \forall y \in \mathcal{E}_E : dep(y, \sigma) \Rightarrow V(y).\sigma$$

Furthermore, the environment must not turn $V(y)$ true or change the value of unobservable interface variables. This results in the condition **NC** (no change):

$$\textbf{NC} \qquad \forall y \in \mathcal{E}_E : \neg V(y).\sigma \Rightarrow \neg V(y).\sigma' \;\wedge\; \neg V(y).\sigma' \Rightarrow y.\sigma = y.\sigma'$$

When these two conditions are fulfilled, E is said to *respect* V.

4 Refinement

The notion of refinement described below is based on the information about the environment, captured by interface protocols.

4.1 Definition of Refinement

Let C denote the concrete and A the abstract design. Let V denote the interface protocol of A and C, and let E denote an environment that respects V. Refinement as defined below ensures that for any trace of $C\|E$ there is a similar trace of $A\|E$, where similar means that there are no differences with respect to the observable variables.

The aim is to compare only the observable part of the traces. This is obtained using $\downarrow_V$ that projects away state variables that are not observable. Given a cell C and $\sigma \in \mathcal{S}_C$

$$\sigma \downarrow_V = \sigma \downarrow \{\, v \in \mathcal{E}_C \mid V(v).\sigma \,\}$$

Applying $\downarrow_V$ to a trace corresponds to applying $\downarrow_V$ to each of the states in the trace.

Example: Multiplier (continued). Below, each column denotes a state, and the succeeding columns make up a part of a trace, t. In the presence of the interface protocol $V(s) = rdy$, $t \downarrow_V$ is the part of the trace between the two lines.

s	$\cdots$	2	2	2	0	3	6	9	12	12	$\cdots$
rdy	$\cdots$	T	T	F	F	F	F	F	F	T	$\cdots$
x	$\cdots$	1	4	4	4	4	4	4	4	4	$\cdots$
y	$\cdots$	2	3	3	3	3	3	3	3	3	$\cdots$

End of example

Having projected away the parts of the traces that are not observable, leaves the traces with adjacent duplicates, where the original traces only differ on non observable state variables. Between two observable state changes the abstract and the concrete designs might make a different number of state transitions. In the definition below this is compensated for by allowing traces to have duplicates (section 2). This leads to the following definition of refinement:

Definition 1. Let A and C denote cells with the same interface protocol, V, and E an environment, such that E respects V. C refines A in E iff

$$\forall t' \in \mathcal{B}_{C\|E}, \ \exists t \in \mathcal{B}_{A\|E} : \ t' \downarrow_V = t \downarrow_V$$

Note that this definition is a generalization of trace inclusion, since leaving out interface protocols implies that $\downarrow_V$ only projects away local variables.

Example: Multiplier (continued). Let A and C denote the abstract and the concrete multiplier, respectively. Let E denote an environment. Consider the two traces $t \in \mathcal{B}_{A\|E}$ and $t' \in \mathcal{B}_{C\|E}$.

t:

s	$\cdots$	2	2	2	2	2	2	2	2	12	$\cdots$
rdy	$\cdots$	T	T	F	F	F	F	F	F	T	$\cdots$
x	$\cdots$	1	4	4	4	4	4	4	4	4	$\cdots$
y	$\cdots$	2	3	3	3	3	3	3	3	3	$\cdots$

t':

s	$\cdots$	2	2	2	0	3	6	9	12	12	$\cdots$
rdy	$\cdots$	T	T	F	F	F	F	F	F	T	$\cdots$
x	$\cdots$	1	4	4	4	4	4	4	4	4	$\cdots$
y	$\cdots$	2	3	3	3	3	3	3	3	3	$\cdots$

While the computation of s takes place, rdy is false and the interface protocol $V(s) = rdy$ ensures that s is projected away from the traces. This means that when rdy is false, there is no difference between $t' \downarrow_V$ and $t \downarrow_V$. When s has reached the value of $x * y$, rdy is set true, and s is again included in the traces.

Still no difference is found between $t' \downarrow_V$ and $t \downarrow_V$, since now s has the same value in both the concrete and the abstract trace. This illustrates the advantage of considering only the observable part of the interface.

End of example

5 Verification

The definition of refinement given in the previous section is not directly useful for formal verification, since it is concerned with infinite traces. This section provides a verification technique that is suitable for mechanizing the verification of refinement. Below, the verification is carried out using a theorem prover. Often when theorem provers are used, the designs are specified directly in the logic of the theorem prover. In the present work another approach is taken. In order to avoid obscuring the descriptions with theorem prover specific details, SYNCHRONIZED TRANSITIONS [11] is chosen as description language.

5.1 SYNCHRONIZED TRANSITIONS

A subset of SYNCHRONIZED TRANSITIONS is used. In SYNCHRONIZED TRANSITIONS, a cell, C, consists of (1) declarations of the local and interface variables, (2) an initialization, $Init(C)$, (3) a set of transitions, $Tr(C)$, (4) an invariant, $Inv(C)$, and (5) an interface protocol, V.

The variable declarations state the name and type of the variables, thereby identifying the set of states, S_C.

The initialization is a predicate on the state variables. It determines the initial values of state variables. The initialization has the form:

$$v_1 = val_1 \ \wedge \ v_2 = val_2 \ \wedge \ v_3 = val_3 \ \wedge \ \cdots$$

where v_1, v_2, v_3 are variables and val_1, val_2, val_3 are suitably typed values. Any state in S_C that fulfills this predicate can be an initial state. The initialization is related to the underlying transition system in the following way

$$\mathcal{I}_C = \{ \ \sigma \in S_C \ | \ Init(C).\sigma \ \}$$

Transitions describe the state changes that the cell can perform. Each transition consists of a precondition, p, and a multi assignment. A precondition is a boolean typed expression. The multi assignment consists of a list of variables and an equally long list of expressions. If p is true in a state, the multi assignment can be executed in that state. Executing the multi assignment, simultaneously assigns the value of the expressions $e_1, e_2, \ldots, e_n$ to the variables $l_1, l_2, \ldots, l_n$. Syntactically, a transition looks like this:

$$\ll p \ \rightarrow \ l_1, \ l_2, \ \ldots, \ l_n \ := \ e_1, \ e_2, \ \ldots, \ e_n \gg$$

Given a transition, t, $exprs(t)$ denotes the list of expressions $< p, e_1, e_2, \ldots, e_n >$, and $exprs(t).\sigma$ denotes the list of values $< p.\sigma, e_1.\sigma, e_2.\sigma, \ldots, e_n.\sigma >$. Given states

σ, σ' and transition t, $\sigma \rightarrow_t \sigma'$ denotes that the precondition evaluates to true in σ, and that executing the multi assignment in σ results in σ'. This is referred to as executing the transition. σ and σ' are referred to as the pre and post states, respectively, for the execution of t. Given a state, σ, several transitions may have a precondition that evaluates to true in σ. In this case, the transition to be executed is selected nondeterministically. This means that the set, T_C, used to define the underlying transition system is defined by

$$T_C = \{ (\sigma, \sigma') \in (S_C \times S_C) \mid \exists t \in Tr(C) : \sigma \rightarrow_t \sigma' \}$$

An invariant, $Inv(C)$, is a predicate on the state variables. It holds in any state of any trace of C. In 5.3 the use of invariants in the verification is explained.

Example: Multiplier. An abstract and a concrete multiplier written in SYN-CHRONIZED TRANSITIONS are listed below.

Since transitions are selected nondeterministically, no assumptions can be made in advance about the order of the execution of transitions. In case of the multiplier, this is handled in the following way: When the environment has supplied the cell with the arguments x and y, rdy is set false to make the computation start, and when the computation has finished, the cell notifies this by setting rdy true.

```
CELL mult(x, y, s : INTEGER ; rdy : BOOLEAN )
INTERFACE PROTOCOL V(s) = rdy
INVARIANT x ≥ 0 ∧ y ≥ 0
BEGIN
   ≪¬rdy → s, rdy := x * y, TRUE ≫
END mult
```

This abstract design states that s is assigned the value of $x * y$ in one state transition. To implement multiplication a number of finer grained operations are combined: s is set to 0, and y is added to s x times.

```
CELL mult(x, y, s : INTEGER ; rdy : BOOLEAN )
INTERFACE PROTOCOL V(s) = rdy
INVARIANT x ≥ 0 ∧ y ≥ 0
STATE
   comp : BOOLEAN
   xl, s' : INTEGER
INITIALLY comp = FALSE
BEGIN
   ≪¬rdy ∧ ¬comp → comp, s, xl, s' := TRUE , 0, x, s≫
   ≪comp ∧ xl > 0 → xl, s := xl − 1, s + y≫
   ≪comp ∧ xl = 0 → rdy, comp := TRUE , FALSE ≫
END mult
```

If rdy and $comp$ are false, the computation can be started by the first transition: s is set to 0, xl gets the value of x and $comp$ gets true. $comp$ being true denotes that the computation is in progress. In the second transition, xl is decremented each time s is increased by y. The third transition deals with the situation where the computation of s has finished ($xl = 0$). This is notified by setting rdy true. s' has no influence on the computation. It is included for verification purposes. This is explained in 5.5.

End of example

5.2   Stepwise Verification

As explained previously, the definition of refinement is expressed in terms of infinite traces. In order to avoid arguing about infinite traces, stepwise verification is used. Stepwise verification breaks the verification task into a number of independent steps. This is done by focusing on the parts of the design that determine the traces, namely the initialization and the transitions. Informally, the stepwise verification of refinement consists of

1. Proving that the abstract and the concrete designs have corresponding initializations.
2. Proving that each state change caused by the concrete design can be explained in terms of the abstract design.

Stepwise verification is used in for example [1, 5, 7, 11], however, due to differences in the view of the interface, their proof obligation for each concrete state transition, differ from the proof obligation given below.

Stepwise verification differ from the approach in [6] and the work on temporal abstraction in [9] in the following way: In their work, the concrete state transitions are not considered one at a time; instead sequences of concrete state transitions corresponding to one abstract state transition are considered.

5.3   Exploiting Invariants in Stepwise Verification

Considering transitions one at a time, it is not possible to utilize information about the history of previously executed transitions. This implies that when a transition is considered, the state in which the transition is executed, is only identified by the precondition being true in that state. This means that unreachable states are also considered. Consequently, a verification could fail because the property that is attempted verified could be violated from an unreachable state.

Unreachable states can be excluded from consideration by establishing an invariant that rules out these states. States are only considered in the verification if they fulfill this invariant. Below, it is assumed that given a cell and its environment, it has been verified that neither the cell nor the environment violates the invariant of the cell. In [10] it is described how this is verified.

Example: Multiplier (continued). In the concrete multiplier the following invariant holds:

$$comp \Rightarrow (\neg rdy \wedge s + xl * y = x * y)$$

Consider for example the execution of the second transition, t_2, of the concrete multiplier:

$$\ll comp \wedge xl > 0 \rightarrow xl, s := xl - 1, s + y \gg$$

Looking at t_2 in isolation, it is impossible to tell the value of rdy in states where t_2 can be executed. Taking the invariant into account, it is easy to conclude that rdy is false in states where t_2 can be executed. Below, invariants are used to provide this kind of information.
End of example

5.4 Refinement Mappings

The verification technique ensures that the changes of the observable part of the interface caused by concrete transitions can also be caused by the abstract transitions. When a transition makes changes to the observable part of the interface, the changes might be based on local variables. Consider for example the abstract transition, t_a, $\ll i := la \gg$ and the concrete transition, t_c, $\ll i := lc \gg$. Assume that i is an observable interface variable, and la and lc are local variables. It is not possible to conclude that the execution of t_c corresponds to the execution of t_a unless it is known that la and lc have corresponding values. This implies that a mechanism is needed for relating the variables of the two designs. For this purpose *refinement mappings* [1, 7] are used.

A refinement mapping, R, is a mapping from the state space of the concrete design to the state space of the abstract design.

$$R : S_C \mapsto S_A$$

Refinement mappings as described in [1, 7] are the identity on interface variables. This requirement can be relaxed by using interface protocols: *It is only required that R is the identity on an interface variable, x, in states where $V(x)$ holds.*

The relaxed requirement to R is used to hide changes of unobservable interface variables. For each interface variable, v, with $V(v) \neq$ TRUE a verification variable, v' is introduced. v' is used to hold the last observable value of v while v is unobservable. The part of the refinement mapping concerning v is constructed in the following way.

$$v.R(\sigma) = \text{IF } cond.\sigma \text{ THEN } v'.\sigma \text{ ELSE } v.\sigma$$

$cond$ is a boolean valued expression, and is expressed in terms of local variables of C. The refinement mapping states that the abstract value of v is interpreted as the value of v' if $cond$ is true and as the value of the concrete v if $cond$ is false. For this to be sound it is required that when the interface protocol for v is true,

R is the identity on v. As explained in 5.3 the requirements to R can be further relaxed by considering only states where the invariant holds. This results in the following requirement to R.

RC $\forall v \in \mathcal{E}_C, \forall \sigma \in \mathcal{S}_C : Inv(C).\sigma \wedge V(v).\sigma \Rightarrow v.\sigma = v.R(\sigma)$

This property is exploited in the verification technique stated below.

Example: Multiplier (continued). The refinement mapping for the multiplier is:

$$
\begin{aligned}
x.R(\sigma) &= x.\sigma \\
y.R(\sigma) &= y.\sigma \\
s.R(\sigma) &= \text{IF } comp.\sigma \text{ THEN } s'.\sigma \text{ ELSE } s.\sigma \\
rdy.R(\sigma) &= rdy.\sigma
\end{aligned}
$$

When the computation of s starts, $comp$ gets true and s' gets the value of s. This causes $s.R(\sigma)$ to remain unchanged. This means that the refinement mapping hides the changes of s until the computation of s has finished. By then, $comp$ gets false, and the value of $s.\sigma$ $(= x.\sigma * y.\sigma)$ is mapped to $s.R(\sigma)$.

For x, y, and rdy, **RC** holds trivially. In the presence of the interface protocol $V(s) = rdy$ and the invariant $comp \Rightarrow \neg rdy$, the proof obligation for s is

$$((comp.\sigma \Rightarrow \neg rdy.\sigma) \wedge rdy.\sigma) \Rightarrow s.\sigma = s.R(\sigma)$$

This is easily proved.
End of example

5.5   Verifying Refinement

This section presents a verification technique that ensures the notion of refinement in definition 1. It is important to note that the verification technique is intended for mechanized verification.

The part of the verification technique concerning the initialization is called **INIT**

INIT $\forall \sigma \in \mathcal{S}_C : Init(C).\sigma \Rightarrow (Init(A).R(\sigma) \wedge \forall v \in \mathcal{E}_C : v.\sigma = v.R(\sigma))$

This ensures that all initial states of the concrete design, when mapped to the state space of the abstract design, fulfill the initialization predicate of the abstract design, and that initially nothing is hidden by the refinement mapping.

The part of the verification technique concerning the transitions is called **STEP**

$$
\begin{aligned}
&\forall t_C \in Tr(C), \sigma, \sigma' \in \mathcal{S}_C : \\
\textbf{STEP} \quad &Inv(C).\sigma \wedge \sigma \rightarrow_{t_C} \sigma' \Rightarrow \\
&R(\sigma) = R(\sigma') \vee \exists t_A \in Tr(A) : R(\sigma) \rightarrow_{t_A} R(\sigma')
\end{aligned}
$$

This ensures that any execution of a concrete transition can be explained either as the execution of an abstract transition ($\exists t_A \ldots$) or as though no change has taken place ($R(\sigma) = R(\sigma')$). **INIT** and **STEP** can be combined to verify that two cells are in accordance with definition 1 using the following verification technique:

Verification technique: *Let A and C denote cells with the same interface protocol, V, and E any environment, such that E respects V. Verifying* **INIT** *and* **STEP** *ensures that C refines A in E.*

As mentioned in 5.3, it is assumed that both C and E maintain the invariant of C. Consequently this assumption does not appear explicitly in the verification technique.

A soundness proof for the verification technique has been carried out, however, it is not given here.

At first glance, the verification technique does not seem only to be concerned with the observable part of the interface. This concern, however, is taken care of by the refinement mapping. When a change of a non observable interface variable is made, the refinement mapping can hide the change. This is illustrated by the following example.

Example: Multiplier (continued). For the multiplier, the verification technique identifies four proof obligations; one for the initialization, and one for each concrete transition.

No initialization is present in the abstract multiplier. This ensures that $Init(C).\sigma \Rightarrow Init(A).R(\sigma)$ in any initial state, σ. For x, y and rdy, the refinement mapping is the identity. In the concrete design $comp$ is initially false. This ensures that $s.\sigma = s.R(\sigma)$ in all initial states.

When the first transition, t_1, is executed, s is changed. Since $comp$ gets true and s' gets the previous value of s, the change of s is hidden by the refinement mapping from the previous example. Consequently, the execution of t_1 is justified by the clause $R(\sigma) = R(\sigma')$ in **STEP**.

Each time the second transition, t_2, is executed, the change of s is still hidden by the refinement mapping, since $comp$ and s' are unchanged. This means that each execution of t_2 is justified by the clause $R(\sigma) = R(\sigma')$ in **STEP**.

When the third transition, t_3, is executed, $comp$ gets false. This means that the refinement mapping no longer hides the change of s. The invariant (from the example in 5.3) ensures that when $xl = 0$, s equals $x * y$. Consequently, the execution of t_3 corresponds to the execution of the transition in the abstract design.
End of example

5.6 Respecting the Interface Protocol

In section 3.2, respecting the interface protocol is defined. In this definition the environment is represented by the set of state transitions that it can make. Respecting the interface protocol is stated in terms of this representation. Given the SYNCHRONIZED TRANSITIONS description, a condition is stated below to ensure that the interface protocol is respected.

Considering the execution of a transition, $t, \sigma \rightarrow_t \sigma'$, condition **NC** can be applied directly to σ and σ'. However, condition **ND** must be rephrased.

A transition, t is dependent on a variable y in σ if the value of y influences any expression of t, i.e.

$$dep(t, y, \sigma) = \exists v \in val(y) : exprs(t).\sigma \neq exprs(t).\sigma[v/y]$$

Consequently, the resulting proof obligation is

$$\forall t_E \in Tr(E), \sigma, \sigma' \in S_E : \sigma \rightarrow_{t_B} \sigma' \Rightarrow$$
$$\forall v \in \mathcal{E}_E : (dep(t_E, v, \sigma) \vee V(v).\sigma') \Rightarrow V(v).\sigma \ \wedge \ \neg V(v).\sigma' \Rightarrow v.\sigma = v.\sigma'$$

Note that respecting the interface protocol is preserved during refinement. This means that if the environment is refined, it still respects the interface protocol. Note also that the verification technique for respecting the interface protocol is independent of the refinement mapping, R. This means that if the concrete design is further refined, probably with a new refinement mapping, R'. If R' does not violate **RC**, the environment still respects the interface protocol. Consequently, it is only necessary to verify that the environment respects the interface protocol once, namely at the most abstract level.

A Modular Approach. In SYNCHRONIZED TRANSITIONS, the designs can be described in a modular way using cells. The modularity can be exploited when proving that the environment maintains the interface protocol. The condition for ensuring that the environment respects an interface protocol quantifies over all transitions. When the environment is extensive, and when several interface protocols are present, it can be laborious to verify that all transitions in the environment respects each of the interface protocols. In [10], an approach is described for verifying certain safety properties. The approach exploits the modularity of the designs to reduce the number of proof obligations. The approach can be generalized to verify that the environment respects an interface protocol.

5.7 Preservation of Safety Properties

In the preceding sections it has been explained how refinement using interface protocols can be used to justify replacing one abstract cell by another concrete cell in an environment. The correctness of this follows from the soundness proof mentioned in section 5.5. Though the correctness focuses on safety properties, the preservation of these is further illustrated below.

Assume that the abstract cell maintains a safety property that is vital for the environment. For example that an arbiter does not grant a privilege to two clients at the same time. If the notion of refinement is trace inclusion on the externally visible behaviors this safety property is also maintained by the concrete design.

If the notion of refinement includes the use of interface protocols as described previously, the concrete cell might cause externally visible behaviors that cannot be caused by the abstract design. However, the refinement using interface protocols implies trace inclusion for the observable part of the interface. For environments that respects the interface protocol, it follows from the proof mentioned in 5.5 that trace inclusion for the observable part of the interface is sufficient. Below, this is illustrated by the multiplier example.

If the multiplier is always provided with non-zero arguments, the safety property $s \neq 0$ holds for the abstract multiplier. For the concrete multiplier, this is not the case. Consider a transition, t, that placed in the environment of the multiplier can detect the difference between the abstract and the concrete multiplier.

$$\ll s = 0 \rightarrow error := \text{TRUE} \gg$$

If t is found in the environment of the abstract multiplier, *error* will never be true, while in the environment of the concrete multiplier, *error* might become true. However, t *does not respect the interface protocol*, so the violation of the safety property does not indicate a lack of soundness in the use of interface protocols.

This illustrates that if a safety property is not preserved, it is too strong for an environment that relies on the safety property to respect the interface protocol. This means that all safety properties that are relevant for the environment are preserved.

5.8 Mechanization

Using the verification technique introduced above, it can be proved that one design refines another. A prototype tool is developed that turns SYNCHRONIZED TRANSITIONS descriptions into an axiomatization in the logic of the theorem prover LP, The Larch Prover [3]. The tool is based on a translator written for proving invariants for SYNCHRONIZED TRANSITIONS [10]. Each of the constituent parts of the designs (i.e. transitions, invariants, initializations, variable declarations, etc.) are translated into corresponding LP constructs. The tool also generates proof obligations for proving refinement. The multiplier used as example in this section and a number of examples from [11] have been verified using the tool and the theorem prover.

The verification technique stated above turns out to be suitable for a theorem prover. The proof is by cases on the concrete transitions. This means that the proof consists of several limited subproofs. Each of these subproofs assumes that one concrete transition has been executed. At any stage of the proof, it is therefore straightforward to relate the current stage of the proof to a specific transition in the concrete design. Consequently, it is easy to interact if manual assistance is needed. Furthermore, changes in a few transitions allow for most of the proof to be reused.

6 Example - The Tamarack Microprocessor

The verification of an implementation of a simple microprocessor is described in [6]. This verification has been redone using the translator and theorem prover described in the previous section.

Below, a part of the verification is described. Partly to illustrate the applicability of the verification technique of the previous section, partly to illustrate the differences between the present approach and the approach in [6].

6.1 Abstract Description

The CPU executes instructions located in a memory (*mem*). A program counter (*pc*) points out the next instruction to be executed. The instructions include addition, subtraction, (conditional) jumping, and loading and storing data in the memory. Below, focus is on the execution of a store instruction.

When *pc* points out a store instruction, the store instruction identifies the location to be changed. The value to be stored in this location is contained in the register *acc*. Let *ST* denote a function that given *pc*, *acc* and *mem* as arguments returns the updated *mem* where the location pointed out by the store instruction contains the value of *acc* and the rest of the locations are unchanged. While *mem* is updated, *pc* is incremented to point out the next instruction to be executed. This means that the abstract description of the store instruction is the following:

$$\ll mem, pc := ST(pc, acc, mem), pc + 1 \gg$$

6.2 Concrete Description

The concrete design describes a microcode implementation of the CPU. In the microcode implementation, it is not possible to update the memory and to increment the program counter in the same state transition. Consequently, *mem* is updated first, and afterwards *pc* is incremented. In order to emphasize on concerns relevant to the use of interface protocols, the part of the concrete design, concerning the store instruction, is less detailed than the original description in [6].

$$\ll mpc = 17 \rightarrow mem, mpc := ST(pc, acc, mem), 18 \gg$$
$$\ll mpc = 18 \rightarrow pc, mpc := pc + 1, 5 \gg$$

mpc denotes the microcode program counter. It ensures the correct order of the update of *mem* and *pc*. The net result of first updating *mem* and afterwards *pc* is the same as executing the abstract transition shown above.

6.3 The Problem

The crucial difference between the abstract and the concrete descriptions is that the changes of *mem* and *pc* in the concrete design do not happen in the same state transition. *mem* and *pc* are both interface variables. This means that the environment might observe different values on the interface of the concrete and the abstract design.

The Solution Using Interface Protocols. In the present approach this is handled by including a flag, *ready*, in the interface of both descriptions. *ready* is set false when *mem* is updated and set true when *pc* is incremented. An interface protocol

$$V(mem) = ready$$

is introduced. This prevents the environment from reading *mem* before *pc* is incremented.

A refinement mapping, R, is constructed. It hides the change of *mem* until *pc* is incremented. When *ready* gets false, the value of *mem* is assigned to a variable, mem', and *mpc* is incremented. Using the refinement mapping described below, this makes $mem.R(\sigma)$ unchanged. When *pc* is incremented, *mpc* is set to 5 and *ready* gets true. At the abstract level it appears as though *mem* and *pc* are changed in the same state transition.

$$
\begin{aligned}
acc.R(\sigma) &= acc.\sigma \\
ready.R(\sigma) &= ready.\sigma \\
pc.R(\sigma) &= pc.\sigma \\
mem.R(\sigma) &= \text{IF } mpc.\sigma = 18 \text{ THEN } mem'.\sigma \text{ ELSE } mem.\sigma
\end{aligned}
$$

Given R and $V(mem) = ready$, the problem degenerates to proving refinement between the following cells:

CELL $A(pc, acc : \text{INTEGER} ; mem : \text{MEMORY}; ready : \text{BOOLEAN})$
INTERFACE PROTOCOL $V(mem) = ready$
BEGIN
 $\ll mem, pc, ready := \text{ST}(pc, acc, mem), pc + 1, \text{TRUE} \gg$
 $\ll ready := \text{FALSE} \gg$
END A

CELL $C(pc, acc : \text{INTEGER} ; mem : \text{MEMORY}; ready : \text{BOOLEAN})$
INTERFACE PROTOCOL $V(mem) = ready$
STATE
 mem' : MEMORY
 mpc : INTEGER
INVARIANT
 $(mpc = 17 \Rightarrow ready)\wedge$
 $(mpc = 18 \Rightarrow (mem = \text{ST}(pc, acc, mem') \wedge \neg ready))$
BEGIN
 $\ll mpc = 17 \rightarrow$
  $mem, mpc, ready, mem' := \text{ST}(pc, acc, mem), 18, \text{FALSE} , mem \gg$
 $\ll mpc = 18 \rightarrow pc, ready, mpc := pc + 1, \text{TRUE} , 5 \gg$
END C

The proof is carried out using the tools described in the previous section.

Note that the variable mem' is a *verification* variable; after the verification has been completed it can be removed since it has no influence on the design (it is never read).

Solution in [6]. In contrast to the asynchronous descriptions above, the descriptions in [6] are synchronous. For both the abstract and the concrete descriptions, the values of the variables are described as functions of time. In the abstract description, the execution of one instruction takes one time unit, whereas in the concrete design several time units are necessary to complete one instruction. This means that the abstract time scale is more coarse grained than the concrete. This is dealt with using temporal abstraction, which is described in more detail in [9].

The concrete design is augmented with a boolean variable, *ready*. The notion of refinement ensures that at points of concrete time where *ready* is true, the concrete design is in a state that could also be caused by the abstract design. The concrete CPU cannot be used directly instead of the abstract CPU. To be able to replace the abstract CPU with the concrete in an environment, it is necessary to ensure that the environment only depends on the interface variables of the CPU when *ready* is true. This issue of composition is not explicitly dealt with in [6].

7   Conclusion

This paper has reported an attempt to generalize the notion of trace inclusion to a more liberal notion of refinement by taking the environment into account.

A verification technique has been stated for proving refinement for designs written in SYNCHRONIZED TRANSITIONS. A prototype tool has been developed for translating SYNCHRONIZED TRANSITIONS descriptions into the logic of an existing theorem prover, and for generating proof obligations for refinement.

Experience with a number of examples has shown the usefulness of both the liberal notion of refinement and the tools for mechanizing the verification.

Acknowledgements

Thanks to J.Frost, H.H.Løvengreen, A.P.Ravn and J.Staunstrup for comments.

References

1. Martin Abadi and Leslie Lamport. The existence of refinement mappings. Technical Report 29, Digital Systems Research Center, 1988.
2. Randal E. Bryant. Can a simulator verify a circuit? In *Formal Aspects of VLSI Design*, pages 125–136. North-Holland, 1985.
3. Stephen J. Garland and John V. Guttag. A guide to LP, the Larch Prover. Technical Report 82, Digital Systems Research Center, 1991.
4. Mike Gordon. Why higher-order logic is a good formalism for specifying and verifying hardware. In *Formal Aspects of VLSI Design*, pages 153–177. North-Holland, 1985.
5. Bengt Jonsson. On decomposing and refining specifications of distributed systems. In *Lecture Notes in Computer Science, 430*, pages 361–385. Springer Verlag, 1990.

6. Jeffrey J. Joyce. Formal verification and implementation of a microprocessor. In Graham Birtwistle and P.A. Subrahmanyam, editors, *VLSI Specification and Synthesis*, pages 129–157. Kluwer Academic Publishers, 1988.
7. Leslie Lamport. The temporal logic of actions. Technical Report 79, Digital Systems Research Center, 1991.
8. Nancy A. Lynch and Mark R. Tuttle. Hierarchical correctness proofs for distributed algorithms. In *Proceedings of the Sixth Annual ACM Symposium on Principles of Distributed Computing*, pages 137–151. ACM, 1987.
9. Thomas F. Melham. Abstraction mechanisms for hardware verification. In Graham Birtwistle and P.A. Subrahmanyam, editors, *VLSI Specification and Synthesis*, pages 267–291. Kluwer Academic Publishers, 1988.
10. Niels Mellergaard. *Mechanized Design Verification*. PhD thesis, Department of Computer Science, Technical University of Denmark, 1994.
11. Jørgen Staunstrup. *A Formal Approach to Hardware Design*. Kluwer Academic Publishers, 1994.

Effective Theorem Proving for Hardware Verification*

D. Cyrluk,[1] S. Rajan,[2] N. Shankar,[3] and M.K. Srivas[3]

{cyrluk, sree, shankar, srivas}@csl.sri.com

[1] Dept. of Computer Science, Stanford University, Stanford CA 94305 and
Computer Science Laboratory, SRI International, Menlo Park, CA 94025
[2] Integrated Systems Design Laboratory, Department of Computer Science,
University of British Columbia, Vancouver, Canada and
Computer Science Laboratory, SRI International, Menlo Park, CA 94025
[3] Computer Science Laboratory, SRI International, Menlo Park CA 94025 USA

Abstract. The attractiveness of using theorem provers for system design verification lies in their generality. The major practical challenge confronting theorem proving technology is in combining this generality with an acceptable degree of automation. We describe an approach for enhancing the effectiveness of theorem provers for hardware verification through the use of efficient automatic procedures for rewriting, arithmetic and equality reasoning, and an off-the-shelf BDD-based propositional simplifier. These automatic procedures can be combined into general-purpose proof strategies that can efficiently automate a number of proofs including those of hardware correctness. The inference procedures and proof strategies have been implemented in the PVS verification system. They are applied to several examples including an N-bit adder, the Saxe pipelined processor, and the benchmark Tamarack microprocessor design. These examples illustrate the basic design philosophy underlying PVS where powerful and efficient low-level inferences are employed within high-level user-defined proof strategies. This approach is contrasted with approaches based on tactics or batch-oriented theorem proving.

1 Introduction

The past decade has seen tremendous progress in the application of formal methods for hardware design and verification. Much of the early work was on applying proof checking and theorem proving tools to the modeling and verification of hardware designs [14, 16]. Though these approaches were quite general, the

* This work was supported in part by the following funding sources: NASA Langley Research Center contract NAS1-18969, ARPA Contract NAG2-891 administered by NASA Ames Research Center, NSF Grant CCR-930044, the Semiconductor Research Corporation contract 92-DJ-295 (to the University of British Columbia), and the Philips Research Laboratories, Eindhoven, The Netherlands.

verification process required a significant human input. More recently, there has been a large body of work devoted to the use of model-checking, language-containment, and reachability analysis to finite-state machine models of hardware [8]. The latter class of systems work automatically but they do not yet scale up efficiently to realistic hardware designs. The challenge then is to combine the generality of theorem proving with an acceptable and efficient level of automation.

Our main thesis is that in order to achieve a balance between generality, automation, and efficiency, a verification system must provide powerful and efficient primitive inference procedures that can be combined by means of user-defined, general-purpose, high-level proof strategies. The efficiency of the inference procedures is important in order to verify complex designs with a greater level of automation. To achieve efficiency, each individual primitive inference procedure must itself perform a powerful and well-defined inference step using well-chosen algorithms and data structures. A number of related deductive operations must be tightly integrated into such a step. The efficiency resulting from a powerful and tightly integrated inference procedure cannot typically be obtained by composing very low-level inference steps by means of tactics. On the other hand, a fully automatic batch-oriented theorem prover has the drawback of being a toolkit with only a single tool. Doing exploratory proof development with such a theorem prover is tedious because of the low bandwidth of interaction. It is difficult to reconcile efficiency with generality in a fully automated theorem prover since a single proof strategy is being applied to all theorems.

The above design philosophy has formed the guiding principle for the implementation of the Prototype Verification System (PVS) [22, 23] developed at SRI. PVS is designed to automate the tedious and obvious low-level inferences while allowing the user to control the proof construction at a meaningful level. Exploratory proofs are usually carried out at a level close to the primitive inference steps, but greater automation can be achieved by defining high-level proof strategies.

In this paper, we present some automatic inference procedures used in the PVS proof checker, show how these inference procedures can be combined into general-purpose proof strategies, and examine the impact of these strategies on the automation of hardware proofs. The primitive inference procedures in PVS include arithmetic and equality decision procedures, an efficient hashing-based conditional rewriter, and a propositional simplifier based on binary decision diagrams (BDDs). The interaction between rewriting, arithmetic, and BDD-based propositional simplification yields a powerful basis for automation. The capabilities of the inference procedures are available to the user of the proof checker as primitive inference steps. These primitive inference steps can either be used as part of an interactive proof attempt or embedded inside a high-level proof strategy. We have developed a basic proof strategy in terms of these inference steps that is particularly effective for automating proofs of microprocessors and inductively defined hardware circuits. The core of this strategy consists of first carrying out a symbolic execution of the hardware and its specification by expanding and simplifying the relevant definitions; the case structure of the symbolic execution is then brought to the surface, and BDD-based propositional simplification is used to generate subgoals that are typically proved by means of the decision procedures. We present the proof strategy and demonstrate its utility on a number of examples including an N-bit ripple carry adder circuit, Saxe's pipelined microprocessor [24] and the Tamarack processor [18]. The point of these examples is to illustrate efficiency and generality that can be derived from the inference capabilities present in PVS. This work is still at a preliminary stage and we feel that there is plenty of scope for obtaining even greater generality, efficiency, and automation by pursuing the line of development indicated in this paper.

The next section gives a brief overview of PVS. In section 3, we describe a general-purpose strategy for hardware proofs in PVS and illustrate this strategy with an N-bit adder circuit. Section 4 describes the use of the PVS inference procedures in the development of verification strategies for microprocessor designs. We present our conclusions in the last section.

2   An Overview of PVS

PVS is an environment for writing specifications and developing proofs [23]. It serves as a prototype for exploring new approaches to mechanized formal methods. The primary goal of PVS is to combine an expressive specification language with a productive, interactive proof checker that has a reasonable amount of theorem proving power. PVS has been strongly influenced by the observation that theorem proving capabilities can be employed to enrich the type system of a typed logic, and conversely, that an enriched type system facilitates expressive specifications and effective theorem proving. PVS has also been guided by the experience that much of the time and effort in verification is in debugging the initial specification or proof idea. A high bandwidth of interaction is useful at the exploratory level whereas more automated high-level proof strategies are desirable at an advanced stage of proof development. PVS has been used to verify several complex fault-tolerant algorithms, real-time and distributed protocols, and several other applications [20].

2.1   The Specification Language

The PVS specification language builds on a classical typed higher-order logic. The base types consist of booleans, real numbers, rationals, integers, natural numbers, lists, and so forth. The primitive type constructors include those for forming function (e.g., [nat -> nat]), record (e.g., [# a : nat, b : list[nat]#]) , and tuple types (e.g., [int, list[nat]]). The type system of PVS includes *predicate subtypes* that consist of exactly those elements of a given type satisfying a given predicate. PVS contains a further useful enrichment to the type system in the form of *dependent* function, record, and tuple constructions where the type of one component of a compound value depends on the value of another component. PVS terms include constants, variables, abstractions (e.g., (LAMBDA (i : nat): i * i)), applications (e.g., mod(i, 5)), record constructions (e.g., (# a := 2, b := cons(1, null) #)), tuple constructions (e.g., (-5, cons(1, null))), function updates (e.g., f WITH [(2) := 7]), and record updates (e.g., r WITH [a := 5, b := cons(3, b(r))]). Note that the application a(r) is used to access the a field of record r, and the application PROJ_2(t) is used to access the second component of a tuple t. PVS specifications are packaged as *theories*.

2.2 The Proof Checker

The PVS proof checker is intended to serve as a productive medium for debugging specifications and constructing readable proofs. The human verifier constructs proofs in PVS by repeatedly simplifying a conjecture into subgoals using inference rules, until no further subgoals remain. A proof goal in PVS is represented by a sequent. PVS differs from most proof checkers in providing primitive inference rules that are quite powerful, including decision procedures for ground linear arithmetic. The primitive rules also perform steps such as quantifier instantiation, rewriting, beta-reduction, and boolean simplification. PVS has a simple strategy language for combining inference steps into more complicated proof strategies. In interactive use, when prompted with a subgoal, the user types in a proof command that either invokes a primitive inference rule or a compound proof strategy. For example, the **skolem** command introduces Skolem constants for universal-strength quantifiers while the **inst** command instantiates an existential-strength quantifier with its witness. The **lift-if** command invokes a primitive inference step that moves a block of conditionals nested within one or more sequent formulas to the top level of the formula. The **prop** command invokes a compound propositional simplification strategy (or tactic) that is a less efficient alternative to the use of a BDD-based simplifier described below. Various other commands are discussed below. Proofs and partial proofs can be saved, edited, and rerun. It is possible to extend and modify specifications during a proof; the finished proof has to be rerun to ensure that such changes are benign.

While a number of other theorem provers do use decision procedures, PVS is distinguishing in the aggressiveness with which it uses them. It is also unique in the manner in which the decision procedures and automatic rewriting are engineered to interact with each other in implementing the primitive inference commands of PVS. The user can invoke the functionality provided by the interacting decision procedures in its full power in a single command (**assert**) or in limited forms by means of a number of smaller commands. Some of these primitive commands are described in the following sections.

2.3 The Ground Decision Procedures

The ground decision procedures of PVS are used to simplify quantifier-free Boolean combinations of formulas involving arithmetic and equality, and to propagate type information. PVS makes extremely heavy use of these decision procedures.

Consider a formula of the form f(x) = f(f(x)) IMPLIES f(f(f(x))) = f(x), where the variable x is implicitly universally quantified. This is really a *ground* (i.e., variable-free) formula since the universally quantified variable x can be replaced by a newly chosen (Skolem) constant, say c. We can then negate this formula and express this negation as the conjunction of literals: f(c) = f(f(c)) AND NOT f(f(f(c))) = f(c). We can then prove the original formula by refuting its negation. To refute the negation we can assert the information in each literal into a data structure until a contradiction is found. In this case, we can use a *congruence closure* data structure to rapidly propagate equality information.

Congruence closure [12] plays a central role in several other systems including the Stanford Pascal Verifier [21] and Ehdm [9]. This basic procedure can be extended in several ways. One basic extension is to the case of ground linear inequalities over the real numbers. In the example, a < 2*b AND b < 3*c AND NOT 3*a < 18*c, the refutation can be obtained by eliminating the variable b in the second inequality in favor of a. Another extension is to the case of ground arrays or functions. This is important in hardware examples where memory can be represented as a function from addresses to data. For example, the decision procedure can deduce (func WITH [(j) := val])(i) to be equal to func(i) under the assumption i < j. The PVS decision procedures combine congruence closure over interpreted and uninterpreted functions and relations with refutation procedures for ground linear inequalities over the real numbers and arrays [26]. This procedure is also extended to integer inequalities in an incomplete though effective manner. The ground decision procedures can, for example, refute i > 1 AND 2*i < 5 AND NOT i = 2.

2.4 The Simplifier

The congruence closure data structure is used to maintain and update contextual information. Any relevant subtype constraints on terms are also recorded in these data structures. The *beta-reduction* of lambda-redexes, and datatype tuple, record and update access are among the automatic simplifications supported by PVS. We do not describe the arithmetic simplifications except to say that they evaluate expressions where the arithmetic operations (+, -, *, and /) are applied to numerical values and reduce any arithmetic expressions to a sum-of-products form. The Boolean simplifications are similarly straightforward and they simplify expressions involving the constants TRUE and FALSE and the operators NOT, OR, AND, IMPLIES, and IFF. In the simplification of conditional expressions, the *test* part of the conditional is used in the simplification of the *then* and *else* parts. These simplifications are shown below where 'l simplifies to r' is shown as '$l \Longrightarrow r$':

1. `(IF A THEN s ELSE t ENDIF)` $\Longrightarrow$ `s`, if `A` $\Longrightarrow$ `TRUE`
2. `(IF A THEN s ELSE t ENDIF)` $\Longrightarrow$ `t`, if `A` $\Longrightarrow$ `FALSE`
3. `(IF A THEN s ELSE s ENDIF)` $\Longrightarrow$ `s`
4. `(IF A THEN s ELSE t ENDIF)` $\Longrightarrow$ `(IF A' THEN s' ELSE t' ENDIF)`, if `A` $\Longrightarrow$ `A'`, `s` $\Longrightarrow$ `s'` assuming `A'`, and `t` $\Longrightarrow$ `t'` assuming `¬A'`

When the **record** command is invoked on a PVS sequent of the form $A_1, \ldots, A_m \vdash B_1, \ldots, B_n$, the simplified form of each atomic A_i (or $\neg B_i$) is recorded in the congruence closure data structures. This information is then used to simplify the remaining formulas in the sequent. The **simplify** command simplifies the formulas using the ground decision procedures and simplifier without recording any new information into the data structures.

2.5 The PVS Rewriter

A (conditional) rewrite rule is a formula of either the form $A \supset p(b_1, \ldots, b_n)$ or $A \supset l = r$. The former case can be reduced to the latter form as $A \supset p(b_1, \ldots, b_n) = \mathtt{TRUE}$. In the latter case, the PVS rewriter then simplifies an instance $\sigma(l)$ of l to $\sigma(r)$ provided the hypothesis instance $\sigma(A)$ simplifies (using simplification with decision procedures and rewriting) to **TRUE** as must any type correctness conditions (TCCs) generated by the substitution σ. The free variables in A and r must be a subset of those in l. The hypothesis can be empty and definitions can also be used as rewrite rules.

There is also a restriction of rewriting where, if the right-hand side $\sigma(r)$ of a rewrite is an **IF-THEN-ELSE** expression, then the rewrite is not applied unless the test part of the conditional simplifies to **TRUE** or **FALSE**. This restricted form of rewriting serves to prevent looping when recursive definitions are used as rewrite rules and to control the size of the resulting expression. The above heuristic restriction on rewriting relies on the effectiveness of the simplifications given by the decision procedures. This heuristic is quite important in the context of processor proofs where the next state of the processor should be computed as long as there is an explicit clock tick available.

For efficiency, PVS maintains a hash-table where corresponding to a term a, the result of the most recent rewriting of a is kept along with the logical *context* at the time of the rewrite. A context consists of the congruence closure data structures and the current set of rewrite rules stored internally at the time of rewrite. This way, if the term a is encountered within the same logical context, the result of the rewrite is taken from this hash-table and the rewriting steps are not repeated. The information that an expression could not be rewritten in a context is also cached. This information is perhaps the more heavily used than the information about successful rewriting.

The context is modified by the proof tree structure and the IF-THEN-ELSE structure of an expression. In case the term a is encountered in a strictly larger logical context (facts have been added to the congruence closure data structures or the set of rewrite rules has been expanded) then the result of the rewrite is taken from the hash-table and further rewritten using the current larger logical context.

The rewriter described above is used automatically in simplification. It can be invoked by the do-rewrite command. The assert command combines the functionality of record, simplify, and do-rewrite.

2.6   The Power of Interaction

The following example illustrates the power that a close interaction between rewriting and the decision procedures can provide for the user in PVS. Such a close interaction is not as easily accomplished if the decision procedures and rewriting were implemented as separate tactics or strategies.

```
t: nat
s: VAR state
MAR_t0: AXIOM t /= 0  => dest(IR(s)) = MAR(s)

MDR5(s): data =
    IF t <= 2
        THEN IF p(t)
            THEN MDR(s)
            ELSE rf(s) WITH [(MAR(s)) := MDR(s)](dest(IR(s)))
            ENDIF
        ELSE somedata
    ENDIF

property: THEOREM t < 3 & p(0) => MDR(s) = MDR5(s)
```

In the PVS specification shown above, the goal is to prove property from the axioms MAR_t0 and the definition of MDR5, where the constants p, MDR, MAR, etc., are declared elsewhere. The constant rf is a function that maps addresses to data. Assuming that the definition of MDR5 and MAR_t0 have been entered as rewrite rules (via the command auto-rewrite), property can be proved (after skolemizing the variable s and flattening the implication) by simply using the command assert on the resulting sequent twice. The first invocation of assert is able to rewrite MDR5(s) to the IF-THEN-ELSE expression beginning at p(t) since, when t is a nat, t <= 2 can be deduced from t < 3 by the decision procedure.

A second invocation of assert attempts to rewrite each branch of the resulting IF-THEN-ELSE expression. The p(t) case is trivially true. In the NOT p(t) case, the decision procedures deduce that t = 0 is false from p(0). This triggers the rewrite rule MAR_t0 so that the goal becomes MDR(s) = rf(s)

`WITH [(MAR(s)) := MDR(s)](MAR(s))`, which the equality procedures simplify to true.

2.7 BDD Simplifier

Binary decision diagrams (BDDs) are widely used in the design, synthesis, and verification of digital logic. They provide an efficient representation for the simplification of propositional formulas. A BDD-based propositional simplifier was recently added to PVS as a primitive inference step. Prior to the introduction of this simplifier, PVS used a propositional simplification tactic called prop that was found to be unsatisfactory since it could generate subgoals that were just permutations of other subgoals.

The basic idea behind the use of BDD-based simplification in PVS is to transform the goal sequent into a Boolean expression where the atomic formulas have been replaced by propositional variables. This formula is given as input to an off-the-shelf BDD package by means of an external function call. Note that PVS is implemented in Common Lisp whereas the BDD package is a C program. The BDD package simplifies the given formula into an equivalent formula in conjunctive normal form that is easily translated into a collection of subgoal sequents by replacing the propositional variables back to the corresponding atomic formulas. It is also possible to provide the BDD simplifier with some contextual information using the *restriction* operator, also known as *cofactoring*, provided by the BDD package. The restriction operation is used to simplify one BDD representation assuming another containing the contextual assumptions.

The BDD simplifier we use is an efficient implementation from EUT [17]. The simplifier uses Reduced Ordered BDD (ROBDD), a canonical representation of boolean expressions, with an associated set of algorithms [5]. The BDD simplifier is invoked in a PVS proof with the command bddsimp.

3   The Nature of Hardware Proofs, and Our Thesis

We have described some of the built-in deductive capabilities of PVS. A PVS proof is constructed by interactively (or automatically) invoking these inference steps to simplify the given goal into simpler subgoals until all the subgoals are trivially true. At the highest level, the user directs the verification process by elaborating and modifying the specification, providing relevant lemmas, and backtracking on the fruitless paths in a proof attempt. At the next level of an interactive PVS proof, particularly a hardware proof, the user provides the following crucial inputs:

Quantifier elimination: Since the decision procedures work on ground formulas, the user must eliminate the relevant universal-strength quantifiers by introducing Skolem constants or suggesting induction schemes. Existential-strength quantifiers are eliminated by suitable instantiation.

Unfolding definitions: The user may have to simplify selected expressions and defined function symbols in the goal by rewriting using definitions, axioms or lemmas.

Case analysis: The user may have to split the proof based on selected boolean expressions in the current goal.

The use of decision procedures for arithmetic and equality yields a significant advantage in that the outcome of a proof attempt is not as logically sensitive to the decisions made in performing the second and third tasks as is the case in provers without decision procedures. However, decisions made during the second and the third tasks critically impact the efficiency of the proof. For example, the extent to which the defined function symbols are unfolded determines the number of cases to be considered during case analysis. Performing case analysis on selected boolean expressions before rewriting can make rewriting more productive and reduce the size of the resulting expressions, whereas a naive case analysis can lead to a needless combinatorial blowup in proof size.

In most of our experiments with hardware proofs, we found that we needed to intervene manually during rewriting and case analysis tasks only to control the complexity of the proof. Our experience suggested that the second and the third tasks could be completely automated for most hardware proofs given an efficient rewriting and propositional simplification engine used in conjunction with the arithmetic decision procedures. In the next section, we illustrate the above thesis on an N-bit ripple-carry adder example.

3.1 An N-bit Adder

The theory **adder** shown below describes the implementation and the correctness statement of the adder. The theory is parameterized with respect to the length of the bit-vectors. It imports the theory **full_adder** which contains a specification of a full adder circuit with output carry bit **fa_cout** and the sum bit **fa_sum**, and the theory **bv** which specifies the bit-vector type (**bvec[N]**) and related bit-vector functions. An N-bit bit-vector is represented as an array, i.e., a function from the type **below[N]** of natural numbers less than **N** to **bool**; the index 0 denotes the least significant bit. Note that the parameter **N** is constrained to be a **posnat** since we do not permit bit-vectors of length 0.

The carry bit that ripples through the full adders is specified recursively by means of the function **nth_cin**[4]. The function **bv_cout** and **bv_sum** define the carry output and the bit-vector sum of the adder, respectively. The theorem **adder_correct** expresses the conventional correctness statement of an adder circuit using **bvec2nat**, which returns the natural number equivalent of the least

---

[4] Recursive function definitions in PVS must have an associated **MEASURE** function to ensure termination. The typechecker automatically generates type correctness proof obligations to show that the measure of the argument to every recursive invocation the function is less than the measure of the original argument.

significant n-bits of a given bit-vector and `bool2bit` converts the boolean constants `TRUE` and `FALSE` into the natural numbers 1 and 0, respectively.

```
adder[N: posnat] : THEORY

BEGIN

  IMPORTING bv[N], full_adder

  n: VAR below[N]
  bv, bv1, bv2: VAR bvec

  nth_cin(n, cin, bv1, bv2): RECURSIVE bool =
      IF n = 0 THEN cin
      ELSE fa_cout(bv_cin(n - 1, cin, bv1, bv2),
                   bv1(n - 1),
                   bv2(n - 1))
      ENDIF
    MEASURE n

  bv_sum(cin, bv1, bv2)(n): bvec =
    fa_sum(bv1(n), bv2(n), nth_cin(n, cin, bv1, bv2))

  bv_cout(n, cin, bv1, bv2): bool =
    fa_cout(nth_cin(n, cin, bv1, bv2), bv1(n), bv2(n))

  full_adder_correct:
    LEMMA
      bool2bit(a) + bool2bit(b) + bool2bit(c)
      =   2 * bool2bit(fa_cout(c, a, b))
        + bool2bit(fa_sum(a, b, c))

  adder_correct: LEMMA (FORALL n:
      bvec2nat(n, bv1) + bvec2nat(n, bv2) + bool2bit(cin)
      =   exp2(n + 1) * bool2bit(bv_cout(n, cin, bv1, bv2))
        + bvec2nat(n, bv_sum(cin, bv1, bv2)))

END adder
```

The proof of `adder_correct` proceeds by induction on the variable n using an induction scheme for the type `below[N]`. This results in a base case that is easily proved by **assert** and an induction case that is displayed below as a sequent containing only one formula.

```
adder_correct2 :


  |-------
{1}   (FORALL (r: below[N]):
         r < N - 1
           AND (FORALL (bv1, bv2: bvec[N]), (cin: bool):
                     bvec2nat(r, bv1) + bvec2nat(r, bv2)
                        + bool2bit(cin)
                 = exp2(r + 1)
                     * bool2bit(bv_cout(r, cin, bv1, bv2))
                     +
                     bvec2nat_rec(r, bv_sum(cin, bv1, bv2)))
               IMPLIES (FORALL (bv1, bv2: bvec[N]), (cin: bool):
                     bvec2nat(r + 1, bv1)
                        + bvec2nat(r + 1, bv2)
                        + bool2bit(cin)
                 = exp2(r + 1 + 1)
                     * bool2bit(bv_cout(r + 1, cin, bv1, bv2))
                     +
                     bvec2nat(r + 1,
                                  bv_sum(cin, bv1, bv2)))))
```

The general strategy to prove the above goal, as in any inductive proof, is to first introduce skolem constants for the universal-strength variables in the goal and flatten the sequent into a form where the inductive hypothesis is in the antecedent. After that, one has to simplify the conclusion to a point where an instance of the induction hypothesis can be used to discharge the conclusion. The simplification of the conclusion can either be done under control or by brute-force automation. We contrast the two approaches for the adder example below. In both approaches, the first step (`skosimp*`) performs the repeated skolemization and flattening of the sequent required at the start of the proof.

Guided Proof	Automatic Proof
(skosimp*)	(skosimp*)
(expand "exp2" 1)	(auto-rewrite-explicit)
(expand "bvec2nat" 1)	(do-rewrite)
(expand "bv_sum" 1 1)	(inst?)
(expand "bv_cout")	(repeat (lift-if))
(expand "nth_cin" 1)	(simplify)
(lemma "full_adder_correct")	(then* (bddsimp)(assert))
(inst?)	
(inst?)	
(assert)	

In the guided proof, shown on the left, we carefully control the rewriting process by selecting a subset of the defined function symbols in the sequent to unfold in order to keep the size of the proof tree under control. The PVS command **expand** is used to expand function definitions in a controlled manner. The optional second and the third argument to **expand** respectively specify the formula and the occurrence of the symbol to be expanded.

At this point, a careful case analysis on the **bool2bit** values of the three most significant boolean bits under consideration would lead to eight subgoals. We construct a shorter proof by using the lemma **full_adder_correct** about the full adder to eliminate the bit-level case analysis required. The **inst?** command attempts to find a suitable set of instantiations for existential-strength quantifiers in the sequent formulas of a subgoal[5] which in this case include the lemma as well as the induction hypothesis. In this case, it manages to instantiate both the inductive hypothesis and the lemma to the desired substitutions. The lemma **full_adder_correct** cannot be successfully applied as a rewriting rule in this proof because the instances of **bool2bit** do not appear contiguously in the expression to be simplified. The last step in the proof, **assert**, invokes the arithmetic and equality decision procedures of PVS to complete the proof.

A More Automatic Proof

We now describe a more automatic proof of the same theorem, shown on the right side of the above table, that takes a brute-force approach by employing automated rewriting and BDD-based propositional simplification. This strategy is part of a general strategy for proofs involving induction and rewriting that has been used on several other examples. Using this strategy we were able to verify the adder in 130 seconds. Every defined function symbol used directly or indirectly in the sequent is set up as a rewrite rule by invoking **auto-rewrite-explicit**. This set of function symbols includes not only those appearing in the **adder** theory but also those in the theories **full_adder** and **bv** imported by **adder**. The automatic rewriter of PVS is then invoked by means of the **do-rewrite** command to rewrite all the expressions in the sequent using the rewrite rules introduced above. The rewriting process simplifies the conclusion into an equation on two nested conditional (**IF-THEN-ELSE**) expressions. Not surprisingly, the size of the expressions resulting from the rewriting is much larger here than in the intelligent proof.

To automate the case analysis, we repeatedly (using **repeat**[6]) lift all the **IF-THEN-ELSE** conditionals to the topmost level (using **lift-if**). The lifting

---

[5] It can be used either in a mode in which all possible instances of the lemma are produced or only a single instance is produced.

[6] **Repeat** and **then∗** are among the *tacticals* provided by PVS for constructing proof strategies from primitive inference steps and other predefined proof strategies. **Repeat** applies a given proof step repeatedly until its application has no change on the current goal; **then∗** applies the first goal from the given list of proof steps to the current goal, and the rest of the steps in the list to each of the subgoals, if any, resulting from the first application.

process transforms the conclusion into a propositional expression in the form of nested IF-THEN-ELSE expressions whose leaf nodes are equalities on unconditional expressions. The propositional expression is simplified using bddsimp. Rewriting and decision procedures (using assert) are applied to any subgoals generated by bddsimp.

The latter proof is automatic in the sense that it applies certain coarse-grain inference steps under a simple control strategy without requiring any specific information from the user to guide the proof. Three elements are crucial to making the automatic proof successful. Firstly, we need efficient rewriting that can rewrite large expressions while exploiting contextual information. Second, we need an efficient propositional simplifier to perform the automatic case analysis on very large formulas. The above automatic proof blows up if bddsimp is replaced with the tactic-based simplifier (prop) of PVS. The third element is the availability of powerful arithmetic decision procedures. This makes the exact syntactic form of the expressions in the simplified sequent less relevant than whether the sequent has enough (semantic) information to complete the proof.

The automatic proof used above can be packaged in a PVS proof strategy and used on other hardware examples. The core of our strategy for automating hardware proofs begins with the user suggesting the initial induction variable and the induction scheme when this is not obvious from the type of induction variable. An automatic strategy takes over from that point and completes the proof in the following manner. First, the PVS rewriter is set up to automatically rewrite every defined function symbol directly or indirectly used in the the theorem to be proved until rewriting is no longer productive. In the next step, all the boolean conditions appearing as the boolean part of conditional expressions in the formulas are lifted to the topmost level. The resulting nested boolean expression is propositionally simplified into a finite number of subgoals. The last step consists of applying arithmetic and equality decision procedures on each of the subgoals resulting from the propositional simplification. We have applied this strategy to an n-bit ALU [7] that executes 12 microoperations. The completely automatic verification took 90 seconds on a SPARC 10. The same strategy is also effective on several non-hardware examples [25].

4 Microprocessor Verification

The automatic inference procedures used in the PVS proof checker have also allowed us to highly automate the task of microprocessor verification. PVS is a relatively new system that has been evolving over the course of our processor verification effort. As more automatic inference procedures have been added to PVS, our effectiveness at automating microprocessor verification has significantly increased. Here we illustrate the usefulness and importance of automatic inference procedures in PVS from the point of view of processor verification. These examples are quite different from the N-bit adder described in Section 3 but the basic idea underlying the proof strategy given there is easily adapted for our present purpose.

We take the approach of describing the specification and implementation of microprocessors in terms of state transition systems. The state of the microprocessor consists of the state of the memory, register file, and internal registers of the processor (these would generally include the program counter, memory address register, and pipeline registers if the processor is pipelined, etc.).

The microprocessor verification problem is to show that the traces induced by the implementation transition system are a *subset* of the traces induced by the specification transition system, where *subset* has to be carefully defined by use of an abstraction mapping. The details of this approach are beyond the scope of this paper (see [2, 11, 24, 27, 29][7]).

In this approach, the proof of correctness makes use of an *abstraction* function that maps an implementation state into a *corresponding* specification state. Correctness can then be reduced to showing that for any execution trace of the implementation machine there exists a *corresponding* execution trace of the specification machine.

The implementation machine may run at a different rate than the specification machine [11, 27]. For example, in the case of the Saxe pipeline example [24], the specification machine takes one state transition to execute each instruction, but the implementation machine might take five cycles to execute branch instructions, but only one cycle for non-branch instructions. In the following we assume that the specification machine always takes one cycle to execute an instruction. We also assume that the number of cycles that the implementation machine takes to execute an instruction can be given as a function of the current state and current input. (This restriction can be slightly relaxed to deal with interrupts which might arrive a bounded number of cycles into the future.)

4.1 A Proof Strategy for Microprocessor Correctness

We denote the function that determines the number of cycles that the implementation machine takes to complete an instruction as num_cycles. We assume that this information is provided by the hardware designer or verifier.

The first step in verifying the correctness of the microprocessor is to split the proof into cases based on the definition of num_cycles. Thus for each case we have a precise number through which we have to cycle the implementation machine.

In the microprocessor verifications we have looked at, the state variables of the specification state are simply a subset of the state variables of the implementation state. The abstraction mapping maps to each specification register the corresponding implementation register, but not necessarily from the exactly

---

[7] The precise details followed in these papers are somewhat different.

corresponding state. For example, the abstraction mapping for the Saxe pipeline is such that the specification program counter is mapped from the corresponding implementation program counter and the specification register file is mapped from the implementation register file, but three cycles into the future. See [11,24] for details. If the abstraction mapping is given this way, then once the proof is split according to the definition of num_cycles, the resulting statement of correctness is usually an instance of a decidable fragment of the theory Ground Temporal Logic (*GTL2*) [10].

The problem is to come up with an effective procedure for deciding this theory. One obvious strategy is to completely rewrite the next-state functions and abstraction mapping until a large IF-THEN-ELSE is generated, then perform a case analysis on the resulting expression and check that each resulting case is valid. This naive strategy has proven to be ineffective for both the Saxe pipeline and Tamarack microprocessors, let alone anything more complex. However, the automatic inference procedures of PVS have allowed us to develop a less naive strategy that is still highly automatic and does succeed in proving the correctness of both the Saxe pipeline and Tamarack microprocessors:

```
(then* (skosimp*)
       (auto-rewrite-all-theories)
       (typepred-impl-state)
       (record)
       (cycle-split)
       (record)
       (rewrite-lift-if-simplify-and-assert)
       (auto-rewrite-all-theories!)
       (rewrite-lift-if-simplify-and-assert))
```

where the rewrite-lift-if-simplify-and-assert strategy is just:

```
(then* (assert) (repeat (lift-if)) (bddsimp) (assert)).
```

The above strategy consists of first skolemizing, then instructing PVS to use the axioms and definitions of the processor as rewrite rules (auto-rewrite-all-theories), then invoking the type predicate of the implementation state-type ((typepred-impl-state)). This is necessary in case there is a pipeline invariant associated with the machine state [11, 27]. The proof goal is then split (cycle-split) according to the num_cycles function. The current case is recorded in the ground decision procedures and assert is called which invokes automatic rewriting. The rewriting here will halt once it is incapable of simplifying a right hand side that is an IF-THEN-ELSE. Any resulting IF-THEN-ELSEs are then lifted and bddsimp is called to generate the resulting cases. The assert command is used to finish up each of these cases. Sometimes this is enough to complete the proof. If it is not then the (assert, lift-if, bddsimp, assert) cycle is repeated, but this time with PVS directed to completely rewrite, even through unsimplifiable IF-THEN-ELSEs.

The intuition behind this strategy is that the first form of rewriting takes care of the simple parts of the proof that require only rewriting and limited amount of case analysis. The second, unrestricted, rewriting takes care of the resulting cases that need to expand to large IF-THEN-ELSEs and require lots of case analysis. In the Saxe pipeline this type of reasoning is needed to verify the correctness of the register bypass logic. Note that this strategy, while not identical to the basic hardware strategy described earlier, has the same core strategy, namely the (do-rewrite, lift-if, bddsimp, assert) cycle.

We have also applied the same strategy to the Tamarack microprocessor first verified by Joyce [18]. This microprocessor is microcoded but not pipelined. Only the first restricted form of rewriting is necessary to finish the Tamarack's proof of correctness. This is because the case splitting generated by the num_cycles function is sufficient to generate all the relevant cases and to direct the rewriter through a single path through the microcode. In the Saxe pipeline more case analysis is needed to deal with the register bypass logic.

Note that prior to adding hashing and bddsimp to PVS we had verified the correctness of the Saxe pipeline, but only with manual assistance. The verification originally done by Saxe et al [24] also required user assistance.

5 Experimental Results

The following table summarizes the performance of PVS's automatic strategy with and without the improvements to PVS's automatic inference procedures. The timings were made on a SPARC 10.

Processor	Hashing and BDDs	No Hashing	Neither
Adder	127 sec.	160 sec.	unfin.
ALU	87 sec.	92 sec.	unfin.
Saxe Pipeline	605 sec.	1400 sec.	unfin.
Tamarack	545 sec.	unfin.	unfin.

Note that hashing was much more important in the microprocessor examples. These examples typically use more rewriting. In the ongoing verification of a simplified version of the MIPS R3000 we find that we get exponential savings due to hashing.

6 Related Work

The HOL system [13] is prototypical of the proof checkers that are based on very simple primitive inference rules combined using tactics. The more powerful primitive inference mechanisms of PVS can, in principle, be developed as tactics in HOL. For example, Boulton [3] has implemented a decision procedure for Presburger arithmetic as a tactic in HOL. However, this procedure does not handle equality over uninterpreted function symbols and, unlike in PVS, is not tightly integrated with the simplification and rewriting procedures. It would be interesting to see if the same degree of integration can be accomplished as effectively in a tactic-based approach and whether individual tactics can match the performance of inference procedures that use specialized algorithms and data structures. The HOL system favors tactics over special-purpose inference procedures since the latter might introduce unsoundness. This is an important consideration: the inference procedures of PVS do need to be scrutinized and tested with great care and rigor, but once this is done, they do not need to be justified down to basic inference steps with each application.

The "super-duper" tactic developed in [1] for hardware proofs is similar to the core strategy described in this paper. The similarity lies in the fact that both combine rewriting, case-splitting and simplifications in a loop for automating hardware proofs. The main differences are in (1) our use of decision procedures for congruence closure, arithmetic, and BDDs, and (2) our conditional rewriter interacts very closely with the decisions procedures and uses several optimizations. This interaction allows rewriting to be more effective, i.e., successful in simplifying more often, and efficient. We have found that the efficiency and effectiveness of rewriting are very crucial in the core strategy being applicable for large examples. The tactic in [1] is also designed to process predicative style of hardware specifications, whereas ours is suited for functional style.

Kumar, Schneider, and Kropf have developed a system MEPHISTO and a sequent calculus prover FAUST [19] which jointly can automatically verify a class of bit-level hardware circuits specified in a relational style popularized by Michael Gordon. Their system cannot automate proofs of complex circuits, such as microprocessors, that use data types since they do not have rewriting and arithmetic capability. This system does incorporate first-order BDD-based techniques that can handle some data types and parameterized hardware. Although our automatic strategy presented in this paper is designed for proving hardware specified in a functional style, we were able to automatically prove all but two of their eight circuits by modifying our strategy slightly to use the heuristic instantiation capability supported by PVS. We had to provide manual instantiations for the other two examples.

The Boyer-Moore theorem prover, Nqthm, is the best known of the batch-oriented theorem proving systems used in hardware verification [4]. Many of its deductive components are quite similar to those in PVS. The system uses a fast propositional simplifier, and also includes a rewriter and a linear arithmetic package. The latest release of the system has been heavily optimized for efficiency. As an experiment, we used Nqthm without any libraries to prove the N-bit adder. The Nqthm formalization of this theorem was slightly different from that of PVS. We found that the theorem could not be proved automatically. It took several hours of effort to fine-tune the definitions and to determine the lemmas needed to help the theorem prover with its proof. Though a significant human effort was required to complete the proof, Nqthm was eventually able to prove the main theorem in about 14 seconds of CPU time (on a Sparc 10/41).

The same example was proved in PVS without any lemmas and very little human input in about 130 seconds.

Burch and Dill [6] report on an automatic stand-alone strategy for microprocessor verification. Although they have not attempted the two examples reported here they report impressive timings for the automatic verification of a small version of the DLX processor [15]. They also describe a method for automating the generation of the abstraction mapping.

7 Conclusions

Automated theorem proving technology clearly has a great deal to contribute to hardware verification since hardware proofs tend to fall into certain systematic patterns. Our contention is that if theorem provers are to be effective in hardware verification, we must employ powerful and efficient deductive components within high-level strategies that capture the patterns of hardware proofs. More specifically, we have argued that:

- Hardware proofs tend to fall into certain patterns so that it is possible to obtain greater automation.
- Effective theorem proving is best achieved by mechanizing the tedious and routine deductive steps so that the human effort can be concentrated on the difficult parts of the proof.
- We can combine automation with efficiency by employing powerful and well-integrated mechanized procedures as can be obtained through the use of decision procedures and BDD-based propositional simplification.
- Batch-oriented theorem provers like Nqthm do contain tightly integrated and highly mechanized inference procedures, but they require a significant amount of tedious human effort in the exploratory phase of proof development.
- PVS strikes a balance between the tactic-based approach and those based on batch-oriented theorem proving. In PVS, efficient mechanization is used to automate the tedious and obvious deductive steps. Proofs can be constructed interactively under human control. Further mechanization can be obtained by defining high-level strategies in terms of tactics.

We have shown how hardware proofs can be automated in PVS through the use of a powerful mechanization of various useful inference steps and the definition of simple proof strategies that invoke these inference steps. We have illustrated the use of PVS an N-bit adder, a pipelined processor, and a simple unpipelined processor. The basic approach shown here is being applied to the mechanization of the correctness proofs of industrial-strength processors including the MIPS R3000 architecture and a commercial avionics processor AAMP5.

AAMP5 is a microcoded pipelined processor built at the Collins Avionics Division of Rockwell International for Avionics applications. It is a complex

CISC processor containing more than half a million transistors and is designed to execute a stack-oriented machine. One of the main purposes in undertaking this project [28], which is sponsored by NASA Langley Research Center and Rockwell International, was to see how well techniques developed and tested on small examples would scale to a commercial processor of significant complexity. We have successfully used the core strategy described in the paper to verify a number of instructions (identified by Rockwell engineers) of AAMP5. The verification revealed several errors some unknown to Rockwell and some planted by Rockwell engineers as a challenge to us.

Acknowledgements. John Rushby provided a great deal of support and encouragement for this work and supplied detailed comments on drafts of this paper. Sam Owre answered a number of questions regarding PVS and also proofread the paper. The N-bit ripple-carry adder example comes from a PVS library for bit-vectors being developed by Rick Butler and Paul Miner of NASA.

References

1. Mark D. Aagard, Miriam E. Leeser, and Phillip J. Windley. Toward a super duper hardware tactic. In *Proceedings of the HOL User's Group Workshop*, pages 401–414, 1993.
2. Martín Abadi and Leslie Lamport. The existence of refinement mappings. In *Third Annual Symposium on Logic in Computer Science*, pages 165–175. IEEE, Computer Society Press, July 1988.
3. R. J. Boulton. The HOL arith library. Technical report, University of Cambridge Computer Laboratory, 1992.
4. R. S. Boyer and J S. Moore. *A Computational Logic Handbook*. Academic Press, New York, NY, 1988.
5. K. S. Brace, R. L. Rudell, and R. E. Bryant. Efficient implementation of a BDD package. In *Proc. of the 27th ACM/IEEE Design Automation Conference*, pages 40–45, 1990.
6. J. R. Burch and D. L. Dill. Automated verification of pipelined microprocessor control. In David Dill, editor, *Computer-Aided Verification '94*, pages 68–80. Volume 818 of *Lecture Notes in Computer Science*, Springer-Verlag, 1994.
7. F. J. Cantu. Verifying an *n-bit* arithmetic logic unit. Blue book note 935, University of Edinburgh, June 1994.
8. E. M. Clarke and O. Grümberg. Research on automatic verification of finite-state concurrent systems. In Joseph F. Traub, Barbara J. Grosz, Butler W. Lampson, and Nils J. Nilsson, editors, *Annual Review of Computer Science, Volume 2*, pages 269–290. Annual Reviews, Inc., Palo Alto, CA, 1987.
9. *User Guide for the EHDM Specification Language and Verification System, Version 6.1*. Computer Science Laboratory, SRI International, Menlo Park, CA, February 1993. Three volumes.
10. D. Cyrluk and P. Narendran. Ground temporal logic—a logic for hardware verification. In David Dill, editor, *Computer-Aided Verification '94*, pages 247–259. Volume 818 of *Lecture Notes in Computer Science*, Springer-Verlag, 1994.

11. David Cyrluk. Microprocessor verification in PVS: A methodology and simple example. Technical Report SRI-CSL-93-12, SRI Computer Science Laboratory, December 1993.

12. P. J. Downey, R. Sethi, and R. E. Tarjan. Variations on the common subexpressions problem. *Journal of the ACM*, 27(4):758–771, October 1980.

13. M. J. C. Gordon and T. F. Melham, editors. *Introduction to HOL: A Theorem Proving Environment for Higher-Order Logic.* Cambridge University Press, Cambridge, UK, 1993.

14. Mike Gordon. Proving a computer correct. Technical Report TR 42, University of Cambridge, Computer Laboratory, 1983.

15. J. L. Hennessy and D. A. Patterson. *Computer Architecture: A Quantitative Approach.* Morgan Kaufmann, 1990.

16. Warren A. Hunt, Jr. Microprocessor design verification. *Journal of Automated Reasoning*, 5(4):429–460, December 1989.

17. G. Janssen. *ROBDD Software.* Department of Electrical Engineering, Eindhoven University of Technology, October 1993.

18. J. Joyce, G. Birtwistle, and M. Gordon. Proving a computer correct in higher order logic. Technical Report 100, Computer Lab., University of Cambridge, 1986.

19. R. Kumar, K. Schneider, and T. Kropf. Structuring and automating hardware proofs in a higher-order therem proving environment. *Formal Methods in System Design*, 2(2):165–223, 1993.

20. Patrick Lincoln, Sam Owre, John Rushby, N. Shankar, and Friedrich von Henke. Eight papers on formal verification. Technical Report SRI-CSL-93-4, Computer Science Laboratory, SRI International, Menlo Park, CA, May 1993.

21. D. C. Luckham, S. M. German, F. W. von Henke, R. A. Karp, P. W. Milne, D. C. Oppen, W. Polak, and W. L. Scherlis. Stanford Pascal Verifier user manual. CSD Report STAN-CS-79-731, Stanford University, Stanford, CA, March 1979.

22. S. Owre, J. M. Rushby, and N. Shankar. PVS: A prototype verification system. In Deepak Kapur, editor, *11th International Conference on Automated Deduction (CADE)*, pages 748–752, Saratoga, NY, June 1992. Volume 607 of *Lecture Notes in Artificial Intelligence*, Springer-Verlag.

23. S. Owre, N. Shankar, and J. M. Rushby. *User Guide for the PVS Specification and Verification System, Language, and Proof Checker (Beta Release).* Computer Science Laboratory, SRI International, Menlo Park, CA, February 1993. Three volumes.

24. James B. Saxe, Stephen J. Garland, John V. Guttag, and James J. Horning. Using transformations and verification in circuit design. *Formal Methods in System Design*, 4(1):181–210, 1994.

25. N. Shankar. Abstract datatypes in PVS. Technical Report SRI-CSL-93-9, Computer Science Laboratory, SRI International, Menlo Park, CA, December 1993.

26. Robert E. Shostak. Deciding combinations of theories. *Journal of the ACM*, 31(1):1–12, January 1984.

27. Mandayam Srivas and Mark Bickford. Formal verification of a pipelined microprocessor. *IEEE Software*, 7(5):52–64, September 1990.

28. Mandayam Srivas and Steve Miller. Formal verification of the AAMP5 microprocessor: A case study in the industrial use of formal methods. Technical report. A Forthcoming NASA Contractor Report.

29. P. Windley and M. Coe. A correctness model for pipelined microprocessors. In *Proceedings of Theorem Provers in Circuit Design*, 1994.

A Formal Framework for High Level Synthesis

Thomas Kropf[*], Klaus Schneider[*] and Ramayya Kumar[**]

*Institut für Rechnerentwurf und Fehlertoleranz (Prof. D. Schmid)
Universität Karlsruhe, Kaiserstr. 12, 76128 Karlsruhe, Germany
**FZI (Prof. D. Schmid), Haid-und-Neu-Str.10-14, 76131 Karlsruhe, Germany
email: {kropf, schneide}@informatik.uni-karlsruhe.de, kumar@fzi.de
WWW: http://goethe.ira.uka.de/hvg/

Abstract. In this paper, we propose a new approach to formal synthesis which focuses on the generation of verification-friendly circuits. Starting from a high-level implementation description, which may result from the application of usual scheduling and allocation algorithms, hardware is automatically synthesized. The target architecture is based on *handshake processes*, modules which communicate by a simple synchronizing handshake protocol. The circuits result from the application of only a few basic operations like synchronization, sequential execution or iteration of base handshake processes. Each process is guided by an abstract theorem that is used to derive proof obligations, to be justified after synthesis. Automation has been achieved to the extend that only those "relevant" proof obligations remain to be proven manually, e.g. theorems for data-dependent loops and lemmata about the used data types. The process-oriented implementation language is enriched by loop invariants. If those are given prior to the synthesis process and the underlying data types are only Booleans, i.e. finite-length bitvectors, then the complete synthesis and verification process runs automatically.

1 Introduction

As high-level synthesis [1] has reached the stage of industrial applications, it has turned out that the implicit assumption, that the generated structures just implement the behavioral specification, does not always hold. Software faults, inevitable in complex design tools, hinder the use of synthesized hardware, especially in safety critical applications, and reduce the cost advantages offered by high-level design tools.

Verifying the *design tools* by software verification techniques fails as a potential remedy to this situation due to the complexity of the algorithms to be checked. Automatically verifying the *synthesized hardware* by post verification techniques is possible when using model checking techniques [2]. However, this approach is limited to only small and medium sized circuits. It already fails if a couple of 32 bit registers are used in the data path. More complex circuits may be verified by proof-based approaches [3]. However, these techniques are interactive in general and require a good understanding of mathematical logic and the functioning of the target hardware. Large designs are verifiable if they are hierarchically composed. Obviously, automatically synthesized hardware is hard to tackle using these approaches, since it is synthesized "flat" and its functioning is, especially if optimized, not always understandable to the human hardware verifier.

In this paper, we propose an approach which makes direct use of high-level synthesis algorithms. The high-level implementation description, resulting from scheduling and allocation is mapped onto a target architecture based on modules which communicate by a simple synchronizing handshake protocol, called *handshake processes* [4]. Handshake processes represent basic operations of the flow graph, resulting from the previous synthesis steps. The processes are composed to get a circuit, satisfying the specification. During the composition process the circuit is verified against the high-level specification in order to check the correctness of scheduling and allocation, which have been performed outside the formal framework. The verification process results in proof obligations, which may not be verified automatically. Usually this comprises lemmata describing properties of the used data types. For data-dependent loops, the implementation description language includes loop invariants. Stating these before the synthesis process is started, only data type related proof goals remain to be verified manually. However, this part is also automated if only bitvectors of finite-length are used [5].

This first approach may result in sub-optimal synthesized circuits with regard to area or speed. However, this is not a fundamental drawback of our approach since many of the well-known optimization techniques may be used afterwards. In general, optimization algorithms are easier to verify, e.g. optimizing transformations used in retiming may be checked for their behavior-preserving properties [6]. Moreover, safety critical applications may lead to the acceptance of a certain hardware overhead in favor of correctness, similar to the additional hardware which is accepted for enhanced testability.

The outline of the paper is as follows. In the next section, our approach is compared to related work. After introducing the formal framework, notation and the used logic, the synthesis approach is presented. There, we first present the basic "protocol" of the handshake processes and then show how base and composed processes may be synthesized, using an HDL. The paper ends with experimental results and conclusions.

2 Related Work

In contrast to post-hoc verification, formal synthesis aims at deriving correct implementations from specifications. Most existing approaches are based on refinement techniques, where a formal specification is successively transformed into hardware implementations. The best known approach is based on the use of the LAMBDA system [7]. Hardware is specified in higher-order logic and synthesized by decomposing the specification into smaller pieces, which can be mapped onto a set of connected modules representing implementations. Although supported by a graphical interface, the synthesis has to be performed completely manually as well as the proof that the resulting structure is behaviorally equivalent to the specification. Hence these designs require a thorough understanding of the used logic and the proof rules to be able to perform the necessary interactive theorem proving. Thus, the overall design methodology is completely different form classical high-level synthesis and many optimization techniques developed in this area cannot be applied.

The formal synthesis of controllers from temporal logic specifications or timing diagrams has been presented in [8]. Due to the underlying propositional logic, these approaches cannot be extended to data path synthesis.

The approaches aiming at the synthesis of self-timed systems [9] are related to our approach with regard to a similar target structure, namely communicating modules. However, we use a synchronous handshake protocol and use higher-order logics as the underlying formal system instead of process algebras. Moreover, we explicitly provide a framework for coping with data dependent loops.

3 Basics

In the following, we use higher order logic as used in the HOL theorem proving environment [10]. In contrast to classical first-order logic, quantification is possible also over functions and predicates. The logic is based on typed lambda calculus, i.e. every term t has a unique type ty, written as $t{:}ty$. We use the usual boolean connectives like $\neg$, $\wedge$, $\vee$, or $\rightarrow$ and the symbol λ for lambda abstraction.

The HOL system provides a set of basic types like ind, the set of individuals, $bool$, the set of booleans and num, the natural numbers. More complex types can be defined using type operators, which take types and produce new ones. Given two types ty_1 and ty_2, the binary type operator for functions $\rightarrow$ gives the type of all functions from ty_1 to ty_2, i.e. $ty_1 \rightarrow ty_2$. The pair operator # forms the cartesian product type of two types $ty_1 \# ty_2$. Another concept, used in the following, is type polymorphism. There are type variables $'ty_{var}$, that can be instantiated with an arbitrary type. In the HOL system, complex types may be defined. Proofs are usually performed interactively in the HOL system. However *tactics* may be defined to automate certain proof tasks. More details can be found in [10]. Proving theorems in the HOL system is usually performed interactively and in "backward" proof style. The goal to be proved is put on top of a goal stack. Repeated rule applications splits the goal into smaller proof goals until axioms are reached and the respective subgoal is removed form the stack. If the goal stack has been completely emptied, the original goal has been proved. Sets of rule applications can be abbreviated as tactics such that partial automation can be achieved.

In the following, modules are represented as predicates and structures of modules as conjunctions of predicates as used in many hardware verification approaches [11]. We use a discrete linear time modelled by natural numbers num and wires are functions from time to the respective data type, e.g. $x{:}num \rightarrow bool$.

Since temporal relationships play an important role for the communication of handshake processes, we use higher-order variants of temporal operators [12, 13, 14].

Specifications of digital circuits can often be described conveniently by the "when" operator [15]. In this paper we use a slightly different semantics and formalization of the when operator which is as follows. The expression $(x \text{ when } f)^{t_0}$ becomes true, if x becomes true for the first time after t_0, at which f becomes true. In figure 3-1, a timing diagram for x and f is given, which makes $(x \text{ when } f)$ true at time instant t_0.

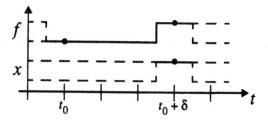

Fig. 3-1. Timing diagram of the when operator

The operator is formally defined in equation 3-1.

$$(x \text{ when } f)^{t_0} := \forall \delta . \left(\forall t . (t < \delta) \rightarrow \neg f^{(t_0 + t)} \right) \wedge f^{(t_0 + \delta)} \rightarrow x^{(t_0 + \delta)} \qquad (3\text{-}1)$$

For the approach, presented in this paper, another definition of the when operator is used (equation 3-2). Intuitively, the when operator can be formalized using a "ready" signal f, a "termination" signal x and an invariant J of a data-dependent loop. The invariant must hold at the beginning of the loop (J^{t_0}). If the loop does not terminate ($\neg f^{(t+t_0)}$) the invariant must also hold at the next loop iteration. Otherwise, the termination signal x must become true in the next time instance. This order of events is visualized by figure 3-2.

$$(x \text{ when } f)^t :=$$

$$\exists J . J^{t_0} \wedge$$

$$\left(\forall t . \neg f^{(t_0 + t)} \wedge J^{(t_0 + t)} \rightarrow J^{(t_0 + t + 1)} \right) \wedge \qquad (3\text{-}2)$$

$$\left(\forall \delta . f^{(t_0 + \delta)} \wedge J^{(t_0 + \delta)} \rightarrow x^{(t_0 + \delta)} \right)$$

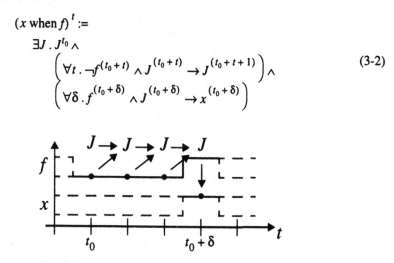

Fig. 3-2. Visualization of the invariant formalization

Note that the above definitions lead to a "weak when", i.e. it is not required that the signal f ever becomes true. Using the weak version of the operator for specifications leads to partial correctness results, i.e. the hardware gives the right result at the right time, but only if it terminates. In circuits, whose termination is not obvious, e.g. when using data-dependent loops, then an additional termination condition $\exists t . f^{t + t_0}$ has to be shown.[1] It has been shown previously that, provided that the invariant J is given, proofs of theorems, which contain the temporal operators may be automated [14]. Proof tactics to perform this task are available for the HOL90 theorem proving environment.

4 Handshake Circuits

The main idea underlying this approach is based on the observation that if it is possible to implement modules with a uniform and consistent input/output behavior, then module composition to perform complex computations is almost trivial and the verification of the overall behavior is considerably eased. In most cases the desired correctness theorem, i.e. that the composed structure behaves as specified, may be derived fully automatically.

The interface of a module is as given in figure 4-1. Note that in the following, all modules are described using only one input *inp* and one output *out*. However, *inp* and *out* are of polymorphic type so that bundles of signals may be easily used as type instances.

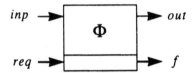

Fig. 4-1. General handshake circuit

Besides input *inp* and output *out* two further signals are used: the request input *req* indicates that a new input value can be read for the next computation and the finish output *f* denotes the end of a computation cycle. Φ represents the specification of the module function, i.e. the behavior to be eventually verified. More formally, the *module function* is expressed by a formula $\Phi(inp, out, t_1, t_2)$, where the starting time t_1 and finish time t_2 is included. These time instants serve as reference time points for the input and output value and may be also used for formalizing restrictions on the duration of a computation cycle. A typical example is given in equation 4-1 where the sum of inp_1 and inp_2 is computed, needing 3 time units.

$$\Phi(inp_1, inp_2, out, t_1, t_2) = \left(out^{t_2} = inp_1^{t_1} + inp_2^{t_1} \right) \wedge (t_2 - t_1 = 3) \qquad (4\text{-}1)$$

The *timing behavior of the module* is formalized in terms of the handshake signals f and *reg* as stated in equation 4-2. The first implication denotes that provided that no computation is active at time t_0 (f^{t_0}), then the outputs are stored until a new task is requested (store mode). The second implication states, that after a request for a computation *req*, the result, computed by Φ, is available at *out* when f is true again.

$$\text{PROTOCOL}\,(req, in, f, out, \Phi) := f^{0} \wedge$$

$$\forall t_0 . \left(f^{t_0} \to [(\lambda t . (out^t = out^{t_0}) \wedge (f^t = f^{t_0})) \text{ until } req]^{t_0} \right) \wedge \qquad (4\text{-}2)$$

$$\left(f^{t_0} \to \left(\left(\lambda t_1 . \left[\lambda t_2 . \Phi(inp, out, t_1, t_2) \text{ when } f \right]^{(t_1 + 1)} \right) \text{ when } req \right)^{t_0} \right)$$

---

1. It is easily possible to define a strong when $\widehat{\text{when}}$ using the weak version as follows:
$$\left(x \,\widehat{\text{when}}\, f \right)^{t} := (x \text{ when } f)^{t} \wedge \neg((\lambda t . \text{F}) \text{ when } f)^{t}$$

5 Verifying Handshake Circuits

In this section a high-level implementation description is presented. It is the result of the scheduling and allocation step of high-level synthesis and its semantics is given directly in terms of handshake circuits.

5.1 Notation

The language distinguishes two levels of abstraction. For creating base modules, there are constructs for synchronous RT-level descriptions using a clock. For compound modules, process composition constructs are provided as given in table 5-1.

Level	construct	description
clock level	=	combinational assignment
	←	synchronous assignment
	|	clock parallel action
	op	arbitrary (combinational) arithmetic or logic operation
	bop	(combinational) boolean operation resulting in a signal of type bool
	const	arbitrary constants (bits, natural numbers)
proces level	;	sequential composition
	,	parallel composition
	:=	assignment

Table 5-1: Base constructs

Variables in capital letters denote bitvectors, i.e. $Out = (out_1, out_2, ..., out_n)$. Sets of registers are abbreviated using an arrow like $\overrightarrow{Out} = (Out_1, Out_2, ..., Out_q)$.

All processes are defined as follows.

```
process <name> (in₁:α₁,..., inₚ:αₚ):(out₁:β₁, ..., outq:βq) =
var
  R₁, ..., Rₖ . register(n)
  W₁, ..., W₁ . wire(m)
  process <subprocess nameᵢ> ...
begin
<body>
end.
```

Each process has a name <name>, a set of inputs in_i of type α_i, resulting in q outputs typed β_i. The types α_i and β are bitvectors of finite length. The process may declare local registers R_j of bitwidth n and wires W_i of bit width m. The <body> depends on the type of the process. There are basic processes, which are either sequential loops or combinational circuits, and composed processes. The latter are nested loops, sequences or synchronized parallel processes.

A key point of this approach is its ability to cope with arbitrary hierarchical structures. This is possible by guaranteeing that composed modules have the same input/output behavior as the base modules, i.e. a uniform interface scheme has to be used. In the following base processes, nested loop processes as well as sequential and parallel composition are treated.

5.2 Base Processes

A base process usually consists of a data dependent loop and is defined as follows.

---

init

$\vec{R} \leftarrow op_{init}(\vec{I}, const)^a$

while bop() do

$\vec{R} \leftarrow op_{loop}(\vec{R})$

end invariant J

output

$\overrightarrow{Out} = op_{out}(\vec{R})$

---

a. abbreviation for
$$R_1 \leftarrow op_{init,1}(\vec{I}, const)|...|$$
$$R_k \leftarrow op_{init,k}(\vec{I}, const)$$

The invariant J is necessary for the automated verification. Currently it is a statement in higher-order logic[1], using the process signals as arguments. Naturally, a similar construct is also available for REPEAT - UNTIL loops.

The corresponding hardware structure of the module is given in figure 5-1.

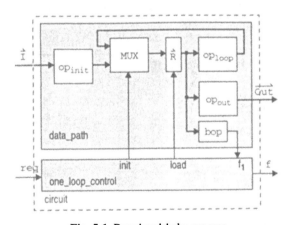

Fig. 5-1. Base handshake process

The controller is realized according to the state transition graph (STG) of figure 5-2 and hence is the same for every realization of a while loop.

Fig. 5-2. STG of one_loop_control

A simple combinational module results from the previous case by omitting the while loop, leading to the simplified structure of figure 5-3. The registers are necessary for maintaining the overall handshake behavior and may be omitted if the input signal remains stable during the whole computation.

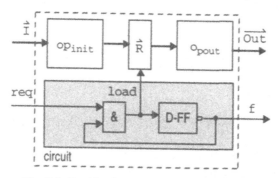

Fig. 5-3. Combinational base handshake process

The register $\vec{R}$ and the controller one_loop_control of figure 5-3 are necessary to store the results of $op_{pout}(op_{init}(\vec{I}))$ to achieve a behavior which conforms to equation 4-2.

A top level formalization of the circuit structure of figure 5-1 in higher-order logic is shown in equation 5-1.

$$circuit(req, \vec{I}, \overrightarrow{Out}, f) :=$$
$$\exists init\ load\ f_1 . data\_path\left(\vec{I}, init, load, f_1, \overrightarrow{Out}\right) \wedge \qquad (5\text{-}1)$$
$$one\_loop\_control(req, init, load, f_1, f)$$

---

1. Currently the syntax of the HOL90 theorem proving environment has to be used for invariants.

To verify the module behavior, it has to be shown that the implementation (equation 5-1) implies a behavioral specification (equation 4-2). This comprises the conformance to the handshake protocol and the computing function Φ. Since the handshake circuits have been synthesized verification-friendly, the timing behavior is automatically verified using appropriate theorems [4]. The remaining proof goal is shown in equation 5-2.

$$f^{t_0}, \mathrm{req}^{t_0} \vdash \left[\lambda t_1 . \Phi(\vec{\mathrm{I}}, \overrightarrow{\mathrm{Out}}, t_0, t_1) \text{ when } f \right]^{t_0 + 1} \tag{5-2}$$

The remaining proof may be further simplified, if the invariant based on the *when* definition of equation 3-2 is used. However, the invariant has to be provided manually either only now or already previously in the implementation description.

In the following, a GCD process is defined to illustrate the definition of base handshake processes. The overall process is written down as follows.

```
process GCD_proc(I₁ :num(16), I₂ :num(16)):(Out : num(16)) =
var
  R₁ , R₂ . register(16)
begin
init
  R₁ ← I₁  |
  R₂ ← I₂
repeat
   if R₁ < R₂
      then R₂ ← R₂ - R₁
      else R₁ ← R₁ - R₂
until (R₁ = 0) ∨ (R₂ = 0)  invariant J_GCD
output
  if R₁ = 0
     then Out ← R₂
     else Out ← R₁
end.
```

The specification Φ to be verified is as follows.

$$\Phi([I_1, I_2], \mathrm{Out}, t_1, t_2) := \left(\mathrm{Out}^{t_2} = GCD\left(I_1^{t_1}, I_2^{t_1} \right) \right) \tag{5-3}$$

Based on a usual definition of the Greatest Common Divisor, GCD, the invariant

$\mathcal{J}_{GCD}$ of equation 5-4 is used. Basically it states that the iterative subtraction performed in the registers of the GCD process do not affect the *GCD* to be computed.

$$\forall t, t > t_0 \cdot \left(GCD\left(R_1^{(t+1)}, R_2^{(t+1)} \right) = GCD\left(R_1^t, R_2^t \right) \right) \tag{5-4}$$

5.3 Compound Processes

5.3.1 Nested Loops

A handshake process may be used to construct a nested data dependent loop. It is called "nested", since most basic handshake processes also contain a data-dependent loop. It thus has a similar function as op_{loop} in figure 5-1 but at a higher level of abstraction.

Given a handshake process₁, which may be defined e.g. as presented in section 5.2.1, nested loops are specified as follows.

```
init
    R  ← op_init,R( I ,const)
    W  ← op_init,W( I ,const)
while bop( R , W ) do
    W  ← process₁( R , W )
end invariant J
output
    Out  = op_out( R , W )
```

The realization of the nested loop using process1 is given in figure 5-1.

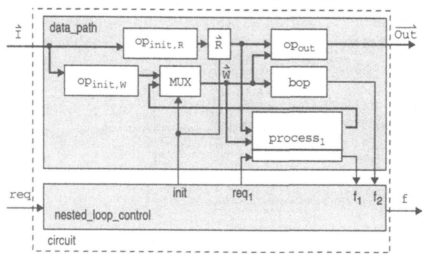

Fig. 5-4. Nested loop handshake process

The STG of the controller nested_loop_control is given in figure 5-2. Similar to the controller of figure 5-2, it has a wait state and a computation state. The controller starts in the wait state and reaches after a request signal the computation state. The inner loop is started by signal req_1. Its termination is signalled to the controller by f_1 and the termination of the outer loop, checked by **bop** is communicated via f_2.

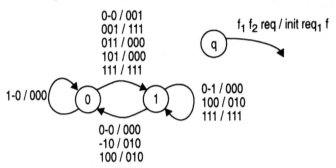

Fig. 5-5. STG of nested_loop_control

As in section 5.2.3, it is sufficient to show, that the following holds.

$$\forall t_0 . \, f^{t_0} \wedge req^{t_0} \rightarrow \left[\lambda t_1 . \Phi \left(\vec{I}, \overrightarrow{Out}, t_0, t_1 \right) \text{ when } f \right]^{t_0 + 1} \tag{5-5}$$

Having proved equation 5-5, the overall behavior PROTOCOL $(req, \vec{I}, \overrightarrow{Out}, f, \Phi)$ may be derived by a preproven lemma [4] if the controller nested_loop_control is used. The remaining proof goals are similar to those of section 5.2.3 and omitted here.

5.3.2 Parallel Composition

The parallel composition or *synchronization* of two processes is stated as follows.

$$\overrightarrow{Out}_1 \leftarrow process_1(\vec{I}_1, \vec{C}_1),$$
$$\overrightarrow{Out}_2 \leftarrow process_1(\vec{I}_2, \vec{C}_2)$$

The corresponding hardware structure is given in figure 5-6.

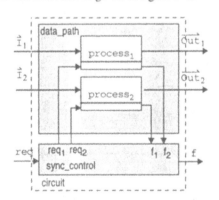

Fig. 5-6. Synchronization of two handshake processes

The controller sync_control essentially is a simple combinational circuit, where f results from the conjunction of f_1 and f_2 and req_i from the conjunction of req and f.

The verification of the overall behavior is easy, since a preproven lemma [4] is also available here, stating that, provided that both processes work as intended, then both results are available at $\overrightarrow{Out}_1$ and $\overrightarrow{Out}_2$, if f indicates the termination of the composed process.

5.3.3 Sequential Composition

The sequential composition of two processes is stated as follows.

```
Out₁ ← process₁(I₁,C₁);
Out₂ ← process₁(I₂,C₂)
```

The corresponding hardware structure is given in figure 5-6.

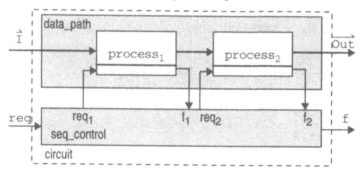

Fig. 5-7. Sequentialization of two handshake processes

The controller seq_control has three states. Its STG is given in figure 5-2. Wait or computation status of both processes are indicated by W and C, i.e. the state CW denotes, that the first process is computing a result and the second waits to be served.

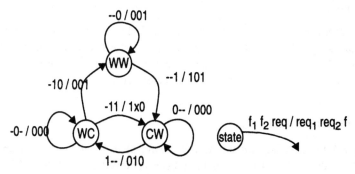

Fig. 5-8. STG of seq_control

Also here a preproven lemma [4] is used to prove, that the whole process provides as a result the functional composition of both processes, if it has terminated.

6 Example

In the following, we will verify a synthesized handshake circuit, which realizes the computation of the euclidean metric, defined in equation 6-1.

$$\Phi(I_1, I_2, Out, t_1, t_2) = \left(Out^{t_2} = \sqrt{\left(I_1^{t_1}\right)^2 + \left(I_2^{t_1}\right)^2}\right) \tag{6-1}$$

The overall process is as follows, where Mult, Add and Sqrt are previously defined processes. For computing the square root, e.g. Heron's algorithms based on the following iteration may be used.

$$x_{n+1} = \frac{1}{2}\left(\frac{a}{x_n} + x_n\right). \tag{6-2}$$

```
process Euclid_Metric (I₁ :num, I₂ :num) : (Out : num) =
var
  W₁ , W₂ , W₃ . wire(16)
begin
  (W₁ ← Mult(I₁ , I₁ ),W₂ ← Mult(I₂ , I₂ ));
  W₃ ← Add(W₁ , W₂ )
output
  Out ← Sqrt(W₃ )
end.
```

The corresponding hardware structure is given in figure 6-1. Note that the control signals have been deliberately omitted in the Add module. This optimization step is possible here, since a combinational module is used in a sequential composition and the overall behavior is not changed if the intermediate storage of the combinational results of Add is missing.

The base processes Mult, Add and Sqrt are verified as shown in section 5.2. The verification of the specification Φ of equation 6-1 is performed as shown in section 5.3.2 and section 5.3.3. Hence besides applying the respective simplification lemmata [4], only some internal lines have to be eliminated [16].

7 Experimental Results

Our approach has been successfully tested on several arithmetic circuits. Runtimes can be found in table 7-1. All runtimes have been achieved on a SUN SPARC 10 with 96MB. Finding the loop invariants has turned out to be the main challenge in synthesizing the circuits.

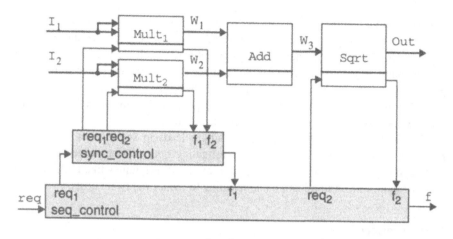

circuit

Fig. 6-1. Synthesized structure of the euclid metric

Circuit Name	Time in seconds
VonNeumann 1	7.19
VonNeumann 2	17.33
IncDecAdder 1	6.22
IncDecAdder 2	15.81
Summation	16.77
Signal Generator	44.52
Greatest Common Divisor	27.78
Russian Multiplier	60.14
Divider	73.60
SquareRoot	66.85

Table 7-1: Runtimes of different benchmark circuits

8 Conclusions and Future Work

In this paper, an approach to the formal synthesis of verification-friendly hardware structures has been presented. Having applied standard scheduling and allocation algorithms, known from high-level synthesis, an implementation description in a HDL-like language results. Its semantics is given directly as a composition of handshake circuits based on a synchronous handshake protocol. During the composition of these handshake circuits, a set of correctness theorems is derived which justifies the correctness of the specified circuit. Hence the correctness of scheduling and allocation, performed outside the formal framework, is checked. If during the specification phase, the designer's key ideas are provided via loop invariants, then all correctness proofs are done automatically, provided that a decidable domain like bitvectors have been used.

In case of more complex data types, proof obligations on the data domain are derived automatically, which have then to be justified otherwise, e.g. by manual proofs. The approach is well suited for data driven applications, which are used in safety critical environments. For pure controller circuits, other approaches are better suited, e.g. by performing a post-verification based on model checking.

Using our approach, no time consuming post verification step is necessary to achieve correctness statements on the resulting circuits.

Obviously, the handshake module paradigm leads in some cases to suboptimal hardware solutions. We are currently investigating optimization strategies at higher abstraction levels like operator folding and low level techniques like multilevel logic minimization, FSM minimization and retiming to reduce hardware costs.

9 References

[1]   M.C. McFarland, A.C. Parker, and R. Camposano. Tutorial on high-level synthesis. In *25th Design Automation Conference*, pages 330–336, 1988.

[2]   J.R. Burch, E.M. Clarke, K.L. McMillan, and D.L. Dill. Sequential circuit verification using symbolic model checking. *27rd ACM/IEEE Design Automation Conference*. IEEE, 1990, pages 46–51.

[3]   A. Gupta. Formal hardware verification methods: A survey. *Journal of Formal Methods in System Design*, pages 151–238, 1992.

[4]   K. Schneider, R. Kumar, and T. Kropf. Automating verification by functional abstraction at the system level. In T. Melham and J. Camilleri, editors, *International Workshop on Higher Order Logic Theorem Proving and its Applications*, pages 391–406, Malta, September 1994. Lecture Notes on Computer Science No. 859, Springer.

[5]   K. Schneider, R. Kumar, and T. Kropf. Modelling generic hardware structures by abstract datatypes. In L. Claesen and M. Gordon, editors, *International Workshop on Higher Order Logic Theorem Proving and its Applications*, pages 419–429, Leuven, Belgium, September 1992. IFIP TC10/WG10.2, Elsevier Science Publishers.

[6]   M. Mutz. Using the HOL theorem proving environment for proving the correctness of term rewriting rules reducing terms of sequential behavior. In K.G. Larsen and A. Skou. *International Workshop on Computer Aided Verification*, volume 575 of *Lecture Notes in Computer Science*. Springer, Aalborg, July 1991, pages 355–366.

[7]   E. Mayger and M. P. Fourman. Integration of formal methods with system design. In A. Halaas and P.B. Denyer. *International Conference on Very Large Scale Integration*, pages 59–70, 1991. Edinburgh, North-Holland

[8]   M. Fujita and H. Fujisawa. Specification, verification and synthesis of control circuits with propositional temporal logic. J.A. Darringer and F.J. Rammig, editors. *Computer Hardware Description Languages and their Applications*, Washington, June 1989. IFIP WG 10.2, North-Holland, pages 265–279.

[9]   S.M. Burns and A.J. Martin. Synthesis of self-timed circuits by program transformation. In G. Milne, editor, *The Fusion of Hardware Design and Verification*, pages 99–116, Glasgow, Scotland, July 1988. IFIP WG 10.2, North-Holland.

[10] M.J.C. Gordon and T. Melham. *Introduction to HOL: A Theorem Proving Environment for Higher Order Logic*. Cambridge University Press, 1993.

[11] F.K. Hanna and N. Daeche. Specification and verification using high-order logic. In C.J. Koomen and T. Moto-oka, editors, *Computer Hardware Description Languages*, pages 418–433. Elsevier Science Publishers, North-Holland, 1985.

[12] E. A. Emerson. *Temporal and Modal Logics*, chapter 16, pages 996–1072. Elsevier Science Publishers, 1990.

[13] F. Kröger. *Temporal Logic of Programs*, volume 8 of *EATCS Monographs on Theoretical Computer Science*. Springer Verlag, 1987.

[14] K. Schneider, T. Kropf, and R. Kumar. Control-path oriented verification of sequential generic circuits with control and data path. In *4th European Design Automation Conference*, pages 648–652, Paris, France, March 1994. IEEE Computer Society Press.

[15] T. Melham. Abstraction mechanisms for hardware verification. In G. Birtwistle and P.A. Subrahmanyam, editors, *VLSI Specification, Verification and Synthesis*, pages 267–291. Kluwer Academic Press, 1988.

[16] K. Schneider, T. Kropf, and R. Kumar. Why hardware verification needs more than model checking. In *International Workshop on Higher Order Logic Theorem Proving and its Applications*, Malta, 1994, Short Paper.

[17] R. Kumar, K. Schneider, and T. Kropf. Structuring and automating hardware proofs in a higher-order theorem-proving environment. *International Journal of Formal System Design*, pages 165–230, 1993.

Tutorial on Design Verification with Synchronized Transitions

Niels Mellergaard and Jørgen Staunstrup

Department of Computer Science, Technical University of Denmark,
DK–2800 Lyngby, Denmark
email: {nm, jst}@id.dtu.dk

Abstract. This tutorial describes a mechanized technique for design verification. The aim is, in the early design phases, to verify selected key properties of a partially specified design. A supporting design language called SYNCHRONIZED TRANSITIONS is used for describing designs. The design verification is mechanized by tools, in particular, a theorem prover called the Larch Prover (LP) used for reasoning about properties of a design, and a translator (called ST2LP) that generates input for the theorem prover from a given design description.

1 Introduction

Design verification aims at the initial phases of a circuit design where descriptions of the design are incomplete and abstract. This tutorial introduces a design verification technique and supporting tools for verifying selected key properties of a partially specified design, e.g., that the design meets a safety requirement, or that its communication with the environment meets a certain protocol. Recently, there has been widespread interest in formalized hardware verification, see [9] for a survey. A major part of this work is oriented towards implementation verification [3, 8] where a high-level description (the specification) is related to another lower level description (the realization). Implementation verification is necessarily aimed at a later stage of design where both a specification and an implementation are available. In contrast, design verification, aims at verifying particular properties of an incomplete design; possibly long before any implementation is started.

To encourage early verification, it is important to enable the designer to carry out the verification on exactly the same design description that is later used for other purposes, e.g., synthesis or simulation. One way to achieve this is to develop tools and transformation techniques that allow such manipulations on design descriptions formulated in a logic intended for verification [6, 11]. The approach described in this tutorial is somewhat different; by supporting verification in a design language that is not specifically aimed at verification. Hence, a design description can be used for a number of purposes, e.g., simulation, synthesis, performance analysis etc.

The design verification technique is supported by mechanical tools: a translator and a theorem prover. The translator transforms design descriptions specified with SYNCHRONIZED TRANSITIONS [14], including formally stated requirements to the design, into verification conditions for a mechanical theorem prover (the LARCH PROVER, LP [10]). The translation is syntax directed which means that the structure of the verification follows the structure of the design description. The verification is broken into a large number of small steps, each syntactical construct yields a constant number of verification conditions (for some zero).

A high-level description of a design is an abstraction of its physical behavior. It is important to realize that formal verification deals with the abstraction and *not* with the physical realization. A formal verification of a requirement is not an absolute guarantee against malfunctioning. For example, in case of a power failure the formal description is no longer a model of the physical realization, and hence, properties verified from the description may no longer hold. The same applies to simulation which is also based on a model of the physical reality. If the model is not an adequate abstraction, a simulation, no matter how exhaustive, does not provide any guarantee against malfunctioning. It is outside the realm of formal methods to ensure that a model adequately reflects physical phenomena such as light or changing voltages. Despite these reservations, formal verification can be a very powerful tool, but it is important to realize that it can never give complete assurance.

2 Synchronized Transitions

A design in SYNCHRONIZED TRANSITIONS describes a computation as a set of independent, concurrently executing transitions. These transitions communicate via (shared) state variables. In addition to these two fundamental concepts: transitions and state variables, SYNCHRONIZED TRANSITIONS has language constructs for expressing initialization, parameterization, protocols, invariants, hierarchy (cells), etc. In Sect. 5, the language is presented through an example. There are strong conceptual similarities between SYNCHRONIZED TRANSITIONS and UNITY, as developed by Chandy and Misra [4]. Both describe a computation as a collection of atomic conditional assignments without any explicit flow of control. Chandy and Misra propose this as a general programming paradigm. Our application of SYNCHRONIZED TRANSITIONS is more specialized, in particular, the development of application specific integrated circuits. The main difference is the structuring concepts of SYNCHRONIZED TRANSITIONS: cells, parameters, statics, etc., in addition, there is a number of syntactical differences.

3 Mechanical Theorem Proving

There are many CAD tools available to help designers master the complexity of large designs. Most of these tools are aimed at a low level, e.g., at the layout or netlist level. There is a similar need for tools to assist in doing high-level design. This tutorial describes such tools that are used for verification.

Currently, research is going on with different approaches for verification, e.g., model checking and theorem proving (see [9] for a survey). The tools described in this tutorial are based on theorem proving, where a proof consists of a number of verification steps. To verify a large design, many verification steps are required. It becomes quite tedious to do this by hand, and practical experience with the approach described here tells that the verification rarely requires much insight or mathematical sophistication. Furthermore, the verification must be redone every time a design description is changed. By using a mechanical theorem prover, it is possible to reduce the routine work needed to do formal verification. There is a number of mechanical theorem provers available. They differ significantly, for example, in the expressive power of the notation allowed for stating a conjecture. However, the notion of proof used in all the currently available mechanical theorem provers is rather similar. A proof is a series of reductions transforming the conjecture into a formula known to be true (or false). The reductions are either built into the theorem prover or supplied by the user. For example, once a conjecture has been verified, it can be used to reduce other conjectures. This mechanical notion of a proof as something consisting of a finite number of applications of reductions is somewhat more restricted than the notion of proof usually found in mathematics. However, verification of most interesting properties of real designs does not require deep mathematical insight and, therefore, the mechanical notion of proof is usually adequate. When this is said, it is also acknowledged that for nontrivial designs, a significant effort is needed to do a formal verification. This is not only a question of finding a successful proof. Often the outcome is the detection of bugs in the design description; either in the explicitly stated requirements of the design, or in the algorithm implemented by the transitions.

The theorem prover LP [7] has been used in experiments with mechanized verification of designs in SYNCHRONIZED TRANSITIONS. LP is a theorem prover for a subset of multi-sorted first-order logic. It is designed to work efficiently on large problems and to be used by relatively naive users. LP is intended primarily for use as an interactive proof assistant, not as a fully automatic theorem prover. Its design is based on the assumption that initial attempts to state conjectures correctly, and then to prove them, usually fail. As a result, LP is designed to carry out routine (and possibly lengthy) steps in a proof automatically and to provide useful information about *why* proofs fail, if and when they do. Unlike the Boyer-Moore prover [1, 2], it does not use heuristics to formulate additional conjectures in a search for a proof that may not exist. Strategic decisions, such as trying induction, must appear as explicit LP commands (either entered by the user or generated by an application-specific front-end). When proof attempts fail, LP provides quick feedback; in fact, if and when time-consuming proof attempts occur, they are generally a sign that something is wrong.

The choice of a first-order logic has not been significant for the overall approach, but merely directed by the theorem prover. LP is used mainly of historical reasons; the axiomatization of conjectures and facts might be more elegant in a higher-order logic, however, the emphasis has been on providing efficient machine assistance, and LP has this far satisfied the needs.

4 Automatic Translation

A translator, called ST2LP, is used for translating design descriptions into verification conditions in the format required by LP. It extracts appropriate declarations for state variables, axioms defining transitions, protocols, invariants, and proof obligations from design descriptions. When formalizing conjectures, the translator often generates fruitful idioms for the proof. For example, when formalizing invariance proofs, the translator generates an LP command to initiate a proof by cases on the transition selected for execution. As a result of these idioms, LP often verifies invariants of designs directly from the output of the translator without any manual assistance. In many cases, when proofs are not completely automatic, there is a number of commands and standard techniques that can push proofs through manually. In the tutorial [16] it is shown how LP is used to do mechanical proofs based on proof scripts generated by ST2LP.

5 Black Jack Dealer

The Black Jack dealer is one of the benchmark examples in the collection distributed for the TPCD[1] conference. This example is used to introduce the notation and key concepts of SYNCHRONIZED TRANSITIONS.

A Black Jack dealer at a casino plays his own hand according to a completely predetermined procedure that leaves no possibility for human choice. This example presents a mechanical Black Jack dealer which plays the dealer's hand. Its inputs are *cardready* (true/false) and *card* (2 of Clubs, ..., Ace of Spades). Its outputs are *hitme, stand,* and *broke* (all truth-valued). The *cardready/hitme* signals are used for a four-phase handshake with the operator. Cards are valued from 2 to 10 (court cards have the value 10), and aces may be valued as either 1 or 11. The Black Jack dealer is repeatedly presented with cards. It must assert *stand* when its accumulated score reaches 17; and it must assert *broke* when its score exceeds 21. In either case the next card starts a new game.

5.1 Transitions and State Variables

A design is modeled in SYNCHRONIZED TRANSITIONS as a collection of transitions operating on a set of state variables, for example:

$$\ll idle \land \neg(broke \lor stand) \land cardready \rightarrow idle := FALSE \gg$$

This transition operates on the state variables *idle, broke, stand,* and *cardready;* it allows *idle* to change (to false) when the precondition $idle \land \neg(broke \lor stand) \land cardready$ holds. All four state variables are boolean. Another transition is:

$$\ll \neg idle \land \neg cardready \rightarrow idle := TRUE \gg$$

This allows *idle* to become true when the precondition holds. Operationally, the computation can be modelled as repeated nondeterministic selection and execution of an enabled transition. In this model, transitions are executed:

---

[1] Theorem Provers in Circuit Design

- *concurrently,*
- *repeatedly,* being ready for another execution immediately upon completion,
- *atomically,* as indivisible operations, and
- *independently* of the order they appear in the design description.

It is not required that a transition is executed immediately after it becomes enabled, because other enabled transitions may be selected. In fact, there is no upper bound on when a transition is selected. This models delays caused by the components in physical circuits. SYNCHRONIZED TRANSITIONS is not intended for specifying and dealing with explicit timing, for example, verifying that a given timing constraint is met. The verification results hold independently of the actual delay.

This is a very concrete and simple model of a computation, and it must be stressed that it is only a model, and a realization may execute transitions differently, as long as the realization does not enter states that cannot be entered by the model. As explained in [14], it is often possible to ensure atomicity, even if several transitions are executed in parallel, thus allowing efficient circuit realizations where many sub-circuits operate simultaneously.

5.2   Initial State

The initial state of a computation, i.e., the initial value of state variables is specified as follows:

```
INITIALLY
    idle = TRUE
    hitme = FALSE
    stand = FALSE
    broke = FALSE
```

It is not required that all state variables are given an initial value.

5.3   Invariants

It is important that the Black Jack dealer never allows *broke* and *hitme* to be true simultaneously (similarly *stand* and *hitme* must not be true simultaneously, neither must *idle* and *hitme*, etc.). These conditions are examples of invariants, and they specify a subset of the state space containing the "legal" states to enter during execution.

Invariants (and protocols introduced in Sect. 5.7) are stated explicitly by the designer to express properties of the design, for example:

```
INVARIANT
    ¬(broke ∧ hitme) ∧¬(idle ∧ hitme) ∧¬(stand ∧ hitme) ∧¬(broke ∧ stand)
```

Invariants do not influence the behavior of a design, and removing them from a design description does not change the computation. This redundancy is quite similar to the use of declarations in high-level languages, for example Pascal: it is used to check consistency between the declaration and use of various quantities. There is, however, an important difference. Normally, declarations are checked syntactically by compilers; this is not possible for invariants and protocols which express dynamic properties. Instead, one could check these

- by an informal, but convincing argument,
- by a simulation, where the design is executed on specific input data,
- by enumerating and checking all possible states and transitions,
- by a formal proof.

The emphasis in this tutorial is on verification based on formal proofs supported by a mechanical theorem prover. One can verify that a given predicate I is an invariant, by showing that:

1. I holds in the initial state, and
2. for each transition T, if I holds before executing T, then I also holds afterwards.

This approach was originally proposed by Floyd for software verification [5]. The second condition is established by assuming that the invariant holds for the values of the state variables prior to executing the transition (called the *pre-state*) and proving that the invariant holds for the values of the state variables afterwards (called the *post-state*). Consider, for example, the transition

$$\ll \neg idle \wedge \neg cardready \wedge hitme \rightarrow hitme := FALSE \gg$$

This transition certainly maintains the invariant: $\neg(idle \wedge hitme)$, because the transition only changes *hitme* to false; this can never violate the invariant.

The approach sketched above suggests that the verification is done by *considering the transitions one at a time*, and showing the implications separately for each transition. Note that the verification of a design is thereby reduced to a number of independent steps, one for each transition. This is important for mechanized verification, because a large design is handled by a number of (relatively) simple proofs instead of a few large and complicated ones. This is essentially what is done by the mechanical tools ST2LP and LP. For each transition, T, the translator generates a verification condition of the form:

$$C: \quad I(pre) \wedge T(pre, post) \Rightarrow I(post)$$

where $T(pre, post)$ is a predicate corresponding to a transition T. LP then tries to verify each of the verification conditions.

5.4   Integers and Subranges

In addition to the boolean state variables used above, a design may contain state variables of type integer or a subrange of integer values, for example:

TYPE cardval = [1..10]

This specifies a type named *cardval* consisting of the integers in the range 1 to 10. Aces are represented by the value 1:

STATIC ace = 1

A type is used to specify the value range of a state variable, for example:

STATE card: cardval

The state variable *card* may hold any of the values indicated by the type *cardval*, in this case the range 1 to 10. This is a transition using such a state variable:

$$\ll card \neq ace \rightarrow count := count + card \gg$$

It is important to check that the value assigned to a state variable is in the domain indicated by the type of the state variable. For example, the state variable *card* must only be assigned values in the range 1 to 10. In many programming languages this check is done at run-time by executing a check every time a new value is assigned. However, for a design language like SYNCHRONIZED TRANSITIONS where many designs are realized as circuits, it is often worthwhile to do the checking before executing the design. When this is possible, there is no need to do the checking during execution. The specification of a range is rather similar to an invariant. The range is also redundant information and it cannot be checked statically. In fact, the range information can be transformed into an invariant and verified together with other invariants. The tools supporting verification of designs in SYNCHRONIZED TRANSITIONS are capable of automatically generating such range invariants and other similar conditions ensuring against indexing arrays out of bounds, division by zero, etc. These invariants are called well-formedness invariants. Well-formedness invariants for the Black Jack design are discussed in Sect. 5.11.

5.5 Records

Boolean, integer, and subrange are simple types. Structured types like records and arrays are used to group associated values, e.g., a record typically describes a number of different attributes of a data item. In the Black Jack example, the dealer's hand is described as a record with two components, a flag that indicates that the hand has at least one ace, and the accumulated score (where aces are counted as 1's):

handtype = RECORD
          anyaces: BOOLEAN
          count: scorerange
     END

If *anyaces* is true the dealer might count one of the aces as 11, thereby obtaining a score of *count* + 10 (he will never count 2 aces as 11).

As it is the case in Pascal and many other high-level programming languages the individual components of a variable, *c*, of type record are obtained by giving the name of the state variable concatenated with the name of the component, e.g., *c.anyaces*, as in the transition

$$\ll card = ace \rightarrow c.anyaces := TRUE \gg$$

5.6   Functions

Functions describe mappings from a list of parameters to a value, for example:

> FUNCTION evallow(h: handtype): scorerange
>   RETURN h.count
> FUNCTION evalhigh(h: handtype): extscorerange
>   RETURN h.count+IF h.anyaces THEN 10 ELSE 0

Functions can be used in all kinds of expressions both in transitions and to describe invariants:

> INVARIANT broke $\Rightarrow$ evallow(hand) > 21

5.7   Protocols

Protocols are predicates on pairs of states, *pre, post*, defining a restriction on the allowable transitions between states (to ones where the pre- and post-state satisfy the predicate). The following is an example of a protocol, stating that x does not change.

> PROTOCOL x.pre = x.post

x.pre denotes the value of x in the pre-state and similarly *x.post* is the value of x in the post-state. The four-phase handshake protocol is specified as follows.

> FUNCTION fourphase(a, b: BOOLEAN): BOOLEAN
>   RETURN ((a.pre$\neq$a.post) $\Rightarrow$ (a.post$\neq$b.pre)) $\wedge$
>   ((b.pre$\neq$b.post) $\Rightarrow$ (b.post=a.pre))

To meet the protocol *fourphase*, a must get the value of $\neg b$ (or be unchanged), and when b changes it must get the value of a. For instance, *idle* and *cardready* follows this protocol:

> PROTOCOL fourphase(idle, cardready)

Like invariants, protocols are verified mechanically. The verification condition, C, given in Sect. 5.3, must be extended to also include verification of protocols. Assume that a design has the protocol P and invariant I, then for each transition, T, the translator ST2LP generates a verification condition of the form:

$$I(pre) \wedge T(pre, post) \Rightarrow I(post) \wedge P(pre, post)$$

5.8   The Asynchronous Combinator

The asynchronous combinator is used to describe the composition of a number of independent transitions. Consider for example:

- a transition for detecting that the dealer is broke

> $\ll$ $\neg idle \wedge$ cardready $\wedge$ evallow(hand) > 21 $\rightarrow$ broke:= TRUE $\gg$

- a transition for detecting that the dealer must stand

$\ll \neg idle \wedge cardready \wedge (standrange(evallow(hand)) \vee$
$\quad standrange(evalhigh(hand))) \rightarrow stand := TRUE \gg$

– a transition for determining that a new card can be dealt

$\ll \neg idle \wedge cardready \wedge evallow(hand) \leq 16 \wedge$
$\quad \neg standrange(evalhigh(hand)) \rightarrow hitme := TRUE \gg$

Each of these transitions (plus another handful) are needed to complete the design. They are combined using the asynchronous combinator $\parallel$.

5.9 The Product Combinator

The product combinator is used to factor a transition into a number of simpler transitions. Let t_1, t_2 be the following two transitions (where v_1 and v_2 are different state variables):

$TRANSITION\ t_1 \ll c_1 \rightarrow v_1 := e_1 \gg \qquad TRANSITION\ t_2 \ll c_2 \rightarrow v_2 := e_2 \gg$

The product, $t_1 * t_2$, is equivalent to the following transition:

$\ll c_1 \wedge c_2 \rightarrow v_1, v_2 := e_1, e_2 \gg$

It is emphasized that the product composition of two transitions is not equivalent to the separate use of the transitions; the two transitions in a product operate simultaneously, as if it was a single transition (formed as shown above).

The product combinator makes it possible to separate transitions dealing with different aspects. The product combinator does not add expressive power, but in many cases it yields simpler and more comprehensible designs.

As an important special case, consider a transition description consisting of a precondition only, for example: $\ll \neg idle \wedge cardready \gg$. The product combinator can be used to factor this precondition out of a number of other transition descriptions, for example, the transitions given in Sect. 5.8:

$\ll \neg idle \wedge cardready \gg * ($
$\quad \ll evallow(hand) > 21 \rightarrow broke := TRUE \gg \parallel$
$\quad \ll standrange(evallow(hand)) \vee standrange(evalhigh(hand))$
$\quad\quad \rightarrow stand := TRUE \gg \parallel$
$\quad \ll evallow(hand) \leq 16 \wedge \neg standrange(evalhigh(hand)) \rightarrow hitme := TRUE \gg)$

There is a third combinator, $+$, the synchronous combinator. It is used for describing a design where two or more enabled transitions are done simultaneously. UNITY has a similar combinator (written as $\parallel$). Section 6.3 describes an example using synchronous composition.

5.10 Design of a Black Jack Dealer

A hand is represented by the sum of the card values (where aces are counted as 1's), and by a flag indicating that the hand has at least one ace. A low and high score of a hand are computed by the two functions *evallow* and *evalhigh*. The low score is computed by counting aces as 1's, i.e., it is just the accumulated score of the hand; the high score is computed by counting one ace (if any) as 11. If either the low or the high score is between 17 and 21 (determined by the function *standrange*) the dealer must stand.

TYPE scorerange = [0..26]
TYPE extscorerange = [0..36]
FUNCTION evallow(c: handtype): scorerange RETURN c.count
FUNCTION evalhigh(c: handtype): extscorerange
  RETURN c.count+IF c.anyaces THEN 10 ELSE 0
FUNCTION standrange(s: extscorerange): BOOLEAN
  RETURN (16 < s) ∧ (s ≤ 21)

The following invariant captures the relationship between the state variables *hand*, *hitme*, *stand*, and *broke*:

INVARIANT
  ¬ idle ⇒ (
  (broke ⇒ (evallow(hand) > 21)) ∧
  (stand ⇒ (standrange(evallow(hand)) ∨ standrange(evalhigh(hand)))) ∧
  (hitme ⇒ (evallow(hand)≤16 ∧ ¬ standrange(evalhigh(hand)))))

Sections 5.1–5.10 have explained most of the design description of the Black Jack dealer. In Fig. 1 all the pieces have been put together.

5.11 Verifying the Black Jack Design

The invariants (I) and protocols (P) specified in the design description can be verified using the translator (ST2LP) and the theorem prover (LP). The verification is divided into a number of separate proofs following the scheme introduced above, i.e., it must be verified that:

- I holds in the initial state, and
- for each transition, T: $I(pre) \land T(pre, post) \Rightarrow I(post) \land P(pre, post)$

The translator generates the corresponding verification conditions: one for the initialization and one for each of the transitions in the design. All of these are verified independently, and it turns out that the theorem prover handles four of the conditions automatically, whereas (a little) manual assistance is needed to verify the remaining four conditions. It is too technical to show the proofs in more detail. In [16] it is shown how proofs for SYNCHRONIZED TRANSITIONS designs look like in LP.

To conclude that the design has the properties stated in the invariants and protocols, it must also be shown that the design description is well-formed. This is partly done by the translator which does the static checks, e.g., that the number of formal and actual parameters to functions match, that types are used consistently, etc. However, as mentioned previously, some checks cannot be done by ordinary static analysis, and in many programming languages such checks are instead done at run-time. The tools supporting verification of designs in SYNCHRONIZED TRANSITIONS automatically extract these dynamic properties and formulate them as (well-formedness) invariants (I_{wf}), ready for verification

```
STATIC ace = 1
TYPE
  cardval = [1..10] scorerange = [0..26] extscorerange = [0..36]
  handtype = RECORD anyaces: BOOLEAN; count: scorerange END
CELL blackjack(cardready, hitme, stand, broke: BOOLEAN; card: cardval; hand: handtype);
  FUNCTION evallow(h: handtype): scorerange RETURN h.count
  FUNCTION evalhigh(h: handtype): extscorerange
    RETURN h.count+IF h.anyaces THEN 10 ELSE 0
  FUNCTION standrange(s: extscorerange): BOOLEAN RETURN (16 < s) ∧ (s ≤ 21)
  FUNCTION fourphase(a, b: BOOLEAN): BOOLEAN
    RETURN ((a.pre≠a.post) ⇒ (a.post≠b.pre)) ∧
           ((b.pre≠b.post) ⇒ (b.post=a.pre))
  STATE idle: BOOLEAN
  INVARIANT
    ¬(broke ∧ stand) ∧ ¬(broke ∧ hitme) ∧ ¬(stand ∧ hitme) ∧ ¬(idle ∧ hitme) ∧
    ¬idle ⇒ (
      ( broke ⇒ (evallow(hand) > 21) ) ∧
      ( stand ⇒ (standrange(evallow(hand)) ∨ standrange(evalhigh(hand))) ) ∧
      ( hitme ⇒ (evallow(hand)≤16 ∧ ¬standrange(evalhigh(hand))) ) ) ∧
    idle ⇒ ( (¬(stand ∨ broke)) ⇒ (evallow(hand)≤16 ∧ ¬standrange(evalhigh(hand))) )
  PROTOCOL
    fourphase(idle, cardready) ∧
    (¬(broke.pre ∨ stand.pre) ⇒ fourphase(cardready, hitme)) ∧
    (¬(hitme.pre ∨ stand.pre) ⇒ fourphase(cardready, broke)) ∧
    (¬(hitme.pre ∨ broke.pre) ⇒ fourphase(cardready, stand))
  INITIALLY
    idle = TRUE hitme = FALSE stand = FALSE broke = FALSE
    hand.anyaces = FALSE hand.count = 0
BEGIN
  ≪ idle ∧ (broke ∨ stand) ∧ ¬cardready →
      hand.anyaces, hand.count, broke, stand:= FALSE, 0, FALSE, FALSE ≫ ||
  ≪ idle ∧ ¬(broke ∨ stand) ∧ cardready → idle:= FALSE ≫ *
    ≪ hand.count := hand.count+card ≫ * (
      ≪ card=ace → hand.anyaces:= TRUE ≫ ||
      ≪ card≠ace ≫ ) ||
  ≪ ¬idle ∧ cardready ≫ * (
    ≪ evallow(hand) > 21 → broke:= TRUE ≫ ||
    ≪ standrange(evallow(hand)) ∨ standrange(evalhigh(hand)) → stand:= TRUE ≫ ||
    ≪ evallow(hand)≤16 ∧ ¬standrange(evalhigh(hand)) → hitme:= TRUE ≫ ) ||
  ≪ ¬idle ∧ ¬cardready → idle := TRUE ≫ * (
    ≪ hitme → hitme:= FALSE ≫ ||
    ≪ broke ∨ stand ≫ )
END blackjack
```

Fig. 1. Black Jack design in SYNCHRONIZED TRANSITIONS.

using the theorem prover. For instance, *hand.count* should keep a value in the integer range specified by its type, *scorerange*:

$$I_{wf} : 0 \leq hand.count \leq 26$$

The invariants for well-formedness initiate a set of proof obligations similar to those for the ordinary invariants and protocols, i.e., they must be verified in the initial state, and each transition should preserve them. Theoretically, they could be verified at the same time as the ordinary invariants by defining the invariant as the conjunction of the ordinary and well-formedness invariants, however, it is more practical to do these proofs independently. In this way it is possible to concentrate the initial efforts in the verification process on the ordinary invariants which often capture the more interesting and essential properties of the design.

For instance, the proof of the well-formedness invariant for the first transition needs manual assistance, whereas the proof of the ordinary invariant is handled automatically. If the invariants were verified at the same time, manual assistance is required, whereas, a separate verification makes it possible to verify the ordinary invariant automatically.

5.12 Summary of Black Jack Design

A number of invariants and protocols were verified for the Black Jack dealer:

- Relationships between the state variables *broke, stand, hitme* as explained in Sect. 5.3, e.g., that *broke* and *hitme* are not true simultaneously,
- relationships between the state variables *broke, stand, hitme* and the score of a hand as explained in Sect. 5.10, e.g., that the low score of a hand is greater than 21 when *broke* is true,
- that the Black Jack dealer obeys the four-phase handshake protocol as explained in Sect. 5.7,
- various well-formedness invariants as explained in Sect. 5.11, e.g., that state variables are not assigned a value outside the range specified by their type.

The verification of the properties listed above was performed with LP version 2.4x on a DEC Alpha 3000/400 with 96 Mb of memory. It takes 1:05 minutes for LP to read in the axiomatization of the Black Jack dealer. During this process the axioms are transformed into rewrite rules and facts are inter-normalized. The proofs of the ordinary (i.e., user supplied) invariants and protocols for the eight transitions take 3:28 minutes, whereas the proofs for the well-formedness invariants take 2:56 minutes. Four of the proofs of the ordinary invariants and protocols need manual assistance. Similarly, two of the proofs of the well-formedness invariants need manual assistance. None of the proofs need more than five user supplied LP commands. The two initialization proofs of the ordinary invariants and the well-formedness invariants take 0:26 minutes and 0:44 minutes, respectively, both proofs need manual assistance (less than ten LP commands).

Besides the CPU time used to do the proofs there is another significant factor: the "human" time used by the designer to supply manual assistance in the proofs; no attempt is made here to quantify this factor. Most of the manual assistance is needed to complete proof steps involving inequalities. This is quite typical, and proofs involving only equalities and booleans tend to require less manual assistance.

The design of the Black Jack dealer illustrates both the SYNCHRONIZED TRANSITIONS notation and the mechanical verification tools. For this example, the most interesting part of the verification is the range checking, i.e., that all assignments of new values to state variables are within the bounds given in the declaration of the state variables. This is a non-trivial and very important aspect of this design, and the tools are able to verify that no value will be out of bounds when this design is executed. Hence, there is no need to do a run-time check.

6   Examples from the Benchmark Collection

This section summarizes the verification of other examples from the TPCD benchmark collection.

6.1   The Traffic Light Controller

This is used as an explanatory example in [16]. It is shown that a design description in SYNCHRONIZED TRANSITIONS maintains the invariant:

$EWlight = red \lor NSlight = red$

Several versions of the design are also discussed in [14].

6.2   The N-Bit Adder

In [4] an efficient adder is described. The design (and efficiency) of this adder is based on a non-trivial invariant. In [13] it is shown how ST2LP and LP are used to verify a very efficient realization of an N-bit adder (without fixing the value of N).

6.3   Min-Max

This is mainly a descriptive benchmark. Below it is shown how to describe the design in SYNCHRONIZED TRANSITIONS. This example illustrates the use of synchronous composition of transitions (indicated by the operator $+$). Synchronous composition is used to specify that two operations are always performed simultaneously (e.g., under control of a global clock); whereas, no such assumptions are made with asynchronous composition.

```
TYPE range = [-256..255]
CELL minmax(in, out: range; clear, enable, reset: BOOLEAN)
STATE
  buffer, min, max: range
BEGIN
```

$\ll$ *clear* $\rightarrow$ *out:=0* $\gg$ +
$\ll$ $\neg$*clear* $\wedge$ $\neg$*enable* $\rightarrow$ *out:= buffer* $\gg$ +
$\ll$ *enable* $\rightarrow$ *buffer:= in* $\gg$ +
$\ll$ *reset* $\wedge$ $\neg$*enable* $\rightarrow$ *buffer, min, max:= 0, 255, -256* $\gg$ +
$\ll$ *(in > max)* $\wedge$ $\neg$*reset* $\rightarrow$ *max:= in* $\gg$ +
$\ll$ *(in < min)* $\wedge$ $\neg$*reset* $\rightarrow$ *min:= in* $\gg$ +
$\ll$ $\neg$*clear* $\wedge$ $\neg$*reset* $\wedge$ *enable* $\rightarrow$ *out:= (min+max) / 2* $\gg$ +
$\ll$ $\neg$*clear* $\wedge$ *enable* $\wedge$ *reset* $\rightarrow$ *out:= in* $\gg$
END minmax

6.4 The Arbiter

This example is described in the paper [15] which also discusses the formal verification of mutual exclusion and a four-phase protocol. This arbiter design has a tree structure and this is reflected in the design description which is recursive. Such regular and other modular structures are verified using a localized verification technique where each cell is verified separately. By using this localized verification technique, it can be avoided that the verification of large designs blows up.

6.5 The Tamarack Processor

This is an example of implementation verification and hence very different from the other examples described in this tutorial. However, it is also possible to do mechanized implementation verification of designs described in SYNCHRONIZED TRANSITIONS. There is a separate translator [12] that generates verification conditions for LP from two design descriptions: an abstract design (the specification) and a concrete design (the realization). This has been used to verify the Tamarack processor.

6.6 The Stop-Watch, GCD, and FIFO

These examples have all been described in SYNCHRONIZED TRANSITIONS and various aspects have been verified using ST2LP and LP, see [14].

7 Larger Designs

It is rarely feasible to view a non-trivial design as one huge collection of transitions. In SYNCHRONIZED TRANSITIONS a design may be modularized by breaking it into cells. For example, the Black Jack dealer is a cell. The Black Jack dealer takes care of the game rules, etc., and it relies on a deck of cards that supplies (legal) cards at the right time. The deck of cards is another cell that works independently of the Black Jack cell. The dealer and the deck of cards communicates through shared variables, e.g., the dealer signals that it wants a new card using the state variable *hitme*, and the deck of cards signals that a

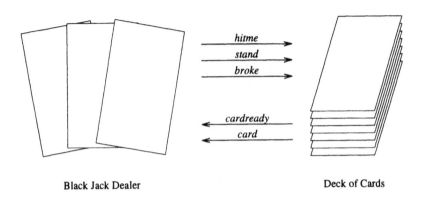

Fig. 2. Shared variable interface.

new card is ready using the state variable *cardready*. The interface consisting of shared variables is illustrated in Fig. 2. Breaking down a large design into cells is not only an aid when describing the design; it also makes the design easier to verify! This section presents a localized technique for verifying modular designs. The technique was originally proposed in [15] and it has been further developed in [17]. Each cell of the design is verified in isolation, showing what is called local correctness. Furthermore, it is verified that each cell does not violate the protocols and invariants of other cells. This is called non-interference, and it is verified solely on the basis of the interface of the cells, without considering their internal details. The localized verification technique is illustrated on the Black Jack dealer extended with a deck of cards cell.

The deck of cards cell must supply cards as the dealer is requesting them. This is the case if the dealer needs more cards for the hand (signalled with the *hitme* state variable), or when a game has finished and a new game must be started (signalled with the state variables *stand* and *broke*). The deck of cards cell uses the state variable *cardready* to tell when a new card is ready. The deck of cards cell may potentially break the four-phase protocol stated in the Black Jack dealer cell because it writes the state variable *cardready*. It is now shown how localized verification is used to verify that this is not the case.

7.1 Local Correctness

Local correctness is one part of the localized verification technique, where a cell is verified independently of its context. To do this, it is shown that all transitions of a cell maintain the invariant and protocol of the cell. For the Black Jack dealer cell this corresponds to the verification condition explained in Sect. 5. However, adding additional cells such as the deck of cards cell may introduce additional invariants and protocols describing properties of the interface. These additional properties are verified locally as described in Sect. 5. For instance, the deck of cards cell complies with the protocol:

INTERNAL PROTOCOL
$$cardready.pre \neq cardready.post \Rightarrow$$
$$cardready.post = \neg(hitme.pre \lor broke.pre \lor stand.pre)$$

This protocol is called an internal protocol because it only holds for the transitions in the deck of cards cell; it need not (and does not) hold when transitions are executed in the Black Jack dealer cell.

The state variables in the two cells are shared; this allows the cells to communicate. But it also means that the transitions of one cell can violate the invariant and protocol of the other cell (interference). For example, the four-phase protocol between *hitme* and *cardready* is verified in the local correctness proof in the Black Jack cell (where *hitme* is changed), but the deck of cards cell writes *cardready*, and this might violate the four-phase protocol. Hence, local correctness is not sufficient to guarantee global correctness. In addition, it must be shown that cells do not interfere.

7.2 Non-interference

Non-interference is the complementary part of the localized verification technique. In this part, it is verified that invariants and protocols involving shared variables (parameters) are preserved. To verify that all invariants and protocols are maintained, it must be shown that the transitions in the deck of cards cell do not interfere with the invariants and protocols in the Black Jack cell, and it must also be shown that the transitions in the Black Jack cell do not interfere with the invariants and protocols in the deck of cards cell.

Instead of showing non-interference for *each transition* separately, it is done once for each cell. The key idea is to assume that the internal protocol, the invariant and the protocol of, e.g., the deck of cards cell capture the essential properties of all transitions in a *single* assertion. This is expressed in the two non-interference conditions presented below.

$$I_{deck}(post) \land P_{deck}(pre, post) \land IP_{deck}(pre, post) \Rightarrow I_{bj}(post) \land P_{bj}(pre, post)$$
$$I_{bj}(post) \land P_{bj}(pre, post) \land IP_{bj}(pre, post) \Rightarrow I_{deck}(post) \land P_{deck}(pre, post)$$

(I, P, and IP are the invariant, protocol and internal protocol in the Black Jack cell (subscript bj) and deck of cards cell (subscript $deck$)). Note that the non-interference conditions avoid considering the individual transitions of the Black Jack dealer and deck of card cells. For example, it is shown from the internal protocol of the deck of cards cell that this cell does not interfere with the four-phase protocol (P_{bj}). This is shown without considering the transitions of the deck of cards cell; as long as they comply with the internal protocol of the cell (this is verified as local correctness), the deck of cards cell will not violate the four-phase protocol.

In general, the cells of a design are arranged in a tree of cell instantiations. It is not necessary to verify the non-interference conditions between any two cells in the design, but only between "neighbor" cells in the instantiation tree. By showing the two non-interference conditions for all the neighbor pairs it is also (implicitly) shown that changes made in accordance with the protocol for one neighboring cell are also in accordance with the protocols for other neighbors. Therefore it circumvents the need to consider separately interferences that might result when parameters are passed through several levels. In [13] the technique is explained in more detail, and it is shown that the technique is sound.

8 Experience

We have used LP and SYNCHRONIZED TRANSITIONS together to verify safety properties of several small, but interesting designs such as an arbiter[16], Chandy and Misra's efficient adder[4], a priority queue[17], and a set of building blocks for asynchronous circuits. We have also used these tools when teaching courses. Students have verified a series of examples such as the Black Jack design.

Designs described with SYNCHRONIZED TRANSITIONS can be manipulated and analyzed in a number of ways:

1. *Design verification:* Proving that transitions maintain properties formulated as invariant assertions.
2. *Implementation verification:* Proving that a concrete program, the realization, is a correct implementation of another program, the abstraction.
3. *Simulation:* Informally verifying a design by executing it with various input data.
4. *Synthesis:* Transforming a design into a circuit description, e.g., a netlist or a layout.
5. *Verification of implementation conditions:* Checking restrictions required for the realization to operate correctly.

Separate experimental tools support (1)–(5). This paper has concentrated on (1), for which LP is used together with a translator that translates a design description into a set of verification conditions. A similar translator for (2) is under development[12].

Our initial experience with LP and SYNCHRONIZED TRANSITIONS is encouraging. We believe that it is currently realistic to design moderate-sized circuits in a high-level language supported by both verification and synthesis tools.

Current techniques and tools for formal verification are not sufficiently refined for widespread application, but the potential benefits are so significant that we think they must be seriously investigated. When formal verification works it corresponds to an exhaustive simulation where a certain property is ensured for all combinations of inputs in all states. Since exhaustive simulation is rarely feasible in practice, formal verification has the potential of improving the quality of integrated circuit design.

8.1 Availability of Tools

LP is written in CLU and runs under Unix. Release 2.4 is currently available free of charge, and without a license, by anonymous ftp from larch.lcs.mit.edu.

Prospective users can retrieve an executable version of LP, along with a supporting run-time library containing sample axiom sets and proofs, for DEC-stations, Sparc-stations, MIPS machines, Vaxes, or Sun-3 workstations, and soon DEC Alpha. Source code is also available.

The tools supporting SYNCHRONIZED TRANSITIONS are still undergoing development. They are written in C and run under Unix and can be obtained by anonymous ftp from ftp.id.dth.dk or by contacting the authors.

Acknowledgements

The work reported in this paper has been supported by The Danish Technical Research Council and by Digital Equipment Corporation through an external research grant. The authors are grateful to Stephen Garland and John Guttag from MIT, Cambridge, USA who have contributed significantly to the approach described in this paper.

References

1. Robert S. Boyer and J. Strother Moore. *A Computational Logic*. Academic Press, 1979.
2. Robert S. Boyer and J. Strother Moore. *A Computational Logic Handbook*. Academic Press, 1988.
3. Randal E. Bryant. Can a simulator verify a circuit? In *Formal Aspects of VLSI Design*, pages 125–136. North-Holland, 1985.
4. K. Mani Chandy and Jajadev Misra. *Parallel Program Design: A Foundation*. Addison-Wesley, 1988.
5. R.W. Floyd. Assigning meanings to programs. In J.T. Schwartz, editor, *Proceedings of the Symposium in Applied Mathematics*, volume 19, pages 19–32. American Mathematical Society, 1967.
6. Michael P. Fourman and Robert L. Harris. Lambda – logic and mathematics behind design automation. In *Proceedings of the 26th ACM/IEEE design Automation Conference*. ACM, 1989.
7. Stephen J. Garland and John V. Guttag. An overview of LP: the Larch Prover. In *Proceedings of the Third International Conference on Rewriting Techniques and Applications*. Springer-Verlag, 1989.
8. Mike Gordon. Why higher-order logic is a good formalism for specifying and verifying hardware. In *Formal Aspects of VLSI Design*, pages 153–177. North-Holland, 1985.
9. Aarti Gupta. Formal hardware verification methods: A survey. *Formal Methods in Systems Design*, 1(2/3):151–238, October 1992.
10. John V. Guttag, James J. Horning with S.J. Garland, K.D. Jones, A. Modet, and J.M. Wing. *Larch: Languages and Tools for Formal Specification*. Springer-Verlag Texts and Monographs in Computer Science, 1993. ISBN 0-387-94006-5, ISBN 3-540-94006-5.
11. Warren A. Hunt. FM8501: A verified microprocessor. In *From HDL Descriptions to Guaranteed Correct Circuit Designs*, pages 85–114. North-Holland, 1986.
12. Niels Maretti. Mechanized verification of refinement. In *Proceedings from TPCD '94*, 1994. To appear.

13. Niels Mellergaard. *Mechanized Design Verification*. PhD thesis, Department of Computer Science, Technical University of Denmark, 1994.

14. Jørgen Staunstrup. *A Formal Approach to Hardware Design*. Kluwer Academic Publishers, 1994.

15. Jørgen Staunstrup, Stephen J. Garland, and John V. Guttag. Localized verification of circuit descriptions. In *Proceedings of the Workshop on Automatic Verification Methods for Finite State Systems, LNCS 407*. Springer Verlag, 1989.

16. Jørgen Staunstrup, Stephen J. Garland, and John V. Guttag. Mechanized verification of circuit descriptions using the Larch Prover. In V. Stavridou and T. Melham, editors, *Proceedings of the IFIP WG 10.2 International Conference on Theorem Provers in Circuit Design: Theory, Practice and Experience*, pages 277–300. Elsevier, 1992.

17. Jørgen Staunstrup and Niels Mellergaard. Localized verification of modular designs. *Formal Methods in System Design*, 1994. accepted for publication.

A Tutorial on Using PVS for Hardware Verification

S. Owre,[1] J. M. Rushby,[1] N. Shankar[1] and M. K. Srivas

{owre, rushby, shankar, srivas}@csl.sri.com

Computer Science Laboratory, SRI International, Menlo Park CA 94025 USA

Abstract. PVS stands for "Prototype Verification System." It consists of a specification language integrated with support tools and a theorem prover. PVS tries to provide the mechanization needed to apply formal methods both rigorously and productively.

This tutorial serves to introduce PVS and its use in the context of hardware verification. In the first section, we briefly sketch the purposes for which PVS is intended and the rationale behind its design, mention some of the uses that we and others are making of it. We give an overview of the PVS specification language and proof checker. The PVS language, system, and theorem prover each have their own reference manuals, [1,2,3] which you will need to study in order to make productive use of the system. A pocket reference card, summarizing all the features of the PVS language, system, and prover is also available.

The purpose of this tutorial is not to describe in detail the features of PVS and how to use the system. Rather, its purpose is to introduce some of the more unique and powerful capabilities that are provided by PVS and demonstrate how these features can be used in the context of hardware verification. We present completely worked out proofs of two hardware examples. One of the examples is a pipelined microprocessor that has been used as benchmark for model checkers and the other is a parameterized implementation of an N-bit ripple-carry adder.

1 Introducing PVS

PVS stands for "Prototype Verification System." It consists of a specification language integrated with support tools and a theorem prover. PVS tries to provide the mechanization needed to apply formal methods both rigorously and productively.

---

[1] S. Owre, N. Shankar, and J. M. Rushby. *The PVS Specification Language (Beta Release).* Computer Science Laboratory, SRI International, Menlo Park, CA, February 1993.

[2] N. Shankar, S. Owre, and J. M. Rushby. *The PVS Proof Checker: A Reference Manual (Beta Release).* Computer Science Laboratory, SRI International, Menlo Park, CA, February 1993.

[3] S. Owre, N. Shankar, and J. M. Rushby. *User Guide for the PVS Specification and Verification System (Beta Release).* Computer Science Laboratory, SRI International, Menlo Park, CA, February 1993.

The specification language of PVS is a higher-order logic with a rich type-system, and is quite expressive; we have found that most of the mathematical and computational concepts we wish to describe can be formulated very directly and naturally in PVS. Its theorem prover, or proof checker (we use either term, though the latter is more correct), is both interactive and highly mechanized: the user chooses each step that is to be applied and PVS performs it, displays the result, and then waits for the next command. PVS differs from most other interactive theorem provers in the power of its basic steps: these can invoke decision procedures for arithmetic and equality, a BDD-based propositional simplifier, efficient hashing-based automatic conditional rewriting, induction, and other relatively large units of deduction; it differs from other highly automated theorem provers in being directly controlled by the user. We have been able to perform some significant new hardware verification exercises quite economically using PVS; we have also repeated some verifications first undertaken in other systems and have usually been able to complete them in a fraction of the original time (of course, these are previously solved problems, which makes them much easier for us than for the original developers).

PVS is the most recent in a line of specification languages, theorem provers, and verification systems developed at SRI, dating back over 20 years. That line includes the Jovial Verification System [13], the Hierarchical Development Methodology (HDM) [25, 26], STP [30], and EHDM [22, 27]. We call PVS a "Prototype Verification System," because it was built partly as a lightweight prototype to explore "next generation" technology for EHDM, our main, heavy-weight, verification system. Another goal for PVS was that it should be freely available, require no costly licenses, and be relatively easy to install, maintain, and use. Development of PVS was funded entirely by SRI International.

The purpose of this tutorial is not to describe in detail the features of PVS and how to use the system. Rather, its purpose is to introduce some of the more unique and powerful capabilities that are provided by PVS and demonstrate how these features can be used in the context of hardware verification. We present completely worked out proofs of two hardware examples. One of the examples is a pipelined microprocessor that has been used as benchmark for testing the capacity of model checkers to handle datapath-oriented circuits. While the size of the datapath is irrelevant in a theorem proving exercise, we wanted to see if the proof would go through just as automatically as in a model checker. The second example is one of the circuits supplied as a TPCD benchmark: a parameterized implementation of an N-bit ripple-carry adder. The second example illustrates proof by induction.

1.1 Design Goals for PVS

The design of PVS was shaped by our experience in doing or contemplating early-lifecycle applications of formal methods. Many of the larger examples we have done concern algorithms and architectures for fault-tolerance (see [23] for an overview). We found that many of the published proofs that we attempted to check were in fact, incorrect, as was one of the important algorithms. We have also found that many of our own specifications are subtly flawed when first written. For these reasons, PVS is designed to help in the detection of errors as well as in the confirmation of "correctness." One way it supports early error detection is by having a very rich type-system and correspondingly rigorous typechecking. A great deal of specification can be embedded in PVS types (for example, the invariant to be maintained by a state-machine can be expressed as a type constraint), and typechecking can generate proof obligations that amount to a very strong consistency check on some aspects of the specification.

Another way PVS helps eliminate certain kinds of errors is by providing very rich mechanisms for conservative extension—that is, definitional forms that are guaranteed to preserve consistency. Axiomatic specifications can be very effective for certain kinds of problem (e.g., for stating assumptions about the environment), but axioms can also introduce inconsistencies—and our experience has been that this does happen rather more often than one would wish. Definitional constructs avoid this problem, but a limited repertoire of such constructs (e.g., requiring everything to be specified as a recursive function) can lead to excessively constructive specifications: specifications that say "how" rather than "what." PVS provides both the freedom of axiomatic specifications, and the safety of a generous collection of definitional and constructive forms, so that users may choose the style of specification most appropriate to their problems.[4]

The third way that PVS supports error detection is by providing an effective theorem prover. The design rationale behind the PVS theorem prover was to provide automatic support for obvious and tedious parts of a proof while giving the user the ability to guide the prover at higher levels of a proof. This goal is accomplished by implementing the primitive inference steps of PVS using automatic rewriting and efficient decision procedures for arithmetic and propositional logic. This approach makes PVS an effective system for hardware verification since most hardware proofs need significant amount of rewriting and case analyses.

Our experience has been that the act of trying to prove properties about specifications is the most effective way to truly understand their content and to identify errors. This can come about incidentally, while attempting to prove a "real" theorem, such as that an algorithm achieves its purpose, or it can be done deliberately through the process of "challenging" specifications as part of a validation process. A challenge has the form "if this specification is right, then the following ought to follow"—it is a test case posed as a putative theorem; we "execute" the specification by proving theorems about it.[5]

---

[4] Unlike EHDM, PVS does not provide special facilities for demonstrating the consistency of axiomatic specifications. We do expect to provide these in a later release, but using a different approach than EHDM.

[5] Directly executable specification languages (e.g., [2, 17]) support validation of specifications by running conventional test cases. We think there can be merit in this

1.2   Uses of PVS

PVS has so far been applied to several small demonstration examples, and a growing number of significant verifications. The smaller examples include the specification and verification of ordered binary tree insertion [28], the Boyer-Moore majority algorithm, an abstract pipelined processor, Fischer's real-time mutual exclusion protocol, and the Oral Messages protocol for Byzantine agreement. Examples of this scale can typically be completed within a day. More substantial examples include the correspondence between the programmer and RTL level of a simple hardware processor [11], the correctness of a real-time railroad crossing controller [29], a variant of the Schröder-Bernstein theorem, and the correctness of a distributed agreement protocol for a hybrid fault model consisting of Byzantine, symmetric, and crash faults [19]. These harder examples can take from several days to a week.

Currently, PVS is being applied to the requirements specification of selected aspects of the control software for NASA's space shuttle project and to verify a commercial pipelined microprocessor, AAMP5, being built for avionics applications at Rockwell International.

2   The PVS Language

The PVS specification language builds on a classical typed higher-order logic. The base types consist of booleans, real numbers, rationals, integers, natural numbers, lists, and so forth. The primitive type constructors include those for forming function (e.g., [nat -> nat]), record (e.g., [# a : nat, b : list[nat]#]), and tuple types (e.g., [int, list[nat]]). PVS departs from simply typed logics by allowing *predicate subtypes*. A predicate subtype consists of exactly those elements of a given type satisfying a given predicate so that, for example, the subtype of positive numbers is given by the type {n : nat | n > 0}. Predicate subtypes are used to explicitly constrain the domains and ranges of operations in a specification and to define partial functions, e.g., division, as total functions on a specified subtype. In general, typechecking with predicate subtypes is undecidable.[6] PVS contains a further useful enrichment to the type system in the form of *dependent* function, record, and tuple constructions where the type of one component of a compound value depends on

---

approach, but that it should not compromise the effectiveness of the specification language as a tool for deductive analysis; we are considering supporting an executable subset within PVS.

[6] PVS does have an algorithmic typechecker that checks for type correctness relative to the simple types. It generates proof obligations corresponding to predicate subtypes. The typical proof obligations can be automatically discharged by the PVS decision procedures. The provability of such proof obligations is the only source of undecidability in the PVS type system so that none of the benefits of decidable typechecking are lost.

the value of another component. PVS terms include constants, variables, abstractions (e.g., `(LAMBDA (i : nat): i * i)`), applications (e.g., `mod(i, 5)`), record constructions (e.g., `(# a := 2, b := cons(1, null) #)`), tuple constructions (e.g., `(-5, cons(1, null))`), function updates (e.g., `f WITH [(2) := 7]`), and record updates (e.g., `r WITH [a := 5, b := cons(3, b(r))]`). PVS specifications are packaged as *theories* that can be parametric in types and constants. Type parametricity (or *polymorphism*) is used to capture those concepts or results that can be stated uniformly for all types. PVS also has a facility for automatically generating abstract datatype theories (containing recursion and induction schemes) for a class of abstract datatypes [28].

3 The PVS Proof Checker

The central design assumptions in PVS are that

- The purpose of an automated proof checker is not merely to prove theorems but also to provide useful feedback from failed and partial proofs by serving as a rigorous skeptic.
- Automation serves to minimize the tedious aspects of formal reasoning while maintaining a high level of accuracy in the book-keeping and formal manipulations.
- Automation should also be used to capture repetitive patterns of argumentation.
- The end product of a proof attempt should be a proof that, with only a small amount of work, can be made humanly readable so that it can be subjected to the *social process* of mathematical scrutiny.

In following these design assumptions, the PVS proof checker is more automated than a low-level proof checker such as AUTOMATH [12], LCF [15], Nuprl [7], Coq [8], and HOL [16], but provides more user control over the structure of the proof than highly automated systems such as Nqthm [3,4] and Otter [21]. We feel that the low-level systems over-emphasize the formal correctness of proofs at the expense of their cogency, and the highly automated systems emphasize theorems at the expense of their proofs.

What is unusual about PVS is the extent to which aspects of the language, the typechecker, and proof checker are intertwined. The typechecker invokes the proof checker in order to discharge proof obligations that arise from typechecking expressions involving predicate subtypes or dependent types. The proof checker also makes heavy use of the typechecker to ensure that all expressions involved in a proof are well-typed. This use of the typechecker can also generate proof obligations that are either discharged automatically or are presented as additional subgoals. Several aspects of the language, particularly the type system, are built into the proof checker. These include the automatic use of type constraints by the decision procedures, the simplifications given by the abstract datatype axioms, and forms of beta-reduction and extensionality.

Another less unusual aspect of PVS is the extent to which the automatic inference and decision procedures involving equalities and linear arithmetic inequalities are employed.[7] The most direct consequence of this is that the trivial, obvious, or tedious parts of the proof are often discharged so that the user can focus on the intellectually demanding parts of the proof, and the resulting proof is also easier to read. PVS also provides an efficient conditional rewriter that interacts very closely with its decision procedures to simplify conditions during rewriting. More details about the rewriting and the decision procedures used in PVS are described in [10]. The capabilities of the inference and decision procedures, which play a central role in almost all proofs in PVS are made available to the user by means of the following primitive inference steps.

1. **Bddsimp** performs efficient BDD-based propositional simplification on the current goal.
2. **Do-rewrite** performs automatic conditional rewriting on expressions in the current goal using rewrite rules stored in the underlying database used by the inference procedures. PVS provides several commands for the user to make rewrite rules out of definitions, lemmas and axioms and enter them in the database. The rewriter invokes the decision procedures to simplify conditions of conditional rewrite rules.
3. **Assert** invokes the arithmetic and equality decision procedures on the current goal. Besides trying to prove the subgoal using the decision procedures, it performs the following tasks
   - it stores the subgoal information in the underlying database, allowing automatic use to be made of it later.
   - it simplifies the subgoal using the decision procedures using rewriting as well as other simplification techniques.

In order to learn how to use the PVS proof checker, one must first understand the sequent representation used by PVS to represent proof goals, the commands used to move around and undo parts of the proof tree, and the commands used to get help. One must then understand the syntax and effects of proof commands used to build proofs. Many of these commands are extremely powerful even in their simplest usage. Several of these commands can be more carefully directed by supplying them with one or more optional arguments. The advanced user will also need to understand how to define proof strategies that capture repetitive patterns of proof commands, and commands used for displaying, editing, and replaying proofs. There are about 20 basic commands and a similar number of commonly used high-level strategies.

---

[7] The Ontic system [20] is a proof checker where decision procedures are ubiquitously used. Nqthm [3,4], Eves [24], and IMPS [14] also rely heavily on the use of decision procedures.

4 Rest of the Tutorial

In the following sections we introduce some of the details of PVS system by working the complete proof of correctness of two examples. This will introduce some of the most useful commands and provide a glimpse into the philosophy behind PVS. PVS uses EMACS as its interface by extending EMACS with PVS functions, but all the underlying capabilities of EMACS are available. Thus the user can read mail and news, edit nonPVS files, or execute commands in a shell buffer in the usual way. All PVS commands are entered as extended EMACS commands. The proof checker runs as a subprocess inside EMACS.

5 A Pipelined Microprocessor

In this section we develop a complete proof of a correctness property of the controller logic of a simple pipelined processor design described at a register-transfer level. The design and the property verified are both based on the processor example given in [5]. The example has been used as a benchmark for evaluating how well finite state-enumeration based tools, such as model checkers, can handle datapath-oriented circuits with a large number of states by varying the size of the datapath. From the perspective of a theorem prover, the size of the datapath is irrelevant because the specification and proof are independent of the datapath size. As a theorem proving exercise, the challenge is to see if the proof can be done just as automatically as a model checker. As we will see in the following, in PVS the proof can be obtained by repeatedly invoking one of its primitive commands **assert**.

5.1 Informal Description

Figure 1 shows a block diagram of the pipeline design. The processor executes instructions of the form (**opcode src1 src2 dstn**), i.e., "destination register **dstn** in the register file **REGFILE** becomes some **ALU** function determined by **opcode** of the contents of source registers **src1** and **src2**. Every instruction is executed in three stages (cycles) by the processor:

1. *Read:* Obtain the proper contents of the register file at **src1** and **src2** and clock them into **opreg1** and **opreg2**, respectively.
2. *Compute:* Perform the ALU operation corresponding to the opcode (remembered in **opcoded**) of the instruction and clock the result into **wbreg**.
3. *Write:* Update the register file at the destination register (remembered in **dstndd**) of the instruction with the value in **wbreg**.

The processor uses a three-stage pipeline to simultaneously execute distinct stages of three successive instructions. That is, the read stage of the current instruction is executed along with the compute stage of the previous instruction

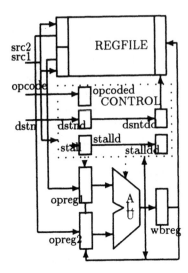

Fig. 1. A Pipelined Microprocessor

and the write stage of the previous-to-previous instruction. Since the **REGFILE** is not updated with the results of the previous and previous-to-previous instructions while a read is being performed for the current instruction, the controller "bypasses" **REGFILE**, if necessary, to get the correct values for the read. The processor can abort, i.e., treat as **NOP**, the instruction in the read stage by asserting the **stall** signal true. An instruction is aborted by inhibiting its write stage by remembering the **stall** signal until the write stage via the registers **stalld** and **stalldd**. We verify that an instruction entering the pipeline at any time gets completed correctly, i.e., will write the correct result into the register file, three cycles later, provided the instruction is not aborted.

5.2 Formal Specification

PVS specifications consist of a number of files, each of which contains one or more theories. A theory is a collection of declarations: types, constants (including functions), axioms that express properties about the constants, and theorems and lemmas to be proved. Theories may import other theories; Every entity used in a theory must be either declared in an imported theory or be part of the prelude (the standard collection of theories built-in to PVS).

The microprocessor specification is organized into three theories, selected parts of which are shown in Figures 2 and 3. (The complete specification can be found in [31].) The theory **pipe** (Figure 2) contains a specification of the design and a statement of the correctness property to be proved. The theories **signal** and **time** (Figure 3) imported by **pipe** declares the types **signal** and **time** used in **pipe**.

The theory **pipe** is parameterized with respect to the types of the register address, data, and the opcode field of the instructions. A theory parameter in

```
pipe[addr: TYPE, data: TYPE, opcodes: TYPE]: THEORY
  BEGIN
   IMPORTING signal, time

   ASSUMING
    addr_nonempty: ASSUMPTION (EXISTS (a: addr): TRUE)
    data_nonempty: ASSUMPTION (EXISTS (d: data): TRUE)
    opcodes_nonempty: ASSUMPTION (EXISTS (o: opcodes): TRUE)
   ENDASSUMING

   t: VAR time

   %% Signal declarations
      opcode: signal[opcodes]
      src1, src2, dstn: signal[addr]
      stall: signal[bool]
      aluout: signal[data]
      regfile: signal[[addr -> data]]
      ...

    %% Specification of constraints on the signals
     dstnd_ax: AXIOM dstnd(t+1) = dstn(t)
     dstndd_ax: AXIOM dstndd(t+1)= dstnd(t)
     .....

     regfile_ax: AXIOM regfile(t+1) =
                       IF stalldd(t) THEN regfile(t)
                       ELSE regfile(t)
                            WITH [(dstndd(t)) := wbreg(t)]
                       ENDIF
     opreg1_ax: AXIOM opreg1(t+1) =
                  IF src1(t) = dstnd(t) & NOT stalld(t)
                     THEN aluout(t)
                  ELSIF src1(t) = dstndd(t) & NOT stalldd(t)
                        THEN wbreg(t)
                  ELSE regfile(t)(src1(t)) ENDIF
      opreg2_ax: AXIOM ...

     aluop: [opcodes, data, data -> data]
     ALU_ax: AXIOM aluout(t) = aluop(opcoded(t), opreg1(t),
                                     opreg2(t))
    correctness: THEOREM (FORALL t:
      NOT(stall(t)) IMPLIES regfile(t+3)(dstn(t)) =
                  aluop(opcode(t), regfile(t+2)(src1(t)),
                                   regfile(t+2)(src2(t))) )
  END pipe
```

Fig. 2. Microprocessor Specification

PVS can be either a type parameter or a parameter belonging to a particular type, such as **nat**. Since **pipe** does not impose any restriction on its parameters, other than the requirement that they be nonempty, which is stated in the **ASSUMING** part of the theory, one can instantiate them with any type. Every entity declared in a parameterized theory is implicitly parameterized with respect to the parameters of the theory. For example, the type **signal** declared in the parameterized theory **signal** is a parametric type denoting a function that maps **time** (a synonym for **nat**) to the type parameter T. (The type **signal** is used to model the wires in our design.) By importing the theory **signal** uninstantiated

```
signal[val: TYPE]: THEORY
  BEGIN
   signal: TYPE = [time -> val]
  END signal

time: THEORY
  BEGIN
   time: TYPE nat
  END signal
```

Fig. 3. Signal Specification

in **pipe**, we have the freedom to create any desired instances of the type **signal**.

In this tutorial, we use a *functional* style of specification to model register-transfer-level digital hardware in logic. In this style, the inputs to the design and the outputs of every component in the design are modeled as signals. Every signal that is an output of a component is specified as a function of the signals appearing at the inputs to the component.

This style should be contrasted with a *predicative* style, which is commonly used in most HOL applications. In the predicative style every hardware component is specified as a predicate relating the input and output signals of the component and a design is specified as a conjunction of the component predicates, with all the internal signals used to connect the components hidden by existential quantification. A proof of correctness for a predicative style specification usually involves executing a few additional steps at the start of the proof to essentially transform the predictative specification into an equivalent functional style. After that, the proof proceeds similar to that of a proof in a functional specification. The additional proof steps required for a predicative specification essentially unwind the component predicates using their definitions and then appropriately instantiate the existentially quantified variables. An automatic way of performing this translation is discussed in [31], which illustrates more examples of hardware design verification using PVS.

Getting back to our example, the microprocessor specification in **pipe** consists of two parts. The first part declares all the signals used in the design—the inputs to the design and the internal wires that denote the outputs of components. The composite state of **REGFILE**, which is represented as a function from **addr** to **data**, is modeled by the signal **regfile**. The signals are declared as uninterpreted constants of appropriate types. The second part consists of a set of AXIOMs that specify the the values of the signals over time. (To conserve space, we have only shown the specification of a subset of the signals in the design.) For example, the signal value at the output of the register **dstnd** at time **t+1** is defined to be that of its input a cycle earlier. The output of the ALU, which is a combinational component, is defined in terms of the inputs at the same time instant.

In PVS, one can use a descriptive style of definition, as illustrated in this example, by selectively introducing properties of the constants declared in a theory as AXIOMs. Or, one can use the definitional forms provided by the language to define the constants. An advantage of using the definitions is that a specification is guaranteed to be consistent, although it might be overspecified. An advantage of the descriptive style is that it gives better control over the degree to which one wants to define an entity. For example, one could have specified dstnd prescriptively by using the conventional function definition mechanism of PVS. PVS's function definition mechanism would have forced us to specify the value of the signal at time t = 0 to ensure that the function is total. In the descriptive style used, we have left the value of the signal at 0 unspecified.

In the present example, the specifications of the signals opreg1 and opreg2 are the most interesting of all. They have to check for any register collisions that might exist between the instruction in the read stage and the instructions in the later stages and bypass reading from the register file in case of collisions. The regfile signal specification is recursive since the register file state remains the same as its previous state except, possibly, at a single register location. The WITH expression is an abbreviation for the result of updating a function at a given point in the domain value with a new value. Note that the function aluop that denotes the operation ALU performs for a given opcode is left completely unspecified since it is irrelevant to the controller logic.

The theorem correctness to be proved states a correctness property about the execution of the instruction that enters the pipeline at t, provided the instruction is not aborted, i.e., stall(t) is not true. The equation in the conclusion of the implication compares the actual value (left hand side) in the destination register three cycles later, when the result of the instruction would be in place, with the expected value. The expected value is the result of applying the aluop corresponding to the opcode of the instruction to the values at the source field registers in the register file at t+2. We use the state of the register file at t+2 rather than t to allow for the results of the two previous instructions in the pipeline to be completed.

5.3 Proof of Correctness

The next step is to typecheck the file, which parses and checks for semantic errors, such as undeclared names and ambiguous types. Typechecking may build new files or internal structures such as *type correctness conditions (TCCs)*. The TCCs represent *proof obligations* that must be discharged before the pipe theory can be considered typechecked. The typechecker does not generate any TCCs in the present example. If, for example, one of the assumptions, say for addr, in the ASSUMING part of the theory was missing, the typechecker would generate the following TCC to show that the addr type is nonempty. The declaration of the signal src1 forces generation of this TCC because a function is nonexistent if its range is empty.

```
% Existence TCC generated (line 17) for src1: signal[addr]
% May need to add an assuming clause to prove this.
  % unproved
src1_TCC1: OBLIGATION (EXISTS (x1: signal[addr]): TRUE);
```

The PVS proof checker runs as a subprocess of Emacs. Once invoked on a theorem to be proved, it accepts commands directly from the user. The basic objective of developing a proof in PVS as in other subgoal-directed proof checkers (e.g., HOL), is to generate a *proof tree* in which all of the leaves are trivially true. The nodes of the proof tree are sequents, and while in the prover you will always be looking at an unproved leaf of the tree. The *current* branch of a proof is the branch leading back to the root from the current sequent. When a given branch is complete (i.e., ends in a true leaf), the prover automatically moves on to the next unproved branch, or, if there are no more unproven branches, notifies you that the proof is complete.

The primitive inference steps in PVS are a lot more powerful than in HOL. So, it is not necessary to build complex tactics to handle tedious lower level proofs in PVS. A user knowledgeable in the ways of PVS can typically get proofs to go through mostly automatically by making a few critical decisions at the start of the proof. However, PVS does provide the user with the equivalent of HOL's tacticals, called *strategies*, and other features to control the desired level of automation in a proof.

The proof of the microprocessor property shown below follows a certain general pattern that works successfully for most hardware proofs. This general proof pattern, variants of which have been used in other verification exercises [1, 18], consists of the following sequence of general proof tasks.

Quantifier elimination: Since the decision procedures work on ground formulas, the user must eliminate the relevant universal quantifiers by skolemization or selecting variables on which to induct and existential quantifiers by suitable instantiation.

Unfolding definitions: The user may have to simplify selected expressions and defined function symbols in the goal by rewriting using definitions, axioms or lemmas. The user may also have to decide the level to which the function symbols have to rewritten.

Case analysis: The user may have to split the proof based on selected boolean expressions in the current goal and simplify the resulting goals further.

Each of the above tasks can be accomplished automatically using a short sequence of primitive PVS proof commands. The complete proof of the theorem is shown below. Selected parts of the proof session is reproduced below as we describe the proof.

```
1: ( then* (skosimp)
2:         (auto-rewrite-theory ''pipe'' :always? t)
3:         (repeat (do-rewrite))
4:         (apply (then* (repeat (lift-if))
5:                       (bddsimp)
6:                       (assert))))
```

In the proof, the names of strategies are shown in *italics* and the primitive inference steps in `type-writer font`. (We have numbered the lines in the proof for reference.) `Then*` applies the first command in the list that follows to the current goal; the rest of the commands in the list are then applied to each of the subgoals generated by the first command application. The `apply` command used in line 5 makes the application of a compound proof step implemented by a strategy behave as an atomic step.

The first goal in the proof session is shown below. It consists of a single formula (labeled {1}) under a dashed line. This is a *sequent*; formulas above the dashed lines are called *antecedents* and those below are called *succedents*. The interpretation of a sequent is that the conjunction of the antecedents implies the disjunction of the succedents.

```
correctness :

  |-------
{1}   (FORALL t: NOT (stall(t))
                  IMPLIES regfile(t + 3)(dstn(t)) =
                        aluop(opcode(t), regfile(t + 2)(src1(t)),
                              regfile(t + 2)(src2(t))))
```

The quantifier elimination task of the proof is accomplished by the command `skosimp`, which skolemizes all the universally quantified variables in a formula and flattens the sequent resulting in the following goal. Note that `stall(t!1)` has been moved to the succedent in the sequent because PVS displays every atomic formula in its positive form.

```
Rule? (skosimp)
Skolemizing and flattening, this simplifies to:
correctness :

  |-------
{1}   (stall(t!1))
{2}   regfile(t!1 + 3)(dstn(t!1))
        =
        aluop(opcode(t!1), regfile(t!1 + 2)(src1(t!1)),
              regfile(t!1 + 2)(src2(t!1)))
```

The next task—unfolding definitions—is performed by the commands in lines 2 through 3. PVS provides a number of ways of unfolding definitions ranging from unfolding one step at a time to automatic rewriting that performs unfolding in a brute-force fashion. Brute-force rewriting usually results in larger expressions than controlled unfolding and, hence, potentially larger number of cases to consider. If a system provides automatic and efficient rewriting and case analysis facilities, then one can dare to use the automatic approach, as we do here. In PVS automatic rewriting is performed by first entering the definitions and AXIOMs that one wants to be used for unfolding as rewrite rules. Once entered, the commands, such as `do-rewrite` and `assert`, that perform rewriting as part of their repertoire repeatedly apply the rewrite rules until none of the rules is applicable. To control the size of the expression resulting from rewriting and the potential for looping, the rewriter uses the following restriction for stopping a rewrite: If the right-hand-side of a rewrite is a conditional expression, then the rule is applied only if the condition simplifies to true or false.

Here our aim is to unfold every signal in the sequent so that every signal expression contains only the start time `t!1`. So, we make a rewrite rule out of every AXIOM in the theory `pipe` by means of the command `auto-rewrite-theory` on line 2. We also force an over-ride of the default restriction for stopping rewriting by setting the tag[8] `always?` to true in the `auto-rewrite-theory` command and embed `do-rewrite` inside a `repeat` loop to force maximum rewriting. In the present example, the rewriting is guaranteed to terminate because every feedback loop is cut by a sequential component.

At the end of automatic rewriting, the succedent we are trying to prove is in the form of an equation on two deeply nested conditional expressions as shown below in an abbreviated fashion. The various cases in conditional expression shown above arise as a result of the different possible conflicts between instructions in the pipeline. The equation we are trying to prove contains two distinct, but equivalent conditional expressions, as in `IF a THEN b ELSE c ENDIF = IF NOT a THEN c ELSE b ENDIF`, that can only be proved to be equal by performing a case-split on one or more of the conditions. While `assert` simplifies the leaves of a conditional expression assuming every condition along the path to the leaves holds, it does not split propositions. One way to perform the case-splitting task automatically is to "lift" all the `IF-THEN-ELSEs` to the top so that the equation is transformed into a propositional formula with unconditional equalities as atomic predicates. After performing such a lifting, one can try to reduce the resulting proposition to true using the propositional simplification command `bddsimp`. If `bddsimp` does not simplify the proposition to true, then it is most likely the case that equations at one or more of the leaves of the proposition need to be further simplified, by `assert`, for instance, using the conditions along the path. If the propositional formula does not reduce to true or false, `bddsimp`

---

[8] Tags are one of the ways in which PVS permits the user to modify the functionality of proof commands.

produces a set of subgoals to be proved. In the present case, each of these goals can be discharged by **assert**. The compound proof step appearing on lines 4 through 6 of the proof accomplishes the case-splitting task.

```
correctness :

  |-------
[1]    (stall(t!1))
{2}    aluop(opcode(t!1),
              IF src1(t!1) = dstnd(t!1) & NOT stalld(t!1)
                THEN aluop(opcoded(t!1), opreg1(t!1), opreg2(t!1))
              ELSIF src1(t!1) = dstndd(t!1) & NOT stalldd(t!1)
              THEN wbreg(t!1)
              ELSE regfile(t!1)(src1(t!1)) ENDIF,
              ....
              ENDIF)
       = aluop(opcode(t!1),
              IF stalld(t!1) THEN IF stalldd(t!1) THEN regfile(t!1)
                ELSE regfile(t!1) WITH [(dstndd(t!1)) := wbreg(t!1)]
                ENDIF
              ELSE ...
              ENDIF(src1(t!1)),
              IF stalld(t!1) THEN IF stalldd(t!1) THEN regfile(t!1)
                ELSE ... ENDIF
              ELSE ...
              ENDIF(src2(t!1)))
```

We have found that the sequence of steps shown above works successfully for proving safety properties of finite state machines that relate states of the machine that are finite distance apart. If the strategy does not succeed then the most likely cause is that either the property is not true or that a certain property about some of the functions in the specification unknown to the prover need to be proved as a lemma. In either case, the unproven goals remaining at the end of the proof should give information about the probable cause.

6   An N-bit Ripple-Carry Adder

The second example we consider is the verification of a parametrized N-bit ripple-carry adder circuit. The theory **adder**, shown in Figure 4, specifies a ripple-carry adder circuit and a statement of correctness for the circuit.

The theory is parameterized with respect to the length of the bit-vectors. It imports the theories (not shown here) **full_adder**, which contains a specification of a full adder circuit (**fa_cout** and **fa_sum**), and **bv**, which specifies the bit-vector type (**bvec[N]**) and functions. An N-bit bit-vector is represented as an

```
adder[N: posnat] : THEORY
BEGIN
  IMPORTING bv[N], full_adder

  n: VAR below[N]
  bv, bv1, bv2: VAR bvec
  cin: VAR bool

  nth_cin(n, cin, bv1, bv2): RECURSIVE bool =
      IF n = 0 THEN cin
      ELSE fa_cout(nth_cin(n - 1, cin, bv1, bv2), bv1(n - 1), bv2(n - 1))
      ENDIF
    MEASURE n

  bv_sum(cin, bv1, bv2): bvec =
    (LAMBDA n: fa_sum(bv1(n), bv2(n), nth_cin(n, cin, bv1, bv2)))

  bv_cout(n, cin, bv1, bv2): bool =
    fa_cout(nth_cin(n, cin, bv1, bv2), bv1(n), bv2(n))

  adder_correct_n: LEMMA
      bvec2nat_rec(n, bv1) + bvec2nat_rec(n, bv2) + bool2bit(cin)
        = exp2(n + 1) * bool2bit(bv_cout(n, cin, bv1, bv2))
          + bvec2nat_rec(n, bv_sum(cin, bv1, bv2))

  adder_correct: THEOREM
      bvec2nat(bv1) + bvec2nat(bv2) + bool2bit(cin)
        = exp2(N) * bool2bit(bv_cout(N - 1, cin, bv1, bv2))
          + bvec2nat(bv_sum(cin, bv1, bv2))
END adder
```

Fig. 4. Adder Specification

array, i.e., a function, from the the type **below[N]**, a subtype of **nat** ranging from 0 through N-1, to **bool**; the index 0 denotes the least significant bit. Note that the parameter **N** is constrained to be a posnat since we do not permit bit vectors of length 0. The **adder** theory contains several declarations including a set of variable declarations in the beginning. PVS allows logical variables to be declared globally within a theory so that the variables can be used later in function definitions and quantified formulas.

The carry bit that ripples through the full adder is specified recursively by means of the function **nth_cin**. Associated with this definition is a *measure* function, following the **MEASURE** keyword, which will be explained below. The function **bv_cout** and **bv_sum** define the carry output and the bit-vector sum of the adder, respectively. The theorem **adder_correct** expresses the conventional correctness statement of an adder circuit using **bvec2nat**, which returns the natural number equivalent of an N-bit bit-vector. Note that variables that are left free in a formula are assumed to be universally quantified. We state and prove a more general lemma **adder_correct_rec** of which **adder_correct** is an instance. For a given n < N, **bvec2nat_rec** returns the natural number equivalent of the least significant n-bits of a given bit-vector and **bool2bit** converts the boolean constants **TRUE** and **FALSE** into the natural numbers 1 and 0, respectively.

6.1 Typechecking

The typechecker generates several TCCs (shown in Figure 5 below) for **adder**. These TCCs represent *proof obligations* that must be discharged before the **adder** theory can be considered typechecked. The proofs of the TCCs may be postponed until it is convenient to prove them, though it is a good idea to view them to see if they are provable.

```
% Subtype TCC generated (line 13) for n - 1
  % unproved
nth_cin_TCC1: OBLIGATION (FORALL n: NOT n = 0 IMPLIES n - 1 >= 0 AND n - 1 < N)

% Subtype TCC generated (line 31) for N - 1
  % unproved
adder_correct_TCC1: OBLIGATION N - 1 >= 0
```

Fig. 5. TCCs for Theory adder

The first TCC is due to the fact that the first argument to **nth_cin** is of type **below[N]**, but the type of the argument (n-1) in the recursive call to **nth_cin** is integer, since **below[N]** is not closed under subtraction. Note that the TCC includes the condition **NOT n = 0**, which holds in the branch of the **IF-THEN-ELSE** in which the expression **n - 1** occurs. A TCC identical to the this one is generated for each of the two other occurrences of the expression **n-1** because **bv1** and **bv2** also expect arguments of type **below[N]**. These TCCs are not retained because they are subsumed by the first one.

The second TCC is generated by the expression **N-1** in the definition of the theorem **adder_correct** because the first argument to **bv_cout** is expected to be the subtype **below[N]**.

There is yet another TCC that is internally generated by PVS but is not even included in the TCCs file because it can be discharged trivially by the typechecker, which calls the prover to perform simple normalizations of expressions. This TCC is generated to ensure that the recursive definition of **nth_cin** terminates. PVS does not directly support partial functions, although its powerful subtyping mechanism allows PVS to express many operations that are traditionally regarded as partial. The measure function is used to show that recursive definitions are total by requiring the measure to decrease with each recursive call. For the definition of **nth_cin**, this entails showing **n-1 < n**, which the typechecker trivially deduces.

In the present case, all the remaining TCCs are simple, and in fact can be discharged automatically by using the **typecheck-prove** command, which attempts to prove all TCCs that have been generated using a predefined proof strategy called **tcc**.

6.2 Proof of Adder_correct_n

The proof of the lemma uses the same core strategy as in the microprocessor proof except for the quantifier elimination step. Since the specification is recursive in the length of the bit-vector, we need to perform induction on the variable n. The user invokes an inductive proof in PVS by means of the command **induct** with the variable to induct on (n) and the induction scheme to be used (**below_induction[N]**) as arguments. The induction used in this case is defined in the PVS prelude and is parameterized, as is the type **below[N]**, with respect to the upper limit of the subrange.

This command generates two subgoals: the subgoal corresponding to the base case, which is the first goal presented to prove, is shown in Figure 6.

```
adder_correct.1 :

  |-------
1   (N > 0
        IMPLIES
        (FORALL
          (bv1: bvec[N], bv2: bvec[N], cin: bool):
            bvec2nat_rec(0, bv1) + bvec2nat_rec(0, bv2)
            + bool2bit(cin)
            = exp2(0 + 1) * bool2bit(bv_cout(0, cin, bv1, bv2))
              + bvec2nat_rec(0, bv_sum(cin, bv1, bv2))))
```

Fig. 6. Base Step

The goal corresponding to the inductive case is shown below.

```
The remaining siblings are:
adder_correct_n.2 :

  |-------
{1}   (FORALL (r: below[N]):
        r < N - 1
        AND (FORALL (bv1, bv2: bvec[N]), (cin: bool):
                bvec2nat_rec(r, bv1) + bvec2nat_rec(r, bv2)
                + bool2bit(cin)
                = exp2(r + 1) * bool2bit(bv_cout(r, cin, bv1, bv2))
                  + bvec2nat_rec(r, bv_sum(cin, bv1, bv2)))
            IMPLIES (FORALL (bv1, bv2: bvec[N]), (cin: bool):
                        bvec2nat_rec(r + 1, bv1)
                        + bvec2nat_rec(r + 1, bv2)
                        + bool2bit(cin)
                        = exp2(r + 1 + 1)
                          * bool2bit(bv_cout(r + 1, cin, bv1, bv2))
                          +
                          bvec2nat_rec(r + 1,
                                        bv_sum(cin, bv1, bv2))))
```

Fig. 7. Inductive Step

The base and the inductive steps can be proved automatically using essentially the same strategy used in the microprocessor proof. A complete proof of **adder_correct_n** is shown in Figure 7.

```
 1: ( spread (induct ''n'' 1 ''below_induction[N]'')
 2:    ( ( then* (skosimp*)
 3:              (auto-rewrite-defs :always? t)
 4:              (do-rewrite)
 5:              ( repeat (lift-if))
 6:              ( apply ( then* (bddsimp)(assert))))
 7:      ( then* (skosimp*)
 8:              (inst?)
 9:              (auto-rewrite-defs :always? t)
10:              (do-rewrite)
11:              ( repeat (lift-if))
12:              ( apply ( then* (bddsimp)(assert)))))))
```

The strategy *spread* used on line 1 applies the first proof step (**induct**) and then applies the ith element of the list of commands that follow to the ith subgoal resulting from the application of the first prof step. Thus, the proof steps listed on lines 2 through 6 prove the base case of induction, the steps on lines 7 through 12 prove the inductive case, and the proof step on line 13 takes care of the third TCC subgoal.

Let us consider the base case first. The **induct** command has already instantiated the variable n to 0. The remaining variables are skolemized away by **skosimp***. To unfold the definitions in the resulting goal, we use the command **auto-rewrite-defs**, which makes rewrite rules out of the definition of every function either directly or indirectly used in the given formula. The rest of the proof proceeds exactly as for the microprocessor.

The proof of the inductive step follows exactly the same pattern except that we need to instantiate the induction hypothesis and use it in the process of unfolding and case-analysis. PVS provides a command **inst?** that tries to find instantiations for existential-strength variables in a formula by searching for possible matches between terms involving these variables with ground terms inside formulas in the rest of the sequent. This command finds the desired instantiations in the present case. The rest of the proof proceeds as in the basis case.

Since the inductive proof pattern shown above is applicable to any iteratively generated hardware designs we have packaged it into a general proof strategy called **name-induct-and-bddrewrite**. The strategy is parameterized with respect to induction scheme to be used and the set of rewrite rules to be used for unfolding. We have used the strategy to prove an N-bit ALU [6] that executes 12 microoperations by cascading N 1-bit ALU slices.

7 Summary

This tutorial gives an overview of some of the unique and important capabilities of PVS. PVS is built to combine a very expressive specification language with effective theorem proving to produce a system to apply formal methods productively. PVS does pay a performance penalty because of the need for the prover to invoke the typechecker more often than in other provers that support a less expressive type system. We are working on reducing the amount of type information that the prover needs to generate and maintain during a proof and also on further optimizing some of our inference procedures. These optimizations should be available with future releases of PVS. In the following, we summarize some of the language and system features that were not covered in the tutorial.

PVS provides a fairly extensive set of commands for determining the status of specification elements such as theories and formulas. For example, the user can inquire whether a theory has been typechecked or a proof has been completed and if the proof is current. It has commands that perform *proof chain analysis* to see the proof status of all the lemmas that a theorem is dependent on.

When a formal specification and verification is complete, it is usually desirable to present it to others in as readable a form as possible. PVS provides commands for generating Latex versions of the specifications and proofs that can be included in typeset documents. The output produced can be controlled by user-supplied tables so that mathematical notation, including infix and mis-fix symbols and sub and superscripts can be created easily.

An important language feature that we haven't illustrated here is the abstract data type feature. This feature is similar to the definitional principle supported by the Boyer-Moore theorem prover, but is generalized to abstract data types with arbitrary constructors. The system provides facility for automatically generating abstract data type theories (containing recursion and induction schemes) from a syntactic definition of the operations of the data type.

7.1 Getting and Using PVS

At the moment, PVS is readily available only for Sun SPARC workstations, although versions of the system do exist for the IBM Risc 6000 (under AIX) and DECSystems (under Ultrix). PVS is implemented in Common Lisp (with CLOS), and has been ported to Lucid and Allegro. All versions of PVS require GNU EMACS, which must be obtained separately.

PVS requires about 30 megabytes of disk space. In addition, any system on which it is to be run should have a minimum of 100 megabytes of swap space and 32 megabytes of real memory (more is better).

To obtain the PVS system, send a request to **pvs-request@csl.sri.com**, and we will provide further instructions for obtaining a tape or for getting the system by FTP. All installations of PVS must be licensed by SRI. A nominal distribution fee is charged for tapes; there is no charge for obtaining PVS by FTP.

References

1. Mark D. Aagard, Miriam E. Leeser, and Phillip J. Windley. Toward a super duper hardware tactic. In *Proceedings of the HOL User's Group Workshop*, pages 401–414, 1993.

2. Heather Alexander and Val Jones. *Software Design and Prototyping using me too*. Prentice Hall International, Hemel Hempstead, UK, 1990.

3. R. S. Boyer and J S. Moore. *A Computational Logic*. Academic Press, New York, NY, 1979.

4. R. S. Boyer and J S. Moore. *A Computational Logic Handbook*. Academic Press, New York, NY, 1988.

5. J. R. Burch, E. M. Clarke, K. L McMillan, D. L. Dill, and L. J. Hwang. Symbolic model checking: 2^{20} states and beyond. In *5th Annual IEEE Symposium on Logic in Computer Science*, pages 428–439, Philadelphia, PA, June 1990. IEEE Computer Society.

6. F. J. Cantu. Verifying an *n-bit* arithmetic logic unit. Blue book note 935, University of Edinburgh, June 1994.

7. R. L. Constable, *et al*. *Implementing Mathematics with the Nuprl*. Prentice-Hall, New Jersey, 1986.

8. T. Coquand and G. P. Huet. Constructions: A higher order proof system for mechanizing mathematics. In *Proceedings of EUROCAL 85, Linz (Austria)*, Berlin, 1985. Springer-Verlag.

9. Costas Courcoubetis, editor. *Computer-Aided Verification, CAV '93*, volume 697 of *Lecture Notes in Computer Science*, Elounda, Greece, June/July 1993. Springer-Verlag.

10. D. Cyrluk, S. Rajan, N. Shankar, and M. K. Srivas. Effective theorem proving for hardware verification. In Ramayya Kumar and Thomas Kropf, editors, *Preliminary Proceedings of the Second Conference on Theorem Provers in Circuit Design*, pages 287–305, Bad Herrenalb (Blackforest), Germany, September 1994. Forschungszentrum Informatik an der Universität Karlsruhe, FZI Publication 4/94.

11. David Cyrluk. Microprocessor verification in PVS: A methodology and simple example. Technical Report SRI-CSL-93-12, Computer Science Laboratory, SRI International, Menlo Park, CA, December 1993.

12. N. G. de Bruijn. A survey of the project Automath. In *To H. B. Curry: Essays on Combinatory Logic, Lambda Calculus and Formalism*, pages 589–606. Academic Press, 1980.

13. B. Elspas, M. Green, M. Moriconi, and R. Shostak. A JOVIAL verifier. Technical report, Computer Science Laboratory, SRI International, January 1979.

14. W. M. Farmer, J. D. Guttman, and F. J. Thayer. IMPS: An interactive mathematical proof system. Technical Report M90-19, Mitre Corporation, 1991.

15. M. Gordon, R. Milner, and C. Wadsworth. *Edinburgh LCF: A Mechanized Logic of Computation*, volume 78 of *Lecture Notes in Computer Science*. Springer-Verlag, 1979.

16. M. J. C. Gordon. HOL: A proof generating system for higher-order logic. In G. Birtwistle and P. A. Subrahmanyam, editors, *VLSI Specification, Verification and Synthesis*, pages 73–128. Kluwer, Dordrecht, The Netherlands, 1988.

17. Sharam Hekmatpour and Darrel Ince. *Software Prototyping, Formal Methods, and VDM*. International Computer Science Series. Addison-Wesley, Wokingham, England, 1988.

18. R. Kumar, K. Schneider, and T. Kropf. Structuring and automating hardware proofs in a higher-order therem proving environment. *Formal Methods in System Design*, 2(2):165–223, 1993.

19. Patrick Lincoln and John Rushby. Formal verification of an algorithm for interactive consistency under a hybrid fault model. In Courcoubetis [9], pages 292–304.

20. D. A. McAllester. *ONTIC: A Knowledge Representation System for Mathematics.* MIT Press, 1989.

21. W. McCune. OTTER 2.0 users guide. Technical Report ANL-90/9, Argonne National Laboratory, 1990.

22. P. Michael Melliar-Smith and John Rushby. The Enhanced HDM system for specification and verification. In *Proc. VerkShop III*, pages 41–43, Watsonville, CA, February 1985. Published as ACM Software Engineering Notes, Vol. 10, No. 4, Aug. 85.

23. Sam Owre, John Rushby, Natarajan Shankar, and Friedrich von Henke. Formal verification for fault-tolerant architectures: Some lessons learned. In J. C. P. Woodcock and P. G. Larsen, editors, *FME '93: Industrial-Strength Formal Methods*, pages 482–500, Odense, Denmark, April 1993. Volume 670 of *Lecture Notes in Computer Science*, Springer-Verlag.

24. W. Pase and M. Saaltink. Formal verification in m-EVES. In G. Birtwistle and P. A. Subrahmanyam, editors, *Current Trends in Hardware Verification and Theorem Proving*, pages 268–302, New York, NY, 1989. Springer-Verlag.

25. L. Robinson, K. N. Levitt, and B. A. Silverberg. *The HDM Handbook.* Computer Science Laboratory, SRI International, Menlo Park, CA, June 1979. Three Volumes.

26. Lawrence Robinson and Karl N. Levitt. Proof techniques for hierarchically structured programs. *Communications of the ACM*, 20(4):271–283, April 1976.

27. John Rushby, Friedrich von Henke, and Sam Owre. An introduction to formal specification and verification using EHDM. Technical Report SRI-CSL-91-2, Computer Science Laboratory, SRI International, Menlo Park, CA, February 1991.

28. N. Shankar. Abstract datatypes in PVS. Technical Report SRI-CSL-93-9, Computer Science Laboratory, SRI International, Menlo Park, CA, December 1993.

29. Natarajan Shankar. Verification of real-time systems using PVS. In Courcoubetis [9], pages 280–291.

30. R. E. Shostak, R. Schwartz, and P. M. Melliar-Smith. STP: A mechanized logic for specification and verification. In D. Loveland, editor, *6th International Conference on Automated Deduction (CADE)*, New York, NY, 1982. Volume 138 of *Lecture Notes in Computer Science*, Springer-Verlag.

31. M.K. Srivas, et. al. Hardware verification using pvs: A tutorial. Technical report, Computer Science Laboratory, SRI International, Menlo Park, CA, 1994. A Forthcoming Technical Report.

A Reduced Instruction Set Proof Environment

Holger Busch

SIEMENS AG, Corporate Research,81730 München,Germany
busch@zfe.siemens.de

Abstract. A general-purpose proof interface has been created on top of the higher-order-logic theorem prover **LAMBDA** in order to improve the efficiency of human interaction and minimize the learning overhead. Users are freed from tedious low-level interactions by way of extended proof automation routines. All essential **LAMBDA** functions for interactive proof development are accessible via a handy set of user commands.

1 Introduction

Higher-order-logic theorem proving has many advantages and potential applications in circuit design and other areas. Its integration in industrial design scenarios, however, has been restricted by a high learning overhead, an insufficient automation, efficiency problems, and the insuitability of proof scripts for the documentation and communication of verification results. Design engineers and even many formal experts therefore consider a higher-order-logic theorem prover to be inadequate for reasoning about commercial designs.

Driven by this discussion and our own experience, we created a prototypical general-purpose proof environment (**Rispe** - *Reduced instruction set proof environment*) which gives non-experts on the proof system **LAMBDA** [4] with minimal learning effort access to all important proof functions through a small number of intuitive proof commands and enables effective proof interaction without scanning large libraries and writing sophisticated tactics.

LAMBDA is supposed as a formal tool for hierarchical top-down system design. It includes the graphical interface **DIALOG**, which provides a schematics editor, various synthesis and hardware-specific proof functions. The kernel **LAMBDA** system is an interactive general-purpose theorem prover for classical higher-order logic. Its meta and implementation language is standard **ML** [5], to which the object language of logic terms has been adapted. **LAMBDA** comprises extended libraries of rules, tactics, and utilities for adding new proof procedures. We use the kernel prover without **DIALOG**.

2 General Approach

General measures are summarized to remedy problems encountered in many years' experience with the system **LAMBDA**. The problems match independent reports from other higher-order-logic systems and reflects the criticism of advocates of automatic first-order provers.

Accessibility. The amount and variety of proof functions are an obstacle not only for novices. In **Rispe**, related proof activities which in **LAMBDA** require a variety of distinct functions are combined. This overloading is internally resolved by analysing the current state of the proof goal along with a simple user-specified parameter. Thus the appropriate internal proof function is invoked and the relevant rules and other parameters are extracted from invisible data bases.

Proof Granularity. Unless users program tactics, trivial preparatory manipulations are often required before available proof tools apply in **LAMBDA**. **Rispe** provides functions which automate these auxiliary transformations. Other proof routines in **Rispe** automate complex proof tasks which are even beyond the capabilities of automatic first-order procedures. Thus the required user interaction is reduced to the essential proof steps.

Goal Handling. Visual analysis of large proof goals is tedious but indispensable for assessing proof tasks and taking decisions. As a side effect of the improved automation, there are many fewer intermediate goals. Most of previously required syntactical analyses of proof goals are included in the internal proof routines. A convenient facility for syntactical abstraction, expansion and reabbreviation of subterms further improves the control of the complexity of goals.

Proof Documentation. Typically, ML-records of interactive **LAMBDA** sessions hardly serve as understandable documentation. In **Rispe** the significant reduction of proof steps and variety of proof comands already leads to better readable proof scripts. Owing to the uniform format of **Rispe** commands, **Rispe** scripts are suitable for automatic documentation generation.

3 Overview of Rispe Commands

A selection of the main **Rispe** commands are displayed in Fig.1. All **Rispe** commands take one string as parameter for specifying subterms to which a proof function is to be applied or supplying other information as guidance. The main groups of **Rispe** functions are characterized in the following.

3.1 Interactive Proof Functions

These commands are intended for transforming a goal in a stepwise manner. They allow the user direct influence on the creation of a subsequent goal state. The distinction to the automatic proof functions discussed later is fluent, though, for significant automation has been added internally.

Rule Application (**APPLY**). Rule schemes are compact and legible portions of

APPLY	: intelligent rule application
INST	: instantiation
GEN	: generalization
IND	: structural and well-founded induction
CASAN	: case analysis
ESIMP	: equational simplification + expansion
FOSC	: first-order calculus
ARITH	: arithmetic conversions
DEF	: adding logic definitions
AX	: definition of (temporary) axioms
PROVE	: start of a proof
POPRL	: store current top rule

Fig. 1. The most important Rispe commands

proof knowledge. Awkward pre-transformations for applying those in LAMBDA are automated by way of unifiability analyses in Rispe.

Instantiation (**INST**). A variety of instantiation functions for free or bound variables exist in LAMBDA. Even obvious instances of quantified variables have to be specified explicitly, often preceded by necessary hypothesis permutations. In Rispe, heuristics permute universally quantified hypotheses and guess instantiation terms fully automatically[3] invoked by one command.

Generalization (**GEN**). This feature allows conveniently abstracting unessential details and creating generic rule schemes [1]. In LAMBDA, basic term generalization utilities are available for programming purposes. In Rispe, a convenient command is offered which includes simultaneous generalization.

Induction (**IND**). The induction command of Rispe supports multiple structural and well-founded induction. Explicit induction rule schemes are constructed for well-founded induction [2]. Auxiliary information is generated for handling tupled induction variables. Partially specified induction schemes may be specialized at a later proof stage. Measure functions are computed if not supplied by the user. Tests for completeness of cases are included; conditions for excluding incomplete cases through all recursions are computed. The induction heuristics find most termination proofs for recursive function definitions automatically.

Case Analysis (**CASAN**). The case analysis command of Rispe combines and extends many different functions of LAMBDA. Its parameter specifies the term(s) to be analysed, which may be a free or bound variable, a term, two expressions to be compared according to a partial order relation, or just the key-word **if** for analysing test expressions. The current goal is used to internally select the appropriate case analysis function. Explicit assignments for analysis expressions are accepted. Incomplete cases are automatically supplemented. Automatic proof routines attempt to discharge the subcases.

3.2 Automatic Proof Functions

These functions allow the user to start various automatic proof mechanisms, which discharge or at least simplify a given goal.

Equational Simplification (**ESIMP**). Conversions in LAMBDA are strategies[1] for recursively replacing subterms by way of conditional equations. The simplifying effect of conditional equations highly depends on the order of subterms. In Rispe various auxiliary transformations including permutations and a subset of a first-order sequent calculus are applied. They yield favourable syntactical representations which greatly increase the effect of equational simplification.

Expansion of definitions (**ESIMP**). In LAMBDA different expansion mechanisms are used. Some of those have to be parameterized with explicit names of equational expansion rules. As frequently an expansion step is followed by equational simplification, it is included in the same Rispe command. It uniformly allows expansions by just specifying the identifier of the abbreviation or function as occurring in the proof goal. Wildcards are supported. It is possible to restrict the expansion to subterms. The expansion function may be applied reversely.

First-Order Automation (**FOSC**). First-order sequent calculus or related procedures are a valuable enhancement of a higher-order logic theorem prover. An according proof function is part of Rispe [3]. It even handles a restricted class of higher-order goals. Combined tactics for instantiating universal quantification, equational simplification, restricted case analysis, and others yield a higher degree of automation than achievable through pure first-order sequent calculus. Look-ahead-functions which avoid unnecessary calls of subtactics and fast conversions yield a good performance. If a goal is not discharged completely, often useful simplifications are obtained. The underlying subtactics are reconfigurable, reusable, and tunable to specific application areas.

Arithmetic Simplifications (**ARITH**). In Rispe, efficient procedures and conversions for arithmetics are combined with subtactics of other automatic Rispe functions including the instantiation of quantifications.

3.3 Proof Management.

Various functions to support the management of proofs starting from the insertion of logic definitions have been provided in Rispe.

Logic Definitions (**DEF**). The functions of LAMBDA for reading logic definitions generate a parser environment and a set of rules about the defined entities. The definition function of Rispe generates auxiliary information which is invisibly referenced by other Rispe functions. This information hiding relieves the user

---

[1] The standard ones are depth-first or breadth-first. Conversionals allow advanced users to program other strategies.

from managing a large amount of proof objects and from much tedious initialisation work before proofs about new definitions are started. For instance, the case analysis function has access to a dynamic list of case analysis rules, which are generated from datatype definitions. An automatic facility for proving the termination of recursive functions can be invoked at definition time or later.

Axioms (**AX**). A useful facility of LAMBDA keeps track of user-defined axioms. This facility has been extended in Rispe for supporting a top-down proof strategy. Dependencies of proofs on axioms can be discharged conveniently once a proof of a preliminary axiom has been given.

Goalstack functions. Auxiliary functions for pushing (**PROVE**) and popping (**POPRL**) goals, rules or axioms to and from the goalstack, saving them in rules, and restoring goal stacks in case of interruptions are called by convenient Rispe commands. An extension for tagging goal stacks with corresponding Rispe proofs is being considered.

Rispe *Procedures.* As Rispe is run in ML, the advanced user has access to all functions and programming facilities of LAMBDA and ML. Nevertheless a simple way of specifying Rispe procedures is supplied. It provides combinators for Rispe commands and accepts formal arguments which are replaced with actual arguments when the procedure is called. A procedure can be applied simultaneously to several specified subgoals.

Example. In order to give a flavour of the use of Rispe commands, in Fig.2 the proof script of a simple hardware proof is displayed. The symbols *FA, BV* here for succinctness replace large subterms of actual intermediate goals. The tactics underlying Rispe are still being extended. As a result scripts will be even shorter. Moreover, Rispe as presented in this paper does not address any specific application domain. Proof functions internally using Rispe subtactics along with hardware-specific extensions would add further automation.

4 Conclusion

Most of the work done so far has centred on the numerous ML routines underlying Rispe rather than on case studies. Comparisons of LAMBDA subproofs of hardware verifications and in other areas redone in Rispe are very encouraging, though. The raised level of user interaction, which is enabled by the powerful proof automation functions, leads to reductions up to a factor of 10 in proof development time and script sizes; Rispe scripts are significantly better readable. Users without knowledge of LAMBDA need some basics of ML for being able to write own specifications. Learning about 20 intuitive and uniform Rispe commands suffices for effectively guiding proofs. A menu provides help information.

The PVS approach [6] employs decision procedures for automated reasoning in higher-order logic. The subtactics and conversions of Rispe are not only

PROVE"..."; $\vdash$ bvVal (n,bv_1) + bvVal (n,bv_2) + bVal cin =
$$2^{n+1} * \text{bVal } (bv_{cout}(n,cin,bv_1,bv_2)) +$$
$$\text{bvVal } (n,bv_{sum}(n,cin,bv_1,bv_2))$$

IND"n";

(base case): $\vdash$ bVal$(bv_1\ 0)$ + bVal$(bv_2\ 0)$ + bVal cin =
$$2 * \text{bVal}(FA_{cout}(\ldots)) + \text{bVal } (FA_{sum}\ (\ldots))$$

ESIMP"{*}"; $\vdash$ bvVal$(bv_1\ 0)$ + bVal$(bv_2\ 0)$ + bVal cin =
$$2 * \text{bVal}(\ldots) + \text{bVal } (cin \text{ xor } bv_1\ 0 \text{ xor } bv_2\ 0)$$

CASAN""; $\vdash$ TRUE

(step case):   bvVal(n,bv_1) + ... + bVal cin =
$$2^{n+1} * \text{bVal } (bv_{cout}(n,\ldots)) + \text{bvVal}(n,bv_{sum}(\ldots)))$$
$$\vdash \text{bvVal}(1+n,bv_1) + \ldots + \text{bVal cin} =$$
$$2^{n+2} * \text{bVal } (bv_{cout}(1+n,\ldots) + \text{bvVal}(1+n,bv_{sum}(\ldots)))$$

ESIMP"{*}";   ...

ARITH""; $\vdash$ bVal$(bv_1\ (1+n))$ + bVal$(bv_2\ (1+n))$ + bVal$(BV_{cout}\ n)$ =
$$2 * \text{bVal}(BV_{cout}\ (1+n) + \text{bVal}(BV_{sum}(1+n))$$

CASAN ""; $\vdash$ TRUE

POPRL "nAddT";

Fig. 2. Rispe proof for an n-bit adder

safe and efficient, but also reconfigurable and customizable. Decision procedures can be made part of conversions and tactics, which in **Rispe** has been done for arithmetic reasoning only. Both approaches demonstrate that the automation achievable in a higher-order system competes well with a first-order prover, while specification and interactive proof steps clearly benefit by higher-order concepts.

Rispe is a first prototype for general-purpose proofs. Application-specific extensions have been started recently, reusing and reconfiguring **Rispe** tactics. First experience indicates further gains in automation.

References

1. H.Busch, 'Transformational Design in a Theorem Prover', in *THEOREM PROVERS IN CIRCUIT DESIGN*, IFIP Transactions A-10, edited by V. Stavridou, T.F. Melham, and R.T. Boute, pp. 175–196, North-Holland, 1992.
2. H. Busch, 'Rule-Based Induction', in *FORMAL METHODS IN SYSTEM DESIGN - Special Issue on HOL'92*, Kluwer, Vol. 5, Issue 1 & 2, July/August 1994.
3. H. Busch, 'First-Order Automation for Higher-Order-Logic Theorem Proving', *HOL 1994 - 7th International Conference on Higher Order Logic Theorem Proving and its Applications*, edited by T. Melham and J. Camilleri, LNCS 859, Springer, September 21-24, 1994, Malta.
4. S. Finn, M. Fourman, M. Francis, B. Harris, R. Hughes, and E. Mayger, Abstract Hardware Limited, **LAMBDA** Documentation, 1993.
5. R. Harper, R. Milner, and M. Tofte, 'The Definition of Standard ML, Version 3', University of Edinburgh, LFCS Report Series, ECS-LFCS-89-81, May 1989. '
6. D. Cyrluk. S. Rajan, N. Shankar, and M.K. Srivas, 'Effective Theorem Proving for Hardware Verification', in *2nd Int. Conf. on THEOREM PROVERS IN CIRCUIT DESIGN*, edited by R. Kumar and T. Kropf, LNCS, Springer, 1994.

Quantitative evaluation of formal based synthesis in ASIC design

G. Bezzi, M. Bombana, P. Cavalloro, S. Conigliaro, G. Zaza

ITALTEL DRSC Settimo Milanese (ITALY)

Abstract. Formal based synthesis allows design space exploration to identify optimized implementations conforming to the initial abstract specification. We propose to exploit the synergies between formal synthesis (at high level of abstraction) and logic synthesis (at lower levels of abstraction). In this way a two–fold goal is reached: quantitative figures are provided as a measure of the applicability of formal reasoning in the design process, and the good integration of the two phases in a unified design flow is demonstrated. Users' benefits include both improved quality of the design process (reduced time–to–market) and improved reliability of the final products (increased competitive profile).

1 Introduction

The industrial interest in the application of formal methods to the design of complex ASICs is noteworthy. Design methodologies have been proposed ([8],[15],[16],[17], [18],[23],[24]) and commercial tools are now available ([9],[19]) to partially support the design practice with the power of formal reasoning. In this paper we focus on the application of high–level formal based synthesis ([3],[14]), that aims at improving the management of the first phases of the design process, addressing both the specification phase and the following partitioning steps at the architectural and/or scheduling/allocation levels. In this way a coherent and sound approach to the initial transformations of the design entities is provided, covering a phase of the design practice where automatic tools are still not widely used.

To apply this approach to real industrial practice ([12],[13]), some pre–requisites must be satisfied. One of them involves the definition of a homogeneous design flow able to link in a coherent way the results coming from the application of high–level formal synthesis with the standard techniques and tools of the lower levels of the design flow. In fact the gap existing between this theoretical approach and the present industrial design environments remains relevant: while the former improves the reasoning capabilities stressing abstraction and generality, the latter are mainly capable of dealing with completely instantiated specifications of RTL descriptions.

In section 2 a formal based synthesis tool (LAMBDA/DIALOG [9]) is introduced for design space exploration as a *'stand–alone' advisor* . In this phase the designer focuses on the search for an 'optimal' solution in terms of architectural choices or scheduling and allocation strategies. In section 3 this design phase is integrated into a homogeneous design flow in accordance with the users' requirements. Experimental results, obtained applying different strategies are compared and design guidelines are extrapolated to increase the efficiency of the global process. Considerations on the still existing problems and some hints on the future of this research are addressed in the conclusion.

2 Formal synthesis: the use of a 'stand–alone' high–level advisor

Formal methods have been applied in the design process at different abstraction levels and with different goals ([1],[7]). LAMBDA/DIALOG allows exploration of the early design space formally guaranteeing the transformations of an abstract behavioral specification of the device into mixed (behavioral/structural) representations [2].

(*) This research activity is part of the ESPRIT II/III Projects n. 5020 and 6128.

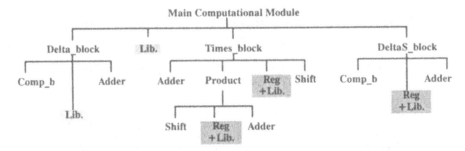

Fig. 1. Hierarchic representation of the modules of the Computational Block.

The tool differs substantially from other high–level behavioral synthesis tools. First, it is the only commercial one applying formal methods to guarantee the correctness of transformations between different description levels, and second it allows extensive user's interaction. The main benefit of the DIALOG interface [9] is to assist the designer in the process of step–wise transforming the starting specification into the chosen implementation. The supported specification language (ML [9]) is a functional language rather different from the most used hardware description languages, as the 'de–facto' standard VHDL [16].

2.1 The initial exploration of the design space

The synthesis process of a complex device is managed applying a 'divide and conquer' strategy: our test case [6] consists of 4 levels of hierarchy including 32 netlists and approximately 330 instances of user–defined modules. The low level modules of this device partially overlap the test cases that have been proposed as benchmarks [20]. The application of the formal techniques are shown to these low–level modules. The same considerations are valid for modules of higher level and for architectural partitionings, without any theoretical difference.

The comparison of implementation alternatives is particularly useful for those modules that are usually critical in terms of area overhead and timing properties, such as those performing the computations. The main Computational Module (Fig. 1) is based on the implementation of three sub–modules and some random logic. A preliminary analysis shows that the efficient combination of N–bit adders (Adder blocks) and N–bit multipliers (Product blocks)strongly influences the values of the physical parameters of the final implementation and the structure of the associated control blocks. As a consequence, the choices applied when designing the Computational Module are crucial.

Different specifications and implementations have been analyzed in LAMBDA in order to identify the most efficient solution for the computational sub–modules focusing on the implementation of the Product blocks (third hierarchic level of partitioning). The considered parameters were synchronous vs. asynchronous specification, parallelism of operations, granularity of the basic operations in terms of bits. The most naïve approach uses the maximum number of resources: the product is implemented through sequential additions and shift operations on the input data.

2.2 The re–use strategy and the search for the best solution

A solution to decrease the area is reached applying the re–use strategy to the sub–elements (Adder and Shift blocks) appearing in the Product netlist. This strategy allows the reduction of the area (the same component is used for more than one computation cycle) and involves the introduction of some registers for synchronization purposes.

Cycles are defined easily in LAMBDA applying a 're–use' tactic followed by a 'scheduling and allocation' tactic to introduce synchronization elements. The selection of the computational blocks to be re–used is done by the designer.

Moreover the computation on a N–bit data can be partitioned into two N/2–bit operations, with the introduction of an adder for the combination of the partial results. In this way the pure sequential procedure applied in the initial specification is substituted by the introduction of a certain level of parallelism, able to reduce the global clock phase of the computation. Applying both techniques and different levels of 're–use' in alternative implementations produces a wide range cases of different characteristics of area and timing: a single adder and a single shift block, two couples of each and finally four couples of each (Table 1, case II–III–IV). The time spent by the designer to produce these alternative representations is very small (a few hours). The control micro–code is also generated automatically by the tool.

The comparison between the different solutions is based on qualitative parameters: the number of components added for the control overhead is given, but no exact indication can be extracted on the relative increase of area on the total, or on the relative variations in the different cases. A more *quantitative* evaluation can be based on the results of logic synthesis.

3 The integration of formal synthesis into a design environment

The integrated approach to design, extensively applied nowadays, doesn't exclude the usefulness of stand–alone advisers on special aspects of the design activity. Anyway, when large–scale exploitation is foreseen, the identification of a global design flow is the winning element for the introduction of new tools into the consolidated design practice.

No single vendor provides tools to cover the whole range of requirements and needs of the design process. As a consequence a multi–vendor design environment is implemented [10] assembling the best tools on the market to solve specific tasks. CAD frameworks [5] provide the unified view and facilities to implement customized features for each proprietary design flow. Different levels of integration into a CAD framework are possible, i.e. a *strong integration* or a *weak integration*, usually called also *encapsulation*. The former implies for a tool the complete accessibility of the internal data–base of the framework, while the latter implies no data–base sharing and the interaction between the proprietary tool and the framework is accomplished in terms of shared files. The encapsulation path is usually followed. The requirements imposed on the tools are rather loose [22]. LAMBDA and the implemented interfaces satisfy these requirements and their encapsulation constitutes the prototype environment used to derive the results presented in the next sections.

3.1 The extended design flow

A design flow including formal synthesis and logic synthesis handles the substantial syntactic/semantic representation gaps existing between the levels of abstraction at which the tools operate. Formal synthesis tools operate abstractly and implementation details are left unspecified on purpose. On the contrary logic synthesis tools need specifications in a restricted subset of VHDL ([4],[11]) where all the implementation details have been correctly instantiated.

The handling of the instantiation mechanism, corresponding to a design level transformation from the most abstract to the most implementation–oriented description, is managed with the introduction of a transformational module supported by libraries of functional blocks, described in VHDL suitable for logic synthesis.

Commercial VHDL data–bases [21] support actively the development of this environment. From an operative point of view, three main design phases are managed by the designer. The first phase prepares the environment definition and the interface instantiation of the entities, as required by the VHDL description. The second phase identifies the standard library blocks in the netlist, for which the VHDL description is included in the functional parametric library. The user specifies the characteristics of each block in order to allow the correct instantiation. The third phase involves the identification of the user–defined modules, for which a behavioral VHDL description is not pre–defined. The resulting design environment has been tested only for the development of prototypes and re–design of already existing devices.

3.2 Experimental results and evaluation of physical parameters

The final result of the first half of the design flow of Fig. 2 is a VHDL netlist. A mixed VHDL simulation controls the correctness of the transformations. The numeric figures obtained applying the second half of the design flow to the examples considered in section 2, are shown in Table 1. The first two rows indicate for each implementation, if the N–bit multiplication has been partitioned, and how many adders and shifts have been inserted to manage the partitioning. The third row gives the number of equivalent gates for each implementation and the fourth the timing parameters. These values can be compared with the initial design constraints. Solution I is satisfactory in terms of equivalent gates, but its timing performances are too elevated; all the sequential components

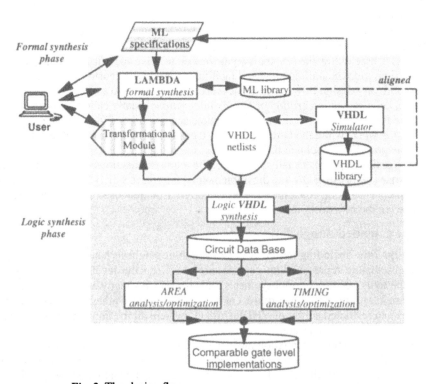

Fig. 2. The design flow.

have been stripped off, with the effect of reducing the area but transforming the circuit into a sort of asynchronous component.

Cases	I	II	III	IV	V	VI	VII
Computational Module	async.	re−use	re−use	re−use	re−use	re−use	re−use
Times_block • n. products (bit dim)	1 (16−bit)	2 (8−bit)	2 (8−bit)	2 (8−bit)	4 (4−bit)	4 (4−bit)	4 (4−bit)
• n. Adder blk/ Product • n. Shift blk/ Product	– –	1 1	2 2	4 4	1 1	2 2	4 3
Mapping (numb. of Equiv. gates)	4712	10927	5612	3696	5623	3824	4645
Area Optim. (numb. of Equiv. gates)	4697	6552	3846	3027	3772	3342	4442
Timing (Clock (ns) & e. gates)	148 4697	72 6552	84 3846	118 3027	90 3772	89 3342	111 4442
Timing Optim. (Clock (ns) & e. gates)	95 5375	44 6736	48 4076	67 3362	61 3984	60 3690	69 5130

Table 1. Summary of the characteristics of the different implementations

Table 1 shows significant differences between the other implementations differing for the value of re–use. Case II is characterized by the greatest amount of area overhead but it is the fastest in terms of clock cycle; on the other hand, case IV has the most reduced area but the timing performances are the worst of this group. Anyway this implementation gives better results than case I for both parameters. The explanation of these results lies in the different delay characteristics of the computation caused by the adopted re–use strategy.

Further refinements are obtained increasing the level of parallelism in decomposing the 16–bit multiplication element into four 4–bit parallel multiplications. The implementation of this case has been done considering, as in the previous case, alternative re–use strategies (cases V–VI–VII). The figures associated with these last implementations show that the parallelism introduced at the multiplication level generates a non–linear dependence between area and timing parameters. The last two rows of Table 1 show the results of the optimization phase on the seven alternative implementations. The comparison of these figures with those of the previous rows of Table 1 shows that in some cases the gain is close to a factor of two.

None of the automatically obtained implementations is more efficient than the one designed and optimized 'by hand' by an expert senior designer.

General guidelines may be abstracted. The synthesis process of computational modules in LAMBDA should be performed reducing as much as possible the number of synchronization components responsible of increased area overhead. This action must be coupled with an optimization of the VHDL functional library. The behavior of components must be written avoiding the use of statements that involve a complex hardware implementation in the logic synthesis phase. The re–use strategy is useful in order to reduce *both* area overhead *and* timing parameters. But to be effective a medium coefficient of re–use must be applied.

4 Conclusions

In this paper we have described the prototype of an innovative design methodology with associated tools supporting both formal reasoning and standard design techniques in a unified design flow. This new approach can be applied to the design of complex ASICs, to increase the global quality of the produced devices coupled with a decreased time–to–market. From this research activity suggestions have been collected to make the tools more user–friendly and more accepted to the users' community.

An extended engineering activity must be pursued in future, before applying this methodology in the current design practice, to make it sound and reliable. An extended test of the involved tools must verify their ability to cope with different classes of applications and with the increasing complexities of the devices to be designed.

References

1. F. Anceau: Formal verification in industrial environment,Workshop on Formal Methods, L'Aquila, (1989)
2. M.Fourman, E. Mayger: Formally Based System Design – Interactive Hardware scheduling, G. Musgrave, U.Lauther (eds), VLSI '89, Elsevier (1989)
3. S. Finn, M. Fourman, M. Francis, R. Harris: Formal System design – Interactive Synthesis based on Computer–Assisted Reasoning, Proc. IFIP WG 10.2, 10.5 Workshop, North Holland, (1990)
4. D. L. Perry: VHDL, McGraw–Hill, Inc. (1991)
5. Mentor Graphics: Getting started with Falcon Framework, (1991)
6. ESPRIT II Project n. 5020: Technical Report WP1.2–1, (1991)
7. F. Anceau: Panel on formal methods in hardware design, 10th Int. Symp. on Computer Hardware Description Languages, Marseille, (1991)
8. L. Claesen, M. Genoe, E. Verlind, F. Proesmans, H. De Man: SFG–Tracing: a methodology of design for Verifiability, Proc. Adv. Res. Workshop on Correct Hardware Design Methodologies, Turin, (1991)
9. AHL: Lambda Reference Manual – Version 4.1, London (1992)
10. M. Miserandino: ITL_TOOLKIT 1.0, Italtel Sit, Milano (1992)
11. C. Costi: A VHDL subset definition for simulation and synthesis in the Italtel environment, ITALTEL Technical Report, (1992)
12. G. Gorla: L'automazione del progetto di sistema nell'industria, Workshop on Specification and Synthesis of Digital Systems, CEFRIEL (1992)
13. M. Bombana, P. Cavalloro, G. Zaza: Specification and formal synthesis of digital circuits, Proc. IFIP TC10/WG10.2 Workshop, North Holland, (1992)
14. R.B. Hughes, G. Musgrave: Design Flow Graph Partitioning, HOL '92, IMEC Leuven Belgium, (1992)
15. R. Schlör, W. Damm: Specification and verification of system–level hardware designs using timing diagrams, EDAC '93, (1993)
16. S. Olcoz, J. M. Colom: Toward a Formal semantics of IEEE Std. VHDL 1076, EURO–DAC '93, Hamburg (1993)
17. C. Bolchini, M. Bombana, P. Cavalloro, C. Costi, F. Fummi, G. Zaza: A design methodology for the correct specification of VLSI systems, Euromicro '93, Barcelona, (1993)
18. T. Robles Valladares, A. Marín López, C. Delgado Kloos, T. de Miguel Moro, G. Rabay Filho: Automatic Hardware Implementation of Formal Specifications, III Jornadas de Concurrencia, Gandía, (1993)
19. CLSI Solutions: VFormal, (1993)
20. Benchmark circuits for hardware verification, Univ. of Karlsruhe, (1993)
21. LEDA: VHDL System, Meylan (1993)
22. ESPRIT III n. 6128: The FORMAT Design Methodology, Tech. Rep., (1994)
23. W. Grass, M. Mutz, W.D. Tiedemann: High Level Synthesis Based on Formal Methods, Euromicro '94, (1994)
24. K. L. McMillan: Fitting Formal methods into the Design Cycle, DAC '94, San Diego (1994)

Formal verification of characteristic properties

Michel Allemand

Laboratoire d'Informatique de Marseille, URA CNRS 1787
CMI de l'Université de Provence - UFR MIM
39, rue Joliot Curie - 13453 Marseille Cedex 13 - France
e-mail: amichel@gyptis.univ-mrs.fr

Abstract. In this paper we introduce a verification methodology well adapted to circuits where the specifications are described in terms of characteristic properties instead of algorithmic procedures. This method avoids most of the interpretation mistakes which could invalidate the proof process. In order to describe implementations, we present a formalism, based on sequences, which is close to HDLs. Then these description and proof methodologies are implemented in the Larch Prover which is adequate for this kind of verifications. This work is illustrated by the verification of the correctness of the nontrivial Minmax circuit.

Introduction

As far as verifying the correctness of digital devices is concerned, an alternative to simulation and test methodologies is the formal proof approach. One aspect of the formal proof of hardware consists in verifying the equivalence between an implementation and its specification (expected behaviour). Both the implementation and the specification are then mathematically modelized. The specification can be either a computational definition (an algorithm or an inductive definition) or an axiomatic definition (a set of characteristic properties). In the second case, a usual approach consists of translating the initial properties into an algorithm which is considered as the new specification. This approach could introduce some mistakes and could invalidate the proof process. Indeed, the translation of informal or formal properties into a formal algorithm is not straightforward. In this paper we present how to deal with such specifications using the semi-automatic Larch Prover [4].

In a first part, we present a proof methodology and a formalism based on temporal sequences which is well adapted to the proof method as well as to LP and to hardware descriptions. In a second part, we apply this work to the verification of the Minmax circuit. We describe its specification in LP according to our formalism and we show how we deal with the modularity of the implementation in order to verify the circuit.

1 Overview of the methodologies

In the field of formal verification of digital devices, two approaches are possible. We can use a specific system like tautology checker or model checker. On the other hand we can use general provers like the Boyer-Moore prover Nqthm, HOL, the Larch Prover and so on. In order to implement our description and proof methodologies, we chose to use LP. The Larch prover is a rewrite-rule based tool which allows induction and works on a subset of multisorted first order logic.

1.1 Description methodology

Our description formalism is based on the approach of [7], of [1] and on the P-Calculus [5] . In [7] the formalism is based on an "execution history" of a circuit which consists of a sequence of values for each data path in the circuit. The approach of [1] is based on streams of values. In a similar way, in the P-calculus, synchronous circuits are modeled by letting the input and output signals be temporal sequences of values (time functions) and by means of functions on such sequences. In order to express temporal features of digital devices, only one operator is needed: a Past functional $\mathcal{P}$ delay operator. In such a formalism the behaviour of a register R (with an input Ri and an output Ro) will be expressed by $Ro = \mathcal{P}(Ri)$. Thus we describe each signal by a sequence of values. Like in the P-calculus, these sequences are associated with a temporal scale. Sequential components are driven by a global clock which corresponds to this temporal scale. Like in [7] we consider an operator "." which takes as inputs a sequence S and an integer t and returns the tth value of S: "$S.t$" (i.e: the value of S at the time step t).

The behaviour of each combinational component C is described by the following first order logic formula: $\forall\ t,\ OUT.t\ =\ \mathcal{C}(IN.t)$ where $\mathcal{C}$ is a function which expresses the combinational behaviour of C and t the time, and the behaviour of each register R is formalized by the formula: $\forall\ t,\ Ro.(t+1)\ =\ Ri.t$

The composition of two components C1 and C2 is performed by a formula which expresses that the input sequence of C2, $C2i$ follows the output sequence $C1o$ of C1:$\forall\ t,\ C2i.t\ =\ C1o.t$

This methodology is well adapted to a hierarchical approach. Indeed, we can describe parts of a circuit as modules with their own inputs and outputs which can be easily connected. The behaviour of a module can be seen at the bit vector level or at the arithmetic level. Descriptions in this formalism can be obtained easily from VHDL ones.

1.2 Proof methodology

In this paper, we only consider the specification of circuits which are described by informal characteristic properties. In order to solve such a problem, several approaches are feasible. The first one consists in deducing a high level algorithm from the various points of such a specification. This algorithm becomes the new specification with respect to which we prove the implementation. The main drawback of such an approach is that the deduced specification can be very far from the initial one. Thus mistakes are likely and the proof process can be falsified.

Our approach consists in transforming the informal properties of the circuit into first order formulas. Such a process cannot prevent all mistakes. Indeed, we are not able to deal with informal specifications and, as soon as we translate informal properties into formal properties, we face mistake risks. However, in case where some ambiguities remain, this transformation can be achieved with designer interactions, thus the safety of this process is sufficient. These formulas are translated into LP descriptions as follows:

- All external universally quantified free variables are translated into LP free variables which are implicitly universally quantified.
- In the same way all external existentially quantified variables could be translated into LP formal constants (i.e. 0-ary operators).
- We translate all expressions such as "$\forall\ t' > t\ P$" by introducing a new free LP variable h and by substituting in P the term "$t + succ(h)$" to t', where $succ$ is the constructor operator for the natural numbers. We use this constructor to refer to a natural number which cannot equal zero.
- We express an expression such as "$(C_1 \wedge \cdots \wedge C_n) \Rightarrow E$", by the LP deduction rule "$when\ C_1, \cdots, C_n\ yield\ E$" if none of the C_i is universally quantified.
- The expression "$(C_1 \wedge \cdots \wedge C_n) \Rightarrow E$" is translated into the LP deduction rule

$$when\ C_1, \cdots, C_{i-1},\ (forall\ t_i)\ C_i', C_{i+1}, \cdots, C_n\ yield\ E$$

if C_i is universally quantified.

Finally the proof process consists in checking with the Larch Prover that the implementation verifies all properties described in the specification. This is done by asserting the description of the implementation and by proving each point of the specification, mainly by induction on time.

The main difficulty of such a verification lies in the strategy of proof. It is independent from the abstraction level of the component descriptions. Thus we chose to describe implementation at the arithmetical level. In case a preverified library of components is not provided, the first step of the proof process will be the verification of the correctness of all of these components.

2 Application to the MinMax circuit

In order to illustrate the application of our description and proof methodologies using the Larch Prover, we will verify the MinMax circuit. This circuit which has been proposed by IMEC [2] has already been verified using symbolic manipulations in [3] and using Nqthm in [6]. The main drawback of these approaches is that the necessity of formalizing the characteristic properties of the circuit as an algorithm takes us far from the initial informal specification. Furthermore, another disadvantage of [3] is that the size of the bit vectors must be fixed.

The Minmax circuit, the implementation of which is given in figure 1, has an input signal IN which consists of a sequence of integers and three boolean

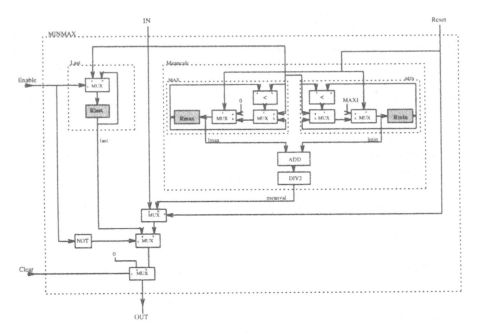

Fig. 1. Implementation of the MinMax circuit

control signals *clear, enable* and *reset*. The unit produces an output sequence *OUT* at the same rate as *IN* in the following way:

- *OUT* is zero if clear is true, independent of the other control signals
- if *clear* is false and *enable* is false then *OUT* equals the last value of *IN* before *enable* became false.
- if *clear* is false and *enable* is true and *reset* is true then *OUT* follows *IN*.
- if *reset* becomes false, then *OUT* equals, on each time point t, the mean value of the maximum and the minimum value of *IN* until that time point.

These informal characteristic properties are rewritten into first order logic. Thus, the formal specification of the second property is:

$$\forall t_0, \forall t_1 > t_0 \ [\ ((clear.t_1 = F) \wedge (enable.t_1 = F) \wedge (enable.t_0 = T) \wedge$$
$$(\forall t' > t_0, t' < t_1 \Rightarrow enable.t' = F)) \Rightarrow$$
$$(OUT.t_1 = IN.t_0)]$$

Although mistakes are possible, it is obvious that the translation into first order formulas is rather straightforward. Then, according to the rules given in section 1.2, we obtain the following LP conjecture:

$$when \ (forall \ l) \ l < h \ => \ enable.(t0 + succ(l)) = false,$$
$$clear.(t0 + succ(h)) \ == \ false,$$
$$enable.(t0 + succ(h)) \ == \ false,$$
$$enable.t0 \ == \ true \ yield$$
$$OUT.(t0 + succ(h)) \ == \ IN.t0$$

2.1 Verification of the implementation

We describe only the verification of the second property. In order to establish this conjecture, we have to prove some intermediate lemmas. These lemmas are deduced from the modular description of the implementation. We generate two subgoals:

- The output *last* of the module "Last" always produces the last value of IN before *enable* became false.

  $when\ (forall\ l)\ l \leq h\ =>\ enable.(t0 + succ(l)) = false,$
  $\quad enable.t0\ ==\ true\ yield$
  $\quad last.(t0 + succ(h))\ ==\ IN.t0$

- If *clear* is false and *enable* is false then OUT follows the output *last* of the module "Last".

  $$((clear.t = false)\&(enable.t = false)) => ((OUT.t) = (last.t))$$

The proof of the second subgoal is as straightforward as the validation of points 1 and 3 of the specification.

On the other hand, we need to introduce a new intermediate lemma to achieve the verification of the first one. This lemma expresses that, as soon as *enable* becomes false the value of *last* is the immediate preceding value of IN, i.e. when *enable* was true. It is formalized by the following LP conjecture:

$$((enable.h = true)\&(enable.succ(h) = false)) => last.succ(h) = IN.h$$

Then, we don't prove directly the first subgoal, but a little more generalized conjecture:

$when\ (forall\ l)\ l \leq h\ =>\ enable.(t0 + succ(l)) = false,$
$\quad enable.t0\ ==\ true\ yield$
$\quad (k \leq h) => last.(t0 + succ(k))\ ==\ IN.t0$

The validation of this conjecture is performed by induction on k. It is obvious that the first subgoal is directly deduced from this conjecture where k equals h. Finally we achieve the proof of the second property using the two subgoals, the verification of which we have achieved.

Conclusion

The main contribution of this work is a formal verification methodology, devoted to circuits whose specification is given in term of characteristic properties. This method has been implemented using the semi-automatic Larch Prover which works on first order logic. Furthermore we introduce a description formalism which is adapted to this prover as well as to the methodology and to hardware descriptions.

As far as the description of the implementations is concerned, the formalism that we have chosen is very legible. It is well adapted to hierarchical implementations. Moreover the LP descriptions in this formalism are very close to the corresponding VHDL descriptions. All of that is a great advantage for the designers.

The separated validation of each point of a specification described in term of characteristic properties is feasible in a relatively short time. Although the mistakes are not completely excluded, this approach preserves from most of them: even in the cases where the proof fails, this approach allows to see the properties which are not verified. This could be the starting point of an error diagnostic procedure.

In order to verify some properties of the circuit we have to introduce several intermediate lemmas according to the modular structure of the implementation. However, due to the readability of LP descriptions and LP conjectures, this work can be achieved by a designer.

The proof methodology concerns only circuits, the specifications of which are described in terms of characteristic properties. On the other hand, our description formalism could also be applied to the description of invariant specifications where we have to state some assumptions on time. Thus, future works will aim at developing this approach.

References

1. A. Bronstein and C. Talcott. Formal verification of synchronous circuits based on String-Functional Semantics: The seven Paillet circuits in Boyer-Moore. In *Workshop on automatic verification methods for finite state systems*, Grenoble, June 1989.
2. L. Claesen, editor. *Internal Workshop on applied Formal Methods for VLSI Design*, Leuven, Belgium, 1990. North-Holland.
3. O. Coudert, C. Berthet, and C. Madre. Verification of sequential machines using boolean functional vectors. In L. Claesen, editor, *Formal VLSI Correctness Verification*. North Holland, 1990.
4. S. J. Garland and J. V. Guttag. A guide to LP, the Larch Prover. Report 82, DEC Systems Research Center, Palo Alto, CA, December 1991.
5. J.-l. Paillet. A functional model for descriptions and specifications of digital devices. In D. Borrione, editor, *IFIP WG 10.2 Workshop From HDL descriptions de guaranted correct circuit designs*. North-Holland, 1987.
6. L. Pierre. The formal proof of the "Min-max" sequential benchmark described in CASCADE using the Boyer-Moore theorem prover. In L. Claesen, editor, *Formal VLSI Correctness Verification*. North Holland, 1990.
7. J. B. Saxe, S. J. Garland, J. V. Guttag, and J. J. Horning. Using transformations and verification in circuit design. In J. Staunstrup and R. Sharp, editors, *International Workshop on Designing Correct Circuits*. North-Holland, IFIP Transactions A-5, 1992. Also published as DEC Systems Research Center Report 78.

Extending Formal Reasoning with Support for Hardware Diagrams

Kathi Fisler

Department of Computer Science
Lindley Hall 215
Indiana University
Bloomington, IN 47405
kfisler@cs.indiana.edu

Abstract. Diagrams have been left as an informal tool in hardware reasoning, thus rendering them unacceptable representations within formal reasoning systems. We demonstrate some advantages of formally supporting diagrams in hardware verification systems via a simple example from the verification of a single-pulser.

1 Introduction

Diagrams have been treated as second-class citizens within the realm of formal reasoning, despite their steady use as informal design tools. The reasons for this appear to be based more on prejudice against diagrams in logic rather than on any inherent properties that render diagrams inappropriate for formal use. Diagrams offer several potential advantages to hardware reasoning: they offer clear, compact and user-transferable representations, and they lack the high learning overhead associated with the formal logics underlying many state-of-the-art sentential reasoning tools. In fact, it would seem that the only thing precluding the rigorous use of diagrams in formal verification is the lack of formalization of diagrammatic representations.

This paper presents initial work in a research project aimed at exploring the interactions of diagrams and sentential representations in the context of hardware design and verification. The goals of this research project are to develop a heterogeneous logic of interacting sentential and diagrammatic representations and to build a proof-checker based upon the logic. Our logic supports four representations: timing diagrams, circuit diagrams, algorithmic state machine (ASM) charts, and higher-order logic. Rules of inference bridge representations, allowing all four representations to interact during the proof process.

2 Previous Work

Using visual representations in hardware design frameworks is not a new idea. Various design tools and description languages have employed diagrammatic representations [5] [7] [13], and systems for reasoning about some aspects of systems using diagrammatic representations have appeared over the past year [3] [2] [10] [12]. Many systems provide formalizations of timing diagrams [2] [10] [12] and some even provide formal definitions of the interaction between timing diagrams and sentential representations [12]; however, none of these support multiple diagrammatic representations. The authors of [3] present a system in which a user can reason about system states using a graphical interval logic, but they translate their visual representations into a sentential logic for purposes of formal manipulation, unlike our logic, which is developed directly at the level of the diagrams. A previous attempt at defining a heterogeneous logic for hardware, along with more complete arguments supporting the use of diagrams in hardware formal methods, is presented in [8]. The logic presented in [8] is less fine-grained than the one presented here; their logic is based only on behavioral relationships while this work allows for reasoning about structural relationships between components. To the best of our knowledge, this work is the first to present examples of fully formal reasoning using diagrammatic logic in the hardware and formal methods communities.

The motivation for our research is based largely upon Hyperproof, the heterogeneous logic reasoning tool developed by Barwise and Etchemendy [1]. Hyperproof consists of a proof-checker for a logic of sentential and diagrammatic representations in a blocks world. We envision constructing an initial tool with much of the same flavor as Hyperproof.

3 Verification and Diagrammatic Representation

We believe that diagrammatic reasoning offers two main advantages in hardware verification: clarity of representation and conciseness of proof. We will use the single-pulser to demonstrate our arguments. The single-pulser is a good choice for this due to its size and clarity; such a study also complements the studies of single-pulser verifications given in these proceedings [9]. We use the work of [9] in this discussion as representative of sentential verification efforts.

To address the issue of clarity, consider the PVS implementation and specification of the single-pulser proposed by [9]. Their implementation is given below and is based upon the accompanying circuit diagram.

$imp(i, O) : bool = (\exists x : (delay(i, x) \land and_{\bullet}^{o}(i, x, O)))$
$delay(i, O) : bool = (\forall t : (O(t + 1) = i(t)))$
$and_{\bullet}^{o}(a, b, c) : bool = (\forall t : (c(t) = (-a(t)) \times b(t)))$

Notice that the PVS representation is nothing but a translation of the diagram into the syntax of PVS — neither representation contains any more information than the other.[1] The behavioral specification of [9] is given in two parts, as follows:

---

[1] We could consider the lengths of wires in the circuit diagram as information not available in the sentential representation, but that information is more fine-grained than our logic is tuned to handle at present.

$$spec1(i, O) : bool = (\forall n, m : Pulse(i, n, m) \supset$$
$$\exists k : n \leq k \wedge k \leq m \wedge O(k) = 1 \wedge$$
$$(\forall j : (n \leq j \wedge j \leq m \wedge O(j) = 1 \supset j = k))$$

$$spec2(i, O) : bool = (\forall k : O(k) = 1 \supset SinglePulse(O, k) \wedge$$
$$(\exists n, m : n \leq k \wedge k \leq m \wedge Pulse(i, n, m)))$$

The same specification could be given in terms of the following:

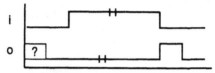

We claim that the timing diagram is a clearer representation of the intended behavior of a single-pulser than the two sentential specifications. The meaning of neither sentential specification is immediately clear, despite the fact that they are written in a straightforward style of higher-order logic. In fact, the average person might construct a diagrammatic depiction of the specifications in the process of understanding their full meanings. The advantages of clear specification are well known in the verification community. Careless interpretation of either specifications or implementations can lead to lost time in establishing proofs, or worse still, invalid proofs of correctness. Of course, the argument can also be made that there are issues of interpretation involved in using diagrams as well; we agree, but claim that the clarity of properly formalized diagrammatic representations minimizes the problem.

We now turn to comparing the sentential single-pulser verification to a possible diagrammatic verification. There are two aspects to consider: the time to develop proofs and the conciseness of the resulting proof. The PVS proof referenced in [9] took an estimated half-hour of proof time for a relatively novice PVS user; the main time expenditure was in properly formulating the specification, which took considerably longer than the actual verification [11]. Although we have no evidence to support this, we believe that specifications may be easier to state and debug using diagrammatic representations that are more familiar to practicing designers.

A brief discussion of our rules of inference is in order; our intent is to design the logic such that diagrammatic rules of inference mimic the informal reasoning steps used by designers in practice. The inference rules relating *and* gates and timing diagrams appear below; other inference rules on *and* gates, such as one where a low input yields a low output, can be derived from these three primitive rules. The portion of the logic necessary to support these rules can be found in [6].

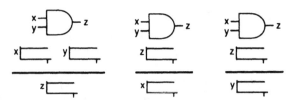

Now consider the following diagrammatic proof that corresponds to the proof of *spec1*. In the timing diagrams, the dashed axis notation denotes that the dashed tick repeats the number of times indicated on the dashed line.

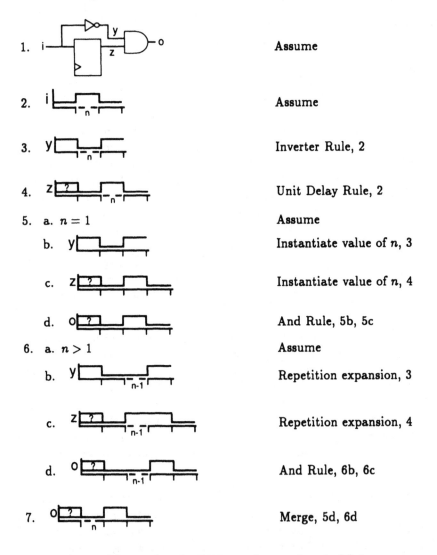

1.	i	Assume
2.	i	Assume
3.	y	Inverter Rule, 2
4.	z	Unit Delay Rule, 2
5.	a. $n = 1$	Assume
	b. y	Instantiate value of n, 3
	c. z	Instantiate value of n, 4
	d. o	And Rule, 5b, 5c
6.	a. $n > 1$	Assume
	b. y	Repetition expansion, 3
	c. z	Repetition expansion, 4
	d. o	And Rule, 6b, 6c
7.	o	Merge, 5d, 6d

Comparing this proof to the PVS proof trace given in [9], it seems reasonable to argue that the diagrammatic proof is easier to follow and quite possibly easier to produce than the one required to verify *spec1* in PVS. The steps taken in the diagrammatic proof are also at a lower granularity (for sake of example) than those we expect to be taken in practice, thus compacting the proof even further. Though we have no measured results to support this, empirical evidence using Hyperproof indicates that proofs are often substantially shorter than their purely sentential equivalents [4].

The above presentation argues the benefits of using diagrammatic representations in formal verification, but it does not adequately address our particular approach of developing a logic of hardware diagrams. Certainly it would seem reasonable to merely provide a diagrammatic interface to an existing sentential logic, thereby allowing us to rely on existing tools for verification. There are certain obvious immediate benefits to such an approach, such as the timeliness with which diagrams could be used to aid in formal verification.

We believe, however, that this approach is not the correct one to take for three reasons. By using diagrams merely as an interface tool, we leave them as second-class citizens to sentential logic in the realm of formal reasoning. We believe that diagrams are as valid a representation as sentential forms in reasoning and we are interested in the creation of logics that put diagrams on equal par as a valid representation. In addition, using diagrams merely in an interface capacity sidesteps our belief that there are logical relationships between different diagrammatic hardware representations. Identifying these relationships may lead us to even stronger frameworks for verification, but to do so requires closer examination of the diagrams themselves as first-class citizens.

Finally, we are not convinced that translation is a desirable approach. Any formal system that uses translation needs to prove that the translation is done correctly; such an argument will require some level of proof that operates on the diagrams, so translation does not save us from needing to formally consider diagrams in proof. In addition, manipulations might be made on the translated representation that do not naturally translate back up to the diagrammatic representation; this problem would be critical for a system that implements inferences on diagrams. Lastly, there may exist natural rules of inference on diagrams that become less natural, perhaps even unwieldy, at the sentential level; this seems at odds with our goal of providing natural diagrammatic inferences.

4 Conclusions and Future Work

We have presented a simple example demonstrating that diagrammatic representations can be formalized and used effectively in proof. Although the actual logic is not presented here for sake of space, the example proof is correct within the logic we have developed, and all inferences used in the example proof have been proven sound. A more complete discussion of the proof and a presentation of the logic that supports it is provided in [6].

We are in the process of completing the definition of the logic presented here; this includes presenting rules of inference and establishing soundness and completeness results for the full logic. The current status of this work is available in [6]. A prototype implementation of a simple proof-checker based upon our logic is in an early development stage. We have yet to apply the logic to a substantial verification effort, but plan to do so in the near future. What is presented here is merely an initial attempt to formalize the interactions between these various diagrammatic representations. There is still much research to do both in understanding the role of diagrams in verification and in developing tools that use such formalizations.

5 Acknowledgements

The author would like to thank Jon Barwise, Steve Johnson, Gerry Allwein, and Shriram Krishnamurthi for their helpful comments in both the research and the preparation of this paper.

References

1. Jon Barwise and John Etchemendy. Hyperproof, CSLI Lecture Notes, University of Chicago Press. To appear, 1994.
2. Viktor Cingel. A graph-based method for timing diagrams representation and verification. In George J. Milne and Laurence Pierre, editors, *Correct Hardware Design and Verification Methods*, pages 1–14. CHARME, Springer-Verlag, 1993.
3. L.K. Dillon, G. Kutty, L.E. Moser, P.M. Melliar-Smith, and Y.S. Ramakrishna. A graphical interval logic for specifying concurrent systems. Technical report, UCSB, 1993.
4. Ruth Eberle, April 1994. Personal communication.
5. Simon Finn, Michael P. Fourman, Michael Francis, and Robert Harris. Formal system design — interactive synthesis based on computer-assisted formal reasoning. In Luc Claesen, editor, *Formal VLSI Specification and Synthesis: VLSI Design-Methods-I*. North-Holland, 1990.
6. Kathi Fisler. A logical formalisation of hardware design diagrams. Indiana University Technical Report TR416, September 1994.
7. Graham Hutton. The Ruby interpreter. Technical Report 72, Chalmers University of Technology, May 1993.
8. Steven D. Johnson, Gerard Allwein, and Jon Barwise. Toward the rigorous use of diagrams in reasoning about hardware. IULG Preprint Series, April 1993.
9. Steven D. Johnson, Paul Miner, and Shyam Pullela. Studies of the single-pulser in various reasoning systems. In *Theorem-Provers and Circuit Design Proceedings*, September 1994.
10. K. Khordoc, M. Dufresne, E. Cerny, P.A. Babkine, and A. Silburt. Integrating behavior and timing in executable specifications. In *CHDL*, pages 385–402, April 1993.
11. Paul S. Miner, July 1994. Personal communication.
12. Rainer Schlör and Werner Damm. Specification and verification of system-level hardware designs using timing diagrams. In *Proc. European Conf. on Design and Automation*, Paris, February 1993.
13. Mandayam Srivas and Mark Bickford. SPECTOOL: A computer-aided verification tool for hardware designs, vol I. Technical Report RL-TR-91-339, Rome Laboratory, Griffiss Air Force Base, NY, December 1991.

Springer-Verlag
and the Environment

We at Springer-Verlag firmly believe that an international science publisher has a special obligation to the environment, and our corporate policies consistently reflect this conviction.

We also expect our business partners – paper mills, printers, packaging manufacturers, etc. – to commit themselves to using environmentally friendly materials and production processes.

The paper in this book is made from low- or no-chlorine pulp and is acid free, in conformance with international standards for paper permanency.

Printing: Weihert-Druck GmbH, Darmstadt
Binding: Theo Gansert Buchbinderei GmbH, Weinheim

Lecture Notes in Computer Science

For information about Vols. 1–822
please contact your bookseller or Springer-Verlag

Vol. 859: T. F. Melham, J. Camilleri (Eds.), Higher Order Logic Theorem Proving and Its Applications. Proceedings, 1994. IX, 470 pages. 1994.

Vol. 860: W. L. Zagler, G. Busby, R. R. Wagner (Eds.), Computers for Handicapped Persons. Proceedings, 1994. XX, 625 pages. 1994.

Vol: 861: B. Nebel, L. Dreschler-Fischer (Eds.), KI-94: Advances in Artificial Intelligence. Proceedings, 1994. IX, 401 pages. 1994. (Subseries LNAI).

Vol. 862: R. C. Carrasco, J. Oncina (Eds.), Grammatical Inference and Applications. Proceedings, 1994. VIII, 290 pages. 1994. (Subseries LNAI).

Vol. 863: H. Langmaack, W.-P. de Roever, J. Vytopil (Eds.), Formal Techniques in Real-Time and Fault-Tolerant Systems. Proceedings, 1994. XIV, 787 pages. 1994.

Vol. 864: B. Le Charlier (Ed.), Static Analysis. Proceedings, 1994. XII, 465 pages. 1994.

Vol. 865: T. C. Fogarty (Ed.), Evolutionary Computing. Proceedings, 1994. XII, 332 pages. 1994.

Vol. 866: Y. Davidor, H.-P. Schwefel, R. Männer (Eds.), Parallel Problem Solving from Nature - PPSN III. Proceedings, 1994. XV, 642 pages. 1994.

Vol 867: L. Steels, G. Schreiber, W. Van de Velde (Eds.), A Future for Knowledge Acquisition. Proceedings, 1994. XII, 414 pages. 1994. (Subseries LNAI).

Vol. 868: R. Steinmetz (Ed.), Multimedia: Advanced Teleservices and High-Speed Communication Architectures. Proceedings, 1994. IX, 451 pages. 1994.

Vol. 869: Z. W. Raś, Zemankova (Eds.), Methodologies for Intelligent Systems. Proceedings, 1994. X, 613 pages. 1994. (Subseries LNAI).

Vol. 870: J. S. Greenfield, Distributed Programming Paradigms with Cryptography Applications. XI, 182 pages. 1994.

Vol. 871: J. P. Lee, G. G. Grinstein (Eds.), Database Issues for Data Visualization. Proceedings, 1993. XIV, 229 pages. 1994.

Vol. 872: S Arikawa, K. P. Jantke (Eds.), Algorithmic Learning Theory. Proceedings, 1994. XIV, 575 pages. 1994.

Vol. 873: M. Naftalin, T. Denvir, M. Bertran (Eds.), FME '94: Industrial Benefit of Formal Methods. Proceedings, 1994. XI, 723 pages. 1994.

Vol. 874: A. Borning (Ed.), Principles and Practice of Constraint Programming. Proceedings, 1994. IX, 361 pages. 1994.

Vol. 875: D. Gollmann (Ed.), Computer Security – ESORICS 94. Proceedings, 1994. XI, 469 pages. 1994.

Vol. 876: B. Blumenthal, J. Gornostaev, C. Unger (Eds.), Human-Computer Interaction. Proceedings, 1994. IX, 239 pages. 1994.

Vol. 877: L. M. Adleman, M.-D. Huang (Eds.), Algorithmic Number Theory. Proceedings, 1994. IX, 323 pages. 1994.

Vol. 878: T. Ishida; Parallel, Distributed and Multiagent Production Systems. XVII, 166 pages. 1994. (Subseries LNAI).

Vol. 879: J. Dongarra, J. Waśniewski (Eds.), Parallel Scientific Computing. Proceedings, 1994. XI, 566 pages. 1994.

Vol. 880: P. S. Thiagarajan (Ed.), Foundations of Software Technology and Theoretical Computer Science. Proceedings, 1994. XI, 451 pages. 1994.

Vol. 881: P. Loucopoulos (Ed.), Entity-Relationship Approach – ER'94. Proceedings, 1994. XIII, 579 pages. 1994.

Vol. 882: D. Hutchison, A. Danthine, H. Leopold, G. Coulson (Eds.), Multimedia Transport and Teleservices. Proceedings, 1994. XI, 380 pages. 1994.

Vol. 883: L. Fribourg, F. Turini (Eds.), Logic Program Synthesis and Transformation – Meta-Programming in Logic. Proceedings, 1994. IX, 451 pages. 1994.

Vol. 884: J. Nievergelt, T. Roos, H.-J. Schek, P. Widmayer (Eds.), IGIS '94: Geographic Information Systems. Proceedings, 1994. VIII, 292 pages. 19944.

Vol. 885: R. C. Veltkamp, Closed Objects Boundaries from Scattered Points. VIII, 144 pages. 1994.

Vol. 886: M. M. Veloso, Planning and Learning by Analogical Reasoning. XIII, 181 pages. 1994. (Subseries LNAI).

Vol. 887: M. Toussaint (Ed.), Ada in Europe. Proceedings, 1994. XII, 521 pages. 1994.

Vol. 888: S. A. Andersson (Ed.), Analysis of Dynamical and Cognitive Systems. Proceedings, 1993. VII, 260 pages. 1995.

Vol. 889: H. P. Lubich, Towards a CSCW Framework for Scientific Cooperation in Europe. X, 268 pages. 1995.

Vol. 890: M. J. Wooldridge, N. R. Jennings (Eds.), Intelligent Agents. Proceedings, 1994. VIII, 407 pages. 1995. (Subseries LNAI).

Vol. 891: C. Lewerentz, T. Lindner (Eds.), Formal Development of Reactive Systems. XI, 394 pages. 1995.

Vol. 892: K. Pingali, U. Banerjee, D. Gelernter, A. Nicolau, D. Padua (Eds.), Languages and Compilers for Parallel Computing. Proceedings, 1994. XI, 496 pages. 1995.

Vol. 893: G. Gottlob, M. Y. Vardi (Eds.), Database Theory – ICDT '95. Proceedings, 1995. XI, 454 pages. 1995.

Vol. 894: R. Tamassia, I. G. Tollis (Eds.), Graph Drawing. Proceedings, 1994. X, 471 pages. 1995.

Vol. 895: R. L. Ibrahim (Ed.), Software Engineering Education. Proceedings, 1995. XII, 449 pages. 1995.

Vol. 896: R. M. Taylor, J. Coutaz (Eds.), Software Engineering and Human-Computer Interaction. Proceedings, 1994. X, 281 pages. 1995.

Vol. 898: P. Steffens (Ed.), Machine Translation and the Lexicon. Proceedings, 1993. X, 251 pages. 1995. (Subseries LNAI).

Vol. 899: W. Banzhaf, F. H. Eeckman (Eds.), Evolution and Biocomputation. VII, 277 pages. 1995.

Vol. 900: E. W. Mayr, C. Puech (Eds.), STACS 95. Proceedings, 1995. XIII, 654 pages. 1995.

Vol. 901: R. Kumar, T. Kropf (Eds.), Theorem Provers in Circuit Design. Proceedings, 1994. VIII, 303 pages. 1995.

Vol. 902: M. Dezani-Ciancaglini, G. Plotkin (Eds.), Typed Lambda Calculi and Applications. Proceedings, 1995. VIII, 443 pages. 1995.

Vol. 903: E. W. Mayr, G. Schmidt, G. Tinhofer (Eds.), Graph-Theoretic Concepts in Computer Science. Proceedings, 1994. IX, 414 pages. 1995.